The Primate Mind

THE PRIMATE MIND

Built to Connect with Other Minds

Edited by Frans B. M. de Waal and Pier Francesco Ferrari

Harvard University Press
Cambridge, Massachusetts, and London, England   2012

Library of Congress Cataloging-in-Publication Data

The primate mind : built to connect with other minds / edited by Frans B. M. de
Waal and Pier Francesco Ferrari
    p. cm.
    Based on a meeting held June 4–7, 2009, at the Ettore Majorana Center in
Erice, Sicily, Italy.
    Includes bibliographical references and index.
    ISBN 978-0-674-05804-0 (alk. paper)
   1. Primates—Psychology.   2. Social psychology.   3. Neuropsychology.
4. Psychology, Comparative.   5. Comparative neurobiology.   I. Waal, F. B. M.
de (Frans B. M.), 1948–   II. Ferrari, Pier Francesco.
   QL737.P9P67259   2011
   599.8'1513—dc23   2011016094

Contents

      Cognitive Development in Humans and Chimpanzees   *288*
      Tetsuro Matsuzawa

      References   *307*
      Contributors   *379*
      Index   *381*

Preface

The Primate Mind seeks to inject the study of primate and human cognition with a new perspective, one less concerned with cognitive "rankings" among humans, apes, monkeys, canines, corvids, and other nonprimates, but rather with the nuts and bolts of cognition: the way it works. The focus is on the way the mind interacts with other minds in terms of emotional and cognitive connections, such as empathy, imitation, cultural learning, and cooperation. Primates are intensely social, hence continuously challenged to compete with, work with, or understand each other. Their brains have been designed for social connection.

From June 4 through 7, 2009, about one hundred participants gathered at the Ettore Majorana Center located in Erice, a city on a 750-meter-high mountaintop in Sicily, Italy, with a gorgeous view of the Mediterranean. A curious location given that we like to look at our new approach as a bottom-up perspective! Sixteen prominent speakers (Figure 0.1), who contributed to the present volume, were asked to cover primate cognition from all possible angles. The challenge was to analyze the cognitive processes underlying social interaction, some of which may be very basic and widespread and emerge early in human development, while others may be more advanced, even uniquely human. How do primates perceive and interpret each other's motor actions, how do they perceive and interpret each other's emotional states and situations, and how does this information in turn influence complex social interaction? What are the developmental landmarks of these processes? Are there neural structures and mechanisms that underpin all of them?

The interaction between neuroscience and primatology (or ethology) proved extremely stimulating, visible in discussions throughout the day as well as over dinners in the local restaurants. Requests for more intellectual gatherings like this have been forthcoming, such as a biannual series at the Ettore Majorana Center. The center was a most appropriate venue, as it is dedicated to higher intellectual endeavors and is a regular meeting point for scientists from all over the world. Professor Antonino Zichichi, its director, and Professor Danilo Mainardi, director of the International School of Ethology at the University of Venice, deserve our gratitude for making possible this memorable meeting of minds.

In order to preserve the element of interdisciplinary exchange, chapters have been ordered in this volume along theoretical topics rather than research methodologies. The first section ("From Understanding of the Actions of Others to Culture") covers themes around social learning, beginning with the neuroscience of how others are perceived and how

Figure 0.1. Speakers at the 2009 Erice conference, including speakers in the junior session and session chairs Top row, from left to right: Erin Hecht, Elisabetta Visalberghi, Tetsuro Matsuzawa, Andrew Whiten, Christophe Boesch, Brian Hare, Charles Menzel, Atsushi Iriki, Simon Baron-Cohen, Marco Iacoboni, Elisabetta Palagi, Teresa Romero, Laurie Santos. Bottom row: Frans de Waal, Ludwig Huber, Filippo Aureli, Pier Francesco Ferrari, Antonino Casile, Stephen Suomi, William Hopkins. Photographer unknown.

this affects an individual's own behavior, followed by the larger themes of imitation and cultural learning. Culture is a hot topic in primatology, fueled by evidence for behavioral variation between groups of the same species. How this variation comes about is informed by how mirror neurons allow an individual to process the behavior of others and make it its own. With all the talk of mirror neurons in the literature, it should never be forgotten that these neurons were first discovered in macaques, hence that many human characteristics now ascribed to them probably also apply to other primates. The chapters in this section range all the way from the working of mirror neurons and the neural processing of others to how primates and other animals are affected by the behavior of others either directly, when they copy them, or indirectly, when they adjust to environmental modifications brought about by members of their own species.

The second section ("Empathy, Perspective Taking, and Cooperation") covers another area of much interest and current debate, from emotional connectedness to "theory of mind." These topics have a long history of study in apes, but there is no reason why being affected by the emotions of others or taking their perspective should not also extend to other species, including monkeys. These capacities promote altruistic behavior and cooperation and are increasingly subjected to controlled experimentation. They obviously relate to both empathy and the absence of empathy in certain human populations, such as people with autism spectrum disorders. In reading these chapters, one may notice definitional issues in that "theory of mind" now usually requires a qualification of what exactly is meant, and that "empathy" is defined rather narrowly by some and more broadly by others. Rather than representing fundamental differences of opinion, these variations in language reflect the struggle to arrive at a common terminology on evolving topics addressed by different disciplines.

The third and final section ("Memory, Emotions, and Communication") concerns emotions and the associated communication, such as body language. It also deals with how apes emotionally develop and how they deal with their own or others' emotions. The perspective is at times historical, and the theme of human uniqueness comes up again, as it did in the other sections. It is easy to see the connection between how apes perceive the intentions and knowledge of others and how they communicate in ways that suggest their appreciation of the attentional states of others. All these capacities are integrated and receive attention in this section along with sensitivity to the emotions of others, which brings us back to the issue of empathy discussed by several authors in the first section. Part of the challenge of the present volume is to explore a new vocabulary that we may use to address the way primate (and human) minds connect with each other. Given the richness of processes involved, it will take time to arrive at a generally agreed-upon terminology. The present volume will hopefully contribute to formation of a field bound for future growth.

The Primate Mind

CHAPTER ONE

A Bottom-Up Approach to the Primate Mind

Frans B. M. de Waal and Pier Francesco Ferrari

THE STUDY of animal cognition traditionally focused on basic processes, such as classical conditioning, orientation, memory, imprinting, and the like, that applied to a wide range of species. Over the years, however, emphasis has shifted to a top-down view in which human accomplishments were placed at the top and the challenge was to find animals that could approach these. Not all scientists pursued this agenda, but it did come to dominate primate cognition studies, which as a result began to favor mentalistic cognitive accounts. We would take language, for example, to see how far apes could go with it. When attempts at teaching apes speech failed, training moved to the gestural domain, and performance exceeded expectations: symbolic communication appeared within grasp of our close relatives (Gardner and Gardner 1969). Definitions of language were quickly changed, however, stressing syntax over reference (Terrace 1979). This move may have kept the supposed species hierarchy intact, but it did little to illuminate basic capacities underlying symbolic communication and the relevance of these capacities for apes or other nonhuman animals. It is the latter type of questions—those concerning basic capacities—that should drive the study of animal cognition, not an obsession with rankings and the human-animal divide.

What if we were to replace interest in complex cognition with an exploration of its foundation? Instead of asking which species can do X, the question would become how does X actually work? What are the necessary ingredients of X, and how did these evolve? Here we propose a

Modified from de Waal, F. B. M., and P. F. Ferrari (2010).

bottom-up perspective that focuses on constituent capacities that underlie larger cognitive phenomena as to be expected from both neuroscience and evolutionary biology. We will briefly review recent research on future planning, imitation, and altruistic behavior in order to show how complex capacities can often be broken down into components that humans share not just with the apes but with a host of other species. This approach, which has been gaining ground in the last few years, will move the field of comparative cognition toward an understanding of capacities in terms of underlying mechanisms, and the degree to which these mechanisms are either widespread or special adaptations.

From Top-Down to Bottom-Up Approaches

Top-down approaches typically focus on all-or-nothing questions. Do only humans have a "theory of mind," or do apes, too? Can animals have culture? What is imitation, and which species are capable of it? Can animals count? The result is a literature of claims and counterclaims regarding complex mental faculties, with a curious lack of references to underlying mechanisms. We believe that this approach suffers from several methodological weaknesses.

Even though Darwin (1871, 105) believed "the difference in mind between man and the higher animals, great as it is, certainly is one of degree and not of kind," the search for continuity is sometimes written off as a pursuit of analogies. The only similarities that matter—homologies—seem impossible to prove, at least to students of pure behavior (Povinelli 2000). Thus, there is the denial of imitation even in apes given that they fail to appreciate others as intentional agents (Tomasello 1999), claims of "humaniqueness" since animals supposedly lack the ability to combine old concepts into new ones (Hauser 2009), and the sweeping conclusion that "the functional discontinuity between human and nonhuman minds pervades nearly every domain of cognition" (Penn and Povinelli 2007, 110).

The distinction between homology (i.e., shared ancestry) and analogy (i.e., independently evolved parallels), however, is far less clear-cut than assumed. Developed by anatomists, this distinction is especially useful in relation to highly defined morphological and behavioral traits with a traceable phylogeny, such as fixed action patterns and facial expressions (Preuschoft and van Hooff 1995). The same distinction is almost impossible to apply to traits that escape precise definition and measurement, such as cognitive capacities. Only if we know their neural underpinnings can these be confidently classified as either analogous or homologous. Recent research on face recognition in humans and other primates, for example, strongly suggests a shared neural background, hence homology

(Tsao et al. 2008; Parr et al. 2009). Until more such evidence is available, the most parsimonious Darwinian assumption is that if closely related species—whether they be squids and octopi or humans and apes—show similar solutions to similar problems, they probably employ similar cognitive mechanisms (de Waal 1999).

A second problem with top-down approaches is the importance they attach to the absence of certain cognitive abilities in certain taxonomic groups. This has led to unwarranted attention to negative evidence. While it is true that both positive and negative evidence risk being false, a profound asymmetry exists: there are many more possible reasons why an existing capacity may not be found than why a nonexisting capacity may be found. Nevertheless, failures to demonstrate certain capacities have led to hasty conclusions about their absence, such as when chimpanzees were said not to care about the welfare of others (Silk et al. 2005), a claim countered by subsequent research (see section on Prosocial Behavior and Empathy). Similarly, a failure to demonstrate that apes understand gravity has been taken to mean that only humans possess such an understanding (Povinelli 2000), even though it would seem a rather adaptive capacity for arboreal primates, which do show signs of it (Hauser et al. 2001). Debates around theory of mind and imitation have gone through similar cycles of initial denial in nonhuman species based on negative evidence, followed by partial or complete acceptance based on experimental paradigms with greater ecological validity (see the section on Imitation).

There are of course also defenders of Darwin's view of mental continuity (e.g., de Waal 2009b), who do not necessarily see analogies everywhere or view cognitive similarities between distant species as a problem for evolutionary theory. Tool use, for example, may have evolved independently in the service of extractive foraging in the great apes, capuchin monkeys, and New Caledonian crows. But even though a sound case for convergent evolution can be made (Emery and Clayton 2004), this does not necessarily imply independent neural mechanisms. The intriguing possibility exists of "deep homologies" in the cognitive domain, much as shared genetic instructions underlie the eyes or limbs of animals as distant as flies and rodents (Shubin et al. 2009). The reevaluation of avian brain evolution implies that even though a layerlike organization is absent from the pallium, this structure may nevertheless derive from the same reptilian telencephalic structure as the mammalian neocortex (Jarvis et al. 2005). If both structures accomplish similar functions, many avian and mammalian cognitive capacities may in fact rest on homologous mechanisms.

A final weakness of top-down approaches is the neglect of mechanistic explanations. This is reflected in entire treatises on animal intelligence devoid of neuroscience (e.g., Tomasello and Call 1997; Penn and Povinelli

2007b), and the assumption that reasoning and cause-effect understanding underlie human mental capacities, often with the implication that only humans can therefore solve certain problems. The latest studies in neuroanatomy and neurophysiology, however, suggest that the situation is more complex than previously thought, as our brain appears merely a linearly scaled-up version of the primate brain (Herculano-Houzel 2009). The excessive attention to so-called "higher" cognitive functions and the corresponding neglect of subcortical processes in cognitive science have been criticized as "corticocentric myopia" (Parvizi 2009). Some have gone so far as labeling the fascination with uniquely human capacities as "nonevolutionary," together with a warning against the belief that a brief evolutionary time interval could have produced a well-integrated set of novel capacities (Margoliash and Nusbaum 2009).

Every species, including ours, comes with an enormous set of evolutionarily ancient components of cognition that we need to better understand before we can reasonably focus on what makes each species' cognition special. Are cognitive specializations due to new capacities or rather to new combinations of old ones? Bottom-up approaches pay attention to these building blocks and represent a new Zeitgeist as reflected in the latest treatments of future planning (Raby and Clayton 2009), reciprocal altruism (Brosnan and de Waal 2002), theory of mind (Call and Tomasello 2008), and comparative cognition in general (Shettleworth 2010). Capacities are never all-or-nothing phenomena. Many nonhuman animals show some but not all aspects of theory-of-mind, self-awareness, culture, reciprocal altruism, planning, and so on. Whereas an outcome-based science stresses differences, a focus on process brings equal attention to commonalities. So, even if humans build cathedrals and produce symphonies, this is no reason to place them beyond comparison: the underlying processes (e.g., social learning, tool use, musical appreciation, sense of rhythm, coordination) are likely shared with other animals.

Remembering the Past and Planning for the Future

Remembering specific personal experiences has been considered uniquely human (Tulving 2005). As Suddendorf and Corballis (1997) put it: "animals other than humans cannot anticipate future needs or drive states, and are therefore bound to a present that is defined by their current motivational state." In this view, the information stored by animals is merely used contingently to react to present stimuli or anticipate immediate events. Experiments with food-caching birds, however, have challenged this view, suggesting that western scrub jays have very precise memories of past caches, including the "what, where, and when" required for episodic memory (Clayton et al. 2003).

Since then, the problem of episodic memory has been dissected with more sophisticated tools by neuroscientists. There is growing evidence from human neuroimaging and brain-damaged patients that remembering past events involves regions of the hippocampus, parahippocampal gyrus, and the prefrontal cortex. But memories of past events may also be pieced together to simulate the future. Brain imaging shows, in fact, that the same neural machinery that serves the recollection of autobiographical events is recruited to make plans, perhaps by piecing together memories of past events so as to simulate the future (Schacter et al. 2007). Thus, episodic memory and planning rely on the same neural structures.

It is therefore not surprising that a species, such as the scrub jay, capable of episodic memory is also capable of future planning as reflected in storing food in anticipation of a future hunger state (Raby and Clayton 2009). This capacity is not only present in this species but also in apes (Mulcahy and Call 2006; Osvath and Osvath 2008). Further descending the outdated *scala naturae* cognitive ladder, even rodents can show some of these capacities (Babb and Crystal 2006). A recent neurophysiological study found activity in a specific cell assembly of the rat hippocampus during memory retrieval. The same assembly also predicts future choices suggesting that rats, like humans, employ a shared neural substrate for memory and action planning (Pastalkova et al. 2008). Instead of viewing episodic memory and future orientation as advanced, language-mediated processes limited to humans (Tulving 2005), therefore, they may be part of general memory and action organization found to varying degrees in a wide range of species.

Imitation

Even though the population-specific traditions of wild primates strongly suggest social learning (e.g., Whiten et al. 1999; van Schaik et al. 2003), the topic of primate imitation has become controversial. The classical definition of imitation as learning an act by seeing it done was replaced by a top-down definition that required a subject to understand the intentional structure of an object's actions, such as the object's goal and specific ways to achieve it. The result has been a plethora of new labels such as "imitative learning," "goal emulation," and "true imitation" (Whiten and Ham 1992). Following the old definition, apes surely imitate, whereas most new definitions tend to exclude them.

However, the majority of studies that failed to find ape imitation employed human behavioral models (e.g., Tomasello, Savage-Rumbaugh, and

Kruger 1993). This is important in light of the alternative view of imitation, which places less emphasis on intentionality and more on body mapping between subject and object, thus erasing the distinction between perception and action (Prinz 2002). Body mapping is facilitated by "identification" with another, which occurs most easily between members of the same species. This has obvious implications for the choice of social learning models (de Waal 2001). Once the extra effort was made to train conspecific models, the issue of ape imitation was quickly settled to the point that major skeptics have come around (e.g., Galef 2005). Exposed to models of their own species, chimpanzees faithfully and reliably copy tool use, foraging techniques, and arbitrary action sequences (e.g., Horner et al. 2006; Whiten et al. 2005; Bonnie et al. 2007).

It remains unclear, however, if this imitation is based on an understanding of the model's intentions. Even for human adults, this may not be the central issue, and simpler processes are likely (Horowitz 2003). Imitation may stem from internal or external mimicry of observed motor movements through shared neural representations (Rizzolatti et al. 2001; Decety and Chaminade 2003). Externally visible mimicry is suggested by reports of "co-action," in which observers place their hand on the model's hand, thus gaining kinesthetic feedback of the other's actions (Horner et al. 2006), or of observers performing hitting movements with an empty hand in precise synchrony with a nut-cracking companion "consistent with a model of imitation in which the imitator codes its observation of the model immediately into a motoric representation" (Marshall-Pescini and Whiten 2008b). Thus, rather than revolving around perceived intentions and goals, imitation more likely revolves around action coding. This would also explain why chimpanzees readily learn solutions to problems from each other but not from repeated demonstrations of these solutions in the absence of a real-life companion (Hopper et al. 2007).

Neuroimaging and neurophysiological studies in humans show that the cortical areas active during the observation of another's action are homologous to those containing mirror neurons in macaques (Iacoboni 2009; Ferrari, Bonini, and Fogassi 2009). These neurons are usually tested in relation to existing behavior, but could they also be involved in complex imitation, such as learning a new motor sequence? Human neuroimaging shows activation of the same areas of the mirror system during simple motor imitation, as during imitation of a novel action sequence (Iacoboni 2009). This suggests that copying of novel behavior recruits neural resources related to the existing behavioral repertoire.

Since mirror neurons have also been found in birds, the evolution of these neurons probably traces back to the common ancestor of birds and mammals (Prather et al. 2008). We should not exclude the possibility, therefore, that all forms of imitation have a shared neural perception-action foundation, from the vocal mimicry of birds to the copying of feeding techniques by primates.

Prosocial Behavior and Empathy

Approaches to altruism are often presented as a quest for "true" altruism, that is, altruism without any obvious benefits for the actor. From this perspective, parental care or aid to kin hardly counts as altruistic (explained by kin selection), and any chance at reciprocation by the beneficiary also disqualifies altruism as genuine. This is a curious approach, however, because motivationally speaking these distinctions are irrelevant unless we assume that actors know about inclusive fitness or are capable of anticipating future return benefits and perform their behavior with these benefits in mind. The evidence that they do so is nonexistent, however. Since one cannot want what one does not know about, evolutionary explanations of altruism are irrelevant at the proximate level. Even animals capable of learning the advantages of reciprocity can only do so if they spontaneously help others in the first place. So, even in their case we still need to explain the helping impulse.

According to the most parsimonious proximate explanation, altruism is a product of socioemotional connections between individuals. Empathy provides such an explanation both for humans (Batson 1991) and other animals (de Waal 2009a). In one chimpanzee study, the role of return benefits was studied by manipulating the availability of rewards. The apes spontaneously assisted humans regardless of whether or not this yielded rewards, and were also willing to assist conspecifics so that these could reach food unavailable to the helper itself. Consistent with predictions from empathy-induced altruism, the decision to help did not seem based on cost/benefit calculations (Warneken et al. 2007).

Spontaneous helping has also been experimentally demonstrated in both marmosets (Burkart et al. 2007) and capuchin monkeys (de Waal, Leimgruber, and Greenberg 2008; Lakshminarayanan and Santos 2008). Familiarity biases prosocial tendencies in the predicted direction: the stronger the social tie between two monkeys, the more they favor prosocial as opposed to selfish options (de Waal, Leimgruber, and Greenberg 2008). The beauty of the empathy mechanism is that it produces a stake in the other's welfare. The behavior comes with an intrinsic reward for improving the other's situation, known in the human literature as the "warm-glow" effect. Humans report feeling good when they do good and show activation of reward-related brain areas (Harbaugh et al. 2007).

Empathy may be widespread among mammals. It has been proposed to rely on an evolutionarily ancient perception-action mechanism (Preston and de Waal 2002), the first sign of which is state matching, also known as emotional contagion, which has been demonstrated in mice (Langford et al. 2006). The same mechanism is being studied through yawn contagion in humans and other animals, which is strongest between bonded individuals, as expected, if empathy is involved (M. Campbell et al. 2009; Palagi et al. 2009).

Human neuroimaging shows that during observation of facial emotions, mirroring activation is present not only in premotor areas but also in insular and cingulated cortices (Wicker et al. 2003; Singer et al. 2004). These areas belong to neural circuits known for their involvement in visceromotor sensations related to unpleasant and painful stimuli. When we observe a facial expression, motorically similar facial expressions are unconsciously triggered, which are associated with corresponding emotions (Dimberg et al. 2000). Thus, instead of empathy starting with a cognitive evaluation of another's situation, it starts, like imitation, with automatic and widespread bodily connections (Preston and de Waal 2002).

Concluding Remarks

If there is one general trend in the field of comparative cognition, it is the expansion of the range of organisms showing signs of any given capacity. In all areas discussed (i.e., episodic memory, planning, imitation, and prosocial behavior), initial claims that humans, or at least hominoids, are special have had to be revised when related capacities were discovered in other, sometimes taxonomically quite distant species. De novo appearances of cognitive capacities are apparently as unlikely as de novo anatomical features.

Rather than focusing on the pinnacles of cognition, the field of comparative cognition seems to be moving toward a bottom-up perspective focused on the nuts and bolts of cognition, including underlying neural mechanisms. Most mechanisms are evolutionarily ancient, tying together phenomena such as the imitation of song learning in birds and tool use in primates or the prosocial tendencies of both humans and other mammals. This does not mean that distinctions between taxonomic groups are irrelevant, or that there is no point to finer-grained classifications. But instead of dividing imitation up into one "true" form and other forms— which apparently do not deserve the name—why not return to the classical definition and include all forms of imitation under one umbrella? The various forms can then be distinguished according to the function they serve in the lives of animals, the stimuli that determine their occurrence (i.e., body actions, perceived goals, environmental changes), and the underlying cognitive processes. The same applies to empathy, which ranges all the way from automatic emotional activation in response to the behavior of others to the perspective taking that allows some large-brained species to gear their helping behavior specifically to the situation and needs of the other (Figure 1.1). All of this falls under empathy, but the cognitive level varies by species, as does the specific context that triggers it.

Figure 1.1. Chimpanzees frequently react to another's need or distress with directed altruism. Here a mother, having heard her son's screams, stretches out a hand to help him out of a tree. Recent experiments confirm that nonhuman primates care about the welfare of others, including non-kin. Photo by Frans de Waal.

In every given domain, functional refinements have evolved as adaptations to a species' ecology, the study of which is critical for an evolutionary cognitive science. The most logical route for comparative cognition, however, is to try to understand the basic processes and common denominators first before exploring species-typical specializations.

From Understanding of the Actions of Others to Culture

The Mirror Neuron System in Monkeys and Its Implications for Social Cognitive Functions

Pier Francesco Ferrari and Leonardo Fogassi

Introduction

One of the most extraordinary features that distinguishes primate species from other mammals is their social complexity. Other taxa have indeed evolved complex societies, but in monkeys and apes we can observe distinctive evolutionary solutions that have shaped their bodies and brains in relation to the social environment. Gestural communication, imitation, empathy, perspective taking, understanding others' intentions, and making coalitions and alliances are part of a range of cognitive abilities that have probably evolved within a social domain and that in recent years have been the target of several investigations and hot debates. Our understanding of the basic neural mechanisms underpinning these cognitive functions may help to shed light on the building blocks from which advanced faculties, presumed to belong uniquely to humans or to apes, might have evolved in the primate brain.

Living in a multifaceted and dynamic social network requires the capacity to understand social interactions. This capacity is crucial for responding adequately to social stimuli, for learning from others, and for the coordination and planning of behaviors with other conspecifics. All these issues raise important questions about the possible neural mechanisms required to allow individuals to understand the actions, intentions, and emotions of conspecifics.

The problem of understanding others' behavior has usually been approached from a traditional perspective based on perception, according to which we understand others' actions through a high-level elaboration of incoming sensory information, which can be subsequently exploited in order to plan and organize behavior. According to this view, perceptual and motor functions are distinct phenomena. However, several neuroanatomical, neurophysiological, and psychological studies challenged this view, showing that action and perception are strictly interconnected at the level of several cortical circuits and that motor cortex possesses cognitive properties. The properties of the mirror neurons are one the best examples of the involvement of motor cortex in higher cognitive processes. The demonstration of the presence of a mirror neuron system in both monkeys and humans has had a large impact on behavioral and cognitive sciences, providing further support to the notion of an evolutionary continuum between monkeys, apes, and humans, not only in morphological traits and genetics, but also in the cognitive domain. Furthermore, this discovery paved the way for a series of investigations aimed at understanding the implications of a dysfunctional mirror neuron system for human psychopathologies.

In this chapter, we will first describe the basic properties of mirror neurons in the ventral premotor cortex and the inferior parietal lobule of the macaque monkey and then the mirror neuron system in humans. In the second part, we will present hypotheses concerning the functions of mirror neurons in monkeys and humans and outline possible evolutionary divergences that differentiated the mirror system in the two species.

Mirror Neurons in the Monkey: Basic Findings

The first description of mirror neurons (MNs) in the convexity of the rostral part of the ventral premotor cortex (area F5) of the macaque monkey was given almost 20 years ago (Di Pellegrino et al. 1992; Gallese et al. 1996; Rizzolatti, Fadiga, Gallese, and Fogassi 1996). These neurons possess both visual and motor properties. In fact, they activate both when a monkey performs a hand goal-directed motor act (e.g., grasping, tearing, or manipulating an object) and when the monkey observes the same, or similar, act performed by the experimenter or by a conspecific (Figure 2.1) (Gallese et al. 1996; Ferrari et al. 2003). The visual response is largely independent of monkey gaze. Pure object presentation does not elicit any activation. Observation of biological movements mimicking the motor act but devoid of a goal (i.e., lacking the target object) is not effective in activating these neurons, suggesting that their discharge is not related to simple body parts displacements, but to the coding of the goal of observed or executed motor acts. This property is not limited to hand actions, since in the lateral part of the ventral premotor cortex MNs have

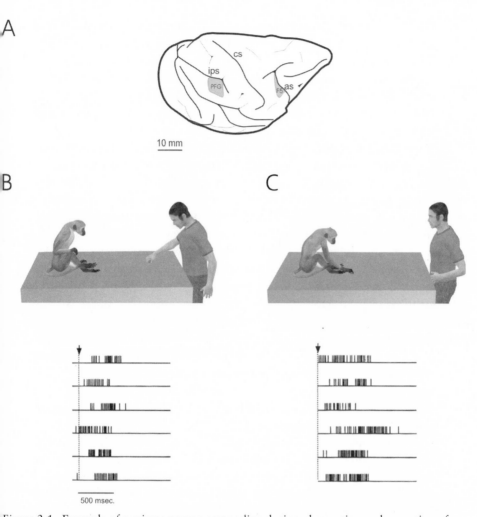

Figure 2.1. Example of a mirror neuron responding during observation and execution of a grasping action. A. Lateral view of the monkey brain. Areas F5 and PFG correspond to the brain regions where mirror neurons have been found (as: arcuate sulcus; cs: central sulcus; ips: intraparietal sulcus). In B and C, on the top is depicted the experimental condition, on the bottom the neuron discharge. B. The experimenter grasps a piece of food. C. The monkey grasps the food in front of the observing experimenter. Six trials are shown for each condition. Each little bar indicates a single action potential. Arrows indicate the grasping onset. Modified from Fogassi and Ferrari (2010).

been described that respond during the observation and execution of goal-directed mouth actions (Ferrari et al. 2003).

What is the function of MNs? Since MNs constitute a system matching observed and executed actions, it has been hypothesized (Gallese et al. 1996; Rizzolatti, Fadiga, Gallese, and Fogassi 1996) that they play a crucial role in understanding others' actions. The basic mechanism through which action understanding occurs is relatively simple: the visual input of an action activates the corresponding motor representation in the motor cortex. Since this motor representation is already available in the motor vocabulary of the observer, its automatic retrieval would give him/her access to the meaning of the observed motor act. MNs in fact show congruence between the effective observed and the effective executed motor act. This congruence may be very strict or broader. In one-third of MNs, defined as "strictly congruent" MNs, the effective executed and observed motor acts correspond both in terms of the goal (e.g., grasping) and the means to achieve the goal (e.g., whole hand grip). In the great majority of MNs ("broadly congruent"), the congruence is more in terms of the goal of the observed and executed motor act, rather than the details of the movements necessary to achieve it (e.g., precision grip on the motor side, whole hand grasping on the visual side). Based on these matching properties, it is plausible to conclude that monkey MNs code motor goals. The hypothesis that MNs have a prominent role in the understanding of motor goals has been confirmed by several neurophysiological investigations.

In a first study (Umiltà et al. 2001), the monkey could see a complete grasping motor act (full vision condition) or only the first phase of the act, because during actual grasping (the closure phase) the hand disappeared behind an opaque screen (hidden condition). The results showed that MNs discharge both when the monkey fully observes a grasping action and when the final part of the action is hidden by the screen, thus suggesting that contextual information allowed MNs to retrieve the motor representation corresponding to the observed motor act, despite the absence of a full visual description of the motor event.

In another study, a subclass of MNs called "audiovisual MNs" was found to discharge not only during execution and observation of a noisy act (i.e., breaking a peanut), but also when the monkey was simply listening to the sound produced by that act (Kohler et al. 2002). This finding clearly indicates that the meaning of an action can be accessed in the motor cortex through different acoustic sensory modalities.

Besides area F5, MNs have been found also within the inferior parietal lobule (IPL), in areas PFG and partly PF (Gallese et al. 2002; Yokochi et al. 2003; Fogassi, Ferrari, et al. 2005; Rozzi et al. 2008; Yamazaki et al. 2010). The properties of parietal MNs are similar to those of F5 MNs, although the percentage of those responding to the observation of mouth actions is lower than in F5. These neurons, like those described in F5, generalize across many aspects of visual stimulus presentation, such as distance, motor act direction, hand preference, type of object to which the motor act is directed, and vantage point. Thus, their discharge does not seem to code a purely visual description of a motor act, but rather its goal.

The presence of mirror neurons in the IPL is not surprising, because of the strong connections of PF and PFG with area F5 (Rozzi et al. 2006). Furthermore, it is very likely that the main source of visual information on biological motion comes, directly or indirectly, from the anterior region of the superior temporal sulcus (STS), which is anatomically connected with IPL. Interestingly, STS has a small percentage of neurons responding to the observation of goal-directed movements performed with the forelimb, but without a motor-related discharge (Perrett et al. 1989; Jellema et al. 2002). Since area F5 does not receive any direct connection from the anterior part of STS, IPL could be the first node of an IPL–F5 circuit processing biologically meaningful movements (Rizzolatti and Luppino 2001).

The data on MNs so far provided are consistent with the idea that their activity enables the monkey to identify *what* an individual is doing and, to a certain extent, *how* he is doing it. More recent neurophysiological findings show that MNs could also enable the monkey to understand others' motor intention, that is, predicting *why* an individual is doing something. The rationale of these experiments has been driven by neurophysiological (Rizzolatti et al. 1988) and psychophysical (Jeannerod 1994; Rosenbaum et al. 2007) experiments showing a distinction between motor act and action. By motor act we mean a movement that has a goal (e.g., grasping a piece of food). By motor act we mean a series of motor acts that, as its final outcome, leads to the achievement of an ultimate goal (e.g., eating a piece of food after reaching it, grasping it, and bringing it to the mouth). According to this distinction, in order to attain the *intended* goal of an action, motor acts must be organized into structured sequences, in which these acts are fluently linked one to the other. Therefore, when an individual *intends* to reach an action goal, this is achieved through the execution of a sequence of motor acts. Taking into account this motor organization of actions, we carried out a series of experiments in which neuronal activity was recorded from grasping MNs of IPL and ventral premotor cortex (PMv) while the monkey executed a motor task and observed the same task performed by an experimenter

(Figure 2.2—Fogassi, Ferrari, et al. 2005; Bonini et al. 2009). The task consisted of two basic conditions: grasping a piece of food, or an object, either to eat it (in the case of food) or to place it (in the case of the object). Thus, the first part of both conditions (grasping the target) was identical, but the motor intentions (action goals) in the two conditions differed. The results showed that the great majority of both parietal and premotor MNs discharged differently during both execution (Figure 2.2A) and observation (Figure 2.2B) of the grasping act, according to the goal of the action in which the act was embedded. This finding indicates that the action goal could be coded well before the beginning of the subsequent motor act that particularly specifies that action (either bringing to the mouth or moving toward another place). In other words, the motor discharge of these neurons seems to reflect the motor intention of the acting individual. In the motor task this may not be surprising because the monkey knows in advance the goal of its own action. However, the differential discharge found in the visual task is more astonishing. Since in the case of the visual task the neuronal preference for the action goal was present already during the initial phase of the observed grasping act, it has been suggested that the differential discharge of MNs would predict the final goal of the observed agent. It has been proposed that these MNs are probably part of a mechanism for understanding others' intentions. Further control experiments demonstrated that contextual information and previous experience of the observer with other individuals acting in that context might provide sufficient clues to activate the correct motor representation of the impending action.

A recent study prompted the idea that beyond coding others' motor acts and intentions, MNs' function could be also related to the potential interaction that might occur between the observer and the observed agent (Caggiano et al. 2009). In fact, this work showed that part of ventral premotor MNs discharged stronger when an experimenter grasped a piece of food within the monkey's peripersonal operant space than when he grasped the same piece of food in the extrapersonal, far space. Other neurons behaved in the opposite way, coding others' actions only when performed in the extrapersonal space. This finding suggests that MNs could code others' actions within the observer's centered spatial framework. This type of coding is likely related to the possibility to interact socially with others.

Mirror Neurons and Learning

A very important issue that in recent years has attracted the attention of neuroscientists and neuropsychologists is whether MN activity can

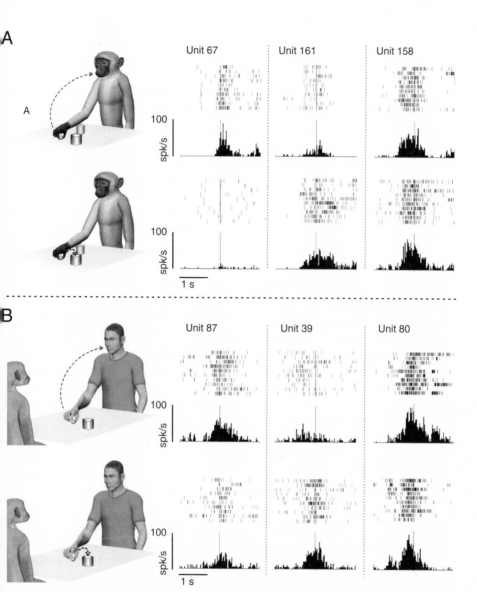

Figure 2.2. Mirror neurons coding motor intention. A. Left: Apparatus and experimental paradigm used for the motor task. Right: Activity of three IPL neurons during grasping in the two basic conditions (grasp-to-eat and grasp-to-place). Rasters and histograms are synchronized with the moment when the monkey touched the object to be grasped. Arrows indicate when the monkey releases the hand from the starting position. Abscissa: time, bin = 20 ms; ordinate: discharge frequency. B. Left: Apparatus and paradigm used for the visual task. Right: Visual responses of IPL mirror neurons during the observation of grasping-to-eat and grasping-to-place done by an experimenter. Modified from Fogassi et al. (2005).

be modified by learning and experience. Although this issue will be probably best addressed by chronic recording experiments, there are already some monkey data showing that mirror neurons can achieve new properties through motor learning or simply through long visual exposure to novel actions (Rochat et al. 2010; Ferrari, Rozzi, and Fogassi 2005). In one study (Rochat et al. 2010), it has been reported that when a novel motor act is incorporated in the motor repertoire, the new motor representation is integrated in the MN system, determining the visual discharge of MNs when this act is observed. In another study, Ferrari, Rozzi, and Fogassi (2005) found a subcategory of MNs (tool-responding mirror neurons) that discharged when the monkey observed motor acts performed by an experimenter with a tool (a stick or a pair of pliers). Unlike the typical response of MNs, the discharge of these neurons was stronger than that obtained when the monkey observed a motor act with similar goal made with the hand. In contrast with Rochat et al., in this latter study monkeys were not trained to use the observed tools, therefore it is possible that the relatively long visual exposure to tool actions (which occurred during long-lasting experiments) created a visual association between the hand and the tool, so that the tool became a sort of extension of the hand. The role of this category of MNs, unlike those described above, could be that of extending the individual's capacity to understand motor acts to incorporate acts that do not strictly correspond with the existing internal motor repertoire but that share similar motor goals.

Mirror Mechanism in Other Cortical Areas and in Other Species

In more recent years, other studies have described sensorimotor neurons with properties very similar to those of MNs.

In one of these investigations (Cisek and Kalaska 2005), it was shown that neurons in the dorsal premotor cortex (PMd) discharged when the monkey was instructed to use a cursor to reach a target on a screen and when it observed the cursor reaching the same target. The main difference between these neurons and MNs is that in order to be activated they do not need visual information about the effector used to reach the target. Thus, they seem to code the goal of the observed cursor movement.

In a similar neurophysiological study, neurons have been described in monkey F1 and PMd that were activated when the monkey performed a movement to move a cursor to reach a target on a screen (Tkach et al. 2007) and when the monkey observed the same replayed movements only when both the cursor and the target were visible. The presence of the replayed cursor movement without the target did not elicit the same effective neuronal response.

Although these studies tested conditioned movements directed to a non-natural target, they clearly suggest that other motor cortical areas can visually code the attainment of a goal even if this is achieved by means of nonbiological effectors (i.e., a cursor reaching a target).

Responses similar to those of MNs have been recently described in the lateral intraparietal area (LIP). This area, located inside the intraparietal sulcus (IPS), is part of a circuit involving the frontal eye field and plays a crucial role in organizing intended eye movements. Most of LIP neurons discharge when the monkey looks in a specific direction. A subset of them has been found to also discharge when the monkey observes another monkey looking in the neuron motor preferred direction (Shepherd et al. 2009). This finding suggests that the motor system involved in the control of eye movements toward targets is endowed with a mechanism similar to that represented by MNs. It is possible that joint attention behaviors and gaze-following responses (here generally referred to as the shift of attention in response to social gaze cues; see Ferrari et al. 2000) might rely on such an MN-like mechanism.

In another study it has been reported that in area VIP, a region buried inside the intraparietal sulcus, there are neurons firing in response to tactile stimuli delivered to the face/body and to visual stimuli delivered in the same peripersonal space area (Ishida et al. 2009). More interesting, some of these neurons responded also when the same visual stimuli were introduced into the peripersonal space of another individual. Since VIP forms an important functional circuit with ventral premotor area F4, which contains neurons active during reaching actions, it is possible that these neurons are involved in encoding not only peripersonal space for pragmatic purposes (i.e., moving the arm in space) but also the body space of others, which becomes important when two or more individuals are interacting.

The notion of a common code for action and perception had been also hypothesized for other species, such as birds (H. Williams and Nottebohm 1985). Neuroanatomical and neurophysiological studies indicate that the same brain structures are involved in song production and perception (Nottebohm et al. 1990). It is plausible that a mechanism similar to that of monkey MNs can have evolved also in other species in order to support interindividual interactions. Very striking has been the recent discovery of MNs in the forebrain areas of songbirds (Prather et al. 2008; Bauer et al. 2008; Prather et al. 2009; G. Keller and Hahnloser 2009). These studies show that the hyperstriatum ventrale pars caudalis (HVC) and the secondary auditory telencephalic region caudal mesopallium (CM) contain neurons that are active during song playback and production. Differences between swamp sparrow populations in neuronal perceptual categorization of songs suggest that these neurons are very likely involved in imitative song learning and in song transmission within natural populations of songbirds.

Mirror Neurons and Their Possible Function in Monkey Social Cognition

The discovery of MNs has provided evidence of a neuronal mechanism in the monkey brain that allows an embodied recognition and automatic understanding of others' motor acts and actions. The question, however, is whether there is any behavioral evidence that monkeys are actually able to recognize others' actions. Direct evidence of MNs' involvement in action understanding would involve inactivating this system through invasive experiments with lesions or pharmacological inactivation. However, these methods are problematic because MNs are present in at least two cortical regions in both hemispheres. Moreover, a lesion of these areas would likely produce other, concomitant cognitive and motor impairments.

An alternative, although not conclusive, strategy is that of assessing the capacity of the monkey to match others' behavior with its own. To this purpose we recently used an experimental approach, classically employed in child development studies (Nadel 2002). We evaluated macaques' responses when facing two human experimenters, one imitating the monkeys' object-directed actions (imitator) and the other performing temporally contingent but structurally different object-directed actions (nonimitator; Paukner et al. 2005). Results clearly show that the macaques gazed more frequently at the imitator than at the non-imitator, thus demonstrating that they recognized when they were being imitated. Similar results have been recently obtained in capuchin monkeys (Paukner et al. 2009). Although indirectly, these data show that monkeys are able to detect contingencies in the social environment structurally matching their own motor behavior and not simply based on temporal synchronies. Using different paradigms, chimpanzees and other apes have also been shown to possess a similar imitation recognition capacity (Nielsen et al. 2005; Haun and Call 2008). Further recent experiments (Rochat et al. 2008) employing a preferential looking paradigm have shown that macaque monkeys can recognize efficiently performed actions, but only when they are directed to achieve goals that have become familiar through previous experience.

In addition, it has been recently reported (see Lyons et al. 2006) that capuchins are able to infer the location of hidden food by observing a human experimenter performing two different actions: 1) looking and attempting to reach a container in a purposeful manner or 2) handling the same container but displaying no goal-directed movement. Capuchins that were then required to choose one of the two containers chose the one handled by the experimenter in a purposeful way, showing not only the ability to recognize others' actions, but also the capacity to discriminate between intentional and accidental ones.

Taken together, these findings show that the ability to match one's own behavior with that of others is a feature shared by several nonhuman primate species, very likely relying on a common neural matching mechanism.

Imitation

By activating the corresponding motor representation during the observation of an action or gesture, MNs would permit translation of the observed action into the correspondent motor plan (Jeannerod, 1994; Rizzolatti et al. 2001; Ferrari, Bonini, and Fogassi 2009), thus providing the potential solution for the "correspondence problem" between one's own and others' movement, which is deemed to be at the basis of imitative behavior (Heyes 2001; Brass and Heyes 2005). Although some forms of imitation have been shown in monkeys and apes, the demonstration of a link between the mirror system (MS) and imitation derives from human studies.

A series of works by Iacoboni and co-workers (see for review Iacoboni 2009 and Chapter 3 in this book) demonstrated that imitation of a simple movement, finger lifting, activated the frontoparietal MS (Iacoboni et al. 1999). Nishitani and Hari (2000), using the event-related neuromagnetic (MEG) technique, showed that in individuals instructed to repeat familiar actions done by another individual, there is an activation of the inferior frontal gyrus (Broca's area 44–IFG), followed in time by the activation of primary motor cortex.

Much wider is the circuit activated during imitation learning (Buccino et al. 2004; see also Chapter 3 in this book). In fact, beyond the activation of the MS, tasks requiring imitation of new actions (playing guitar chords) revealed activation of additional areas, including the middle prefrontal cortex (area 46). This latter area was strongly active during a period in which the subject, after observing the action performed by a model, was internally planning his reproduction of the observed action. Interestingly, this prefrontal activation disappeared during actual imitation. It is very likely that this cortical sector is implicated in the recombination of the motor representations corresponding to the different motor acts, recognized by the MS, in order to reconstruct the model's action. This process would occur after observation of the model's action, in order to prepare actual imitation, and would disappear during execution.

Taken together, these findings suggest that different forms of imitation, whether or not involving learning, share a core mirror circuit (Ferrari, Bonini, and Fogassi 2009) that, in the case of complex actions, recruits additional areas that are probably necessary to maintain in memory the observed motor act sequences, recombine them into the appropriate sequence, and determine the timing in which imitation has to be actually performed.

Recently we proposed that MNs can exert an influence on motor output through two distinct anatomo-functional pathways (Figure 2.3A). One pathway (*direct* pathway) exerts a direct influence on the motor output *during* action observation mediated by parietal-premotor MNs (Figure 2.3A, 2.3B, and 2.3C). The existence of a motor output due to MN activity during action observation has been recently demonstrated in an elegant electrophysiological study in the adult monkey (Kraskov et al. 2009). MNs projecting to the spinal cord were recorded during action execution and observation (Figure 2.3B). Half of these neurons showed activation, the other half suppression of the discharge during action observation (for a possible circuit see Figure 2.3B). Interestingly, the population of MNs with inhibitory properties probably contributes to inhibition of self-movement during action observation. This inhibitory action on the motor output is very important to prevent automatic imitation of actions that might occur during action observation and which could represent an impediment to normal social interactions. In specific neurological syndromes, such as echopraxia, such inhibition is compromised due to prefrontal lesions, and patients tend to spontaneously imitate body gestures of another individual even in the cases in which they are explicitly requested to inhibit their behavior (Lhermitte et al. 1986).

At behavioral level there is ample evidence that action observation often leads to the reenactment of the observed action (see Ferrari, Bonini, and Fogassi 2009 for a review). Probably the most convincing phenomenon that seems to imply a direct influence of MNs on behavior is neonatal imitation (Ferrari, Visalberghi, et al. 2006; Lepage and Théoret 2007). First demonstrated in humans (Meltzoff and Moore 1977, 1983), it has been subsequently shown in apes (Myowa 1996; Myowa-Yamakoshi et al. 2004) and, more extensively, in rhesus macaques (Ferrari, Visalberghi, et al. 2006; Ferrari, Paukner, Ionica, and Suomi 2009).

It has been hypothesized that this phenomenon may involve a mirror mechanism present at birth and capable of matching some facial features with a hardwired internal motor representation of these features (Ferrari,

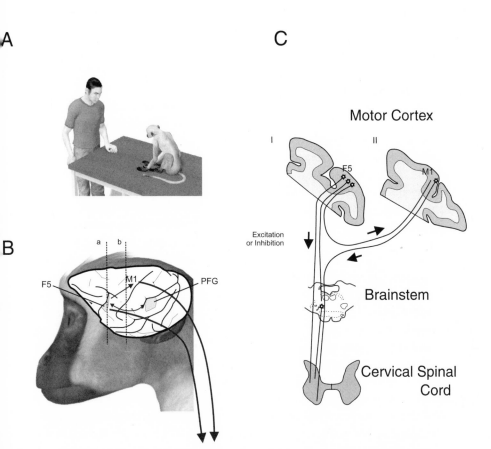

Figure 2.3. Schematic view of the "direct" mirror pathway described in the text. A. The figure illustrates the typical experimental setting in which MNs are recorded from the monkey brain. B. Arrows indicate anatomo-functional connections and information flow during the observation of a hand action. The visual information related to actions is activating the mirror areas. These areas have their main access to the descending motor pathways involving mainly the primary motor cortex (M1) and F5. The resulting motor output from these areas could be excited or inhibited according to whether the matched behaviors have to be facilitated or suppressed. Dotted lines refer to the location of coronal sections shown in Figure 2.3C. C. Coronal sections taken at the level of F5 and M1 showing the anatomical connections between F5 and M1 and the descending pathways from the hand sectors of the motor cortex (F5 and M1) to the brainstem and cervical spinal cord.

Visalberghi, et al. 2006). Recent electroencephalography (EEG) studies in human infants support the idea that an MN system is present in the first months after birth. However, the issue of whether this is present at birth has been only recently addressed in infant macaques (Ferrari et al. 2008). One-week-old macaques showed significant suppression of the alpha rhythm (falling probably around 5–7 Hz frequency band in infant macaques) when compared to control stimuli (Ferrari et al. 2008). Since this inhibition probably reflects the activation of areas recorded in the central-parietal motor regions, this finding, although preliminary, would indicate that the MN system is selectively sensitive to biological meaningful stimuli already at a very early age.

More recently we demonstrated in macaques that infants respond to their mothers' lip smacking by lip smacking back at them (Ferrari, Paukner, Ionica, and Suomi 2009). These behaviors most likely correspond to the neonatal imitative responses that we reported under more-controlled laboratory conditions and suggest that the mirror mechanisms underlying this behavior could be activated at birth for emotional communication (Figure 2.4).

In several imitative phenomena there is a time lag between action observation and execution. First of all, it must be considered that there are several distinct forms of imitative response that are delayed in time, and that they can be distinguished based on both the complexity of the observed behavior and on its presence in the observer's motor repertoire. Contagion (Thorpe 1963), response facilitation (Byrne and Russon 1998), emulation (Nagell et al. 1993), and true imitation (Thorpe 1963; Tomasello and Call 1997) constitute the best-studied categories. Furthermore, we propose that besides requiring a basic form of recognition and understanding of the observed behavior such as those provided by MNs, all these imitative phenomena rely on additional neural systems working in coordination with the "core" MN brain regions.

A second pathway (*indirect* pathway) linking parietal and premotor areas with ventro-lateral prefrontal cortex (see Tanji and Hoshi 2008) could exploit the sensory-motor representations provided by the mirror regions of the "direct pathway" for more-complex cognitive and behavioral functions, such as those required for delayed imitative behaviors (Figure 2.3B). Based mainly on fMRI experiments on imitation learning in humans, we have recently proposed that delayed imitative phenomena can rely on the activity of prefrontal cortical regions. MNs provide information about action goals and, in humans, movements (actions nonrelated to a target) and kinematics. During imitation, often the behavioral responses of the observer are delayed in time with respect to the observed

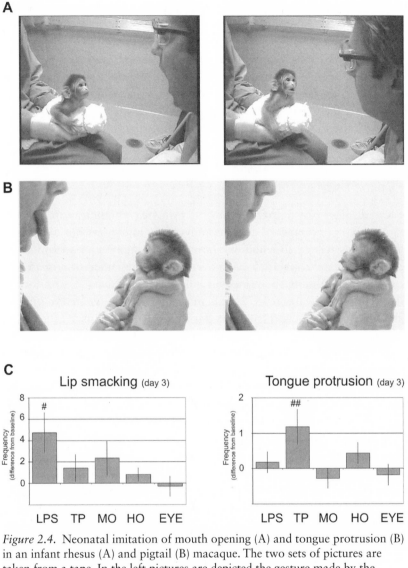

Figure 2.4. Neonatal imitation of mouth opening (A) and tongue protrusion (B) in an infant rhesus (A) and pigtail (B) macaque. The two sets of pictures are taken from a tape. In the left pictures are depicted the gesture made by the model. The pictures on the right have been taken about within three seconds after those on the left. The gesture made by the model was repeated seven or eight times in a period of 20 seconds. C. Averaged scores (difference of frequency of behaviors between stimulus period and baseline) +/– SEM of lip-smacking (left) or tongue-protrusion (right) gesture in different experimental conditions. Data are reported for infants' responses tested on postpartum day 3. Experimental conditions. LPS: lip smacking; TP: tongue protrusion; MO: mouth opening; HO: hand opening; Eye: repeated eye opening and closing. Modified from Ferrari, Visalberghi, et al. (2006).

behaviors; thus visual information about the actions to be copied is no longer available. Since mirror neurons firing is dependent on the availability of the sensory information, how then can mirror neurons mediate delayed imitative behavioral responses? In these instances prefrontal regions have been shown to come into play by (a) parsing the behavior and reconstructing it in a novel sequence, (b) maintaining motor programs active after the visual information is not available anymore, and (c) removing the inhibition from specific neuronal motor representations and thus allowing the motor program to run.

Emotion Understanding and Empathy

Facial displays play a prominent role in primate communication, not only in conveying messages about the individual's intentions but also about inner states related to emotion. Darwin was a pioneer in attempting to characterize and compare facial movements among different primate species and in outlining the principles linking specific facial expressions to their emotions (Darwin 1872; Ekman 1992). Darwin wanted to know if animals had facial expressions similar to those of humans. At the neurophysiological level we now have more advanced knowledge on how emotions are linked to specific functional brain circuits located in the limbic system and in other cortical and subcortical structures. These structures are responsible for the affective states, body modifications (e.g., hair erection, skin color changes, etc.) and for the facial expressions accompanying emotions and their visceromotor reactions.

As for action understanding, emotions can be understood through a sensory elaboration of body postures and of facial changes during displays, followed by an inferential process (a type of expression means fear, another type means happiness, etc.) or through a direct mapping of emotion-related external information onto the motor structures producing the same emotion. Only in the second case, however, is there an automatic and first-person recognition of others' emotions. The motor structures involved in the matching process would be, in this case, areas that control both somatomotor and visceromotor reactions. Visceromotor structures would be those that allow the observer to experience part of the other's affective state.

Several studies in the monkey temporal cortex and amygdala have shown neurons that code visual information related to faces and emotions (Gothard et al. 2007). The activation of some of these structures has been shown to produce clear autonomic responses. Although the presence has been reported, in monkey area F5, of a small percentage of MNs responding to affiliative communicative gestures (Ferrari et al. 2003), there is still no evidence of MNs responding selectively to emotional expressions involving limbic or cortical structures. Interestingly, a recent preliminary work (Caruana et al. 2008) reported that, in the monkey, electrical stimulation of the insula can produce facial expressions and visceromotor reactions. From this work it appears that the organization of the monkey insula is similar to that of humans, suggesting that a system matching expression and perception of emotional states could exist, grounded on areas controlling visceromotor responses.

As will be reviewed in Chapters 3 and 10, several studies in humans have demonstrated that the observation of facial expressions produces activation of limbic structures, and that parts of these structures are activated both when the subject feels a given emotion and when he/she observes the same emotion felt by another individual either by observing emotional facial expressions or cognitive cues indicating a painful event (Carr et al. 2003; Wicker et al. 2003; Singer, Seymour, O'Doherty, H. Kaube, et al. 2004; see also Chapter 10 in this book). More-extensive literature in humans has addressed the issue of the neural basis of understanding emotions and empathy (see Chapters 3 and 10), and therefore it will be not considered further here.

Most of the brain imaging and other physiological studies in humans seem to converge on the fact that humans can understand emotions by means of a direct mapping mechanism, through which emotions activate those structures that give a first-person experience of the same emotions. This allows the sharing of affective states between individuals.

The Evolution of Mirror Mechanisms—Concluding Remarks

The MS represents a neural model that, through a matching mechanism, enables the emergence of important cognitive functions such as the understanding of others' behavior. This mechanism is automatic and prereflexive, that is, does not imply any inferential mechanism or verbal processing. Thus, in its basic properties the MS constitutes a relatively simple action-perception mechanism that could have been exploited several times in the course of organisms' evolution, and it should be not necessarily considered as a uniquely primate specialization.

The fact that this mechanism is present not only in humans and monkeys but also in songbirds for song production/perception indicates that action-perception matching mechanisms are probably parsimonious solutions the brain of vertebrates has evolved for the purpose of sensorimotor transformation. The recent reevaluation of avian brain evolution (Jarvis et al. 2005) implies also the concept that some of the primate cortical regions involved in cognitive functions may not be as different from birds' brain centers as previously theorized, although in the latter a cortical layer-like organization is almost totally absent. In fact, even though structurally different, the avian pallial and mammalian cerebral cortex seem to have a shared evolutionary background (Jarvis et al. 2005). Advanced cognitive functions of birds, such as object categorization and vocal learning, often rely on pallial structures in the same way that those of primates are highly dependent on cortical areas. We can speculate that the mirror mechanisms present in primates share with birds most of their basic properties due to a common evolutionary origin. Mimicry phenomena, so well documented in birds and primates, may thus have a shared action-perception mechanism foundation (de Waal and Ferrari 2010).

Even in the case in which analogy could be considered the most parsimonious explanation, we can see that some functions, such as the capacity for vocal mimicry in birds, could have emerged, similarly to the speech system in humans, from preexisting constraints of the motor system originally evolved for the voluntary control of movement (Feenders et al. 2008). Thus, in order to accomplish similar functions, the motor system has coupled sensory and motor information not only to better control song or speech but also to process acoustic information and to learn new motor sequences.

The presence of mirror mechanisms in different cortical areas and brain structures has several implications for our understanding of their functional role. The MS has been found in humans' and nonhuman primates' parieto-frontal cortical areas closely involved in action organization, in face motor areas involved in speech and facial gesture production, and in brain structures implicated in emotional and visceromotor responses such as the insula and the anterior cingulate cortices, while in birds it is located in brain centers controlling the song production.

The widespread distribution of brain areas endowed with mirror properties suggest that the motor system has been exploited in evolution to accomplish social functions. Starting from a purely pragmatic system oriented to interact with the physical world, the motor system evolved and extended its original function, becoming a system crucially involved in the analysis and interpretation of the social world.

In the absence of information about possible mirror mechanisms in other species, we can only provide a hypothetical account about the possible evolution of mirror systems in vertebrates. Evolution often retains core mechanisms even when new functions (e.g., language in humans) are generated (de Waal and Ferrari 2010). For example, the human Broca's area is part of the inferior frontal gyrus and is involved in speech production and perception, and part of it is involved in the human mirror system. Even though monkeys do not have a corresponding area accomplishing the same function, it has been proposed that this cortical region might have evolved starting from a primate ancestor in which the motor areas were already endowed with mirror properties and with the capacity to control and coordinate mouth gestures and larynx movements (Fogassi and Ferrari 2007).

To understand the evolution of mirror neurons it is therefore important to search in different species for the homologous elements of mirror neurons. The methodological tools available to investigate the presence of mirror neurons in other species are often not practicable, and this may limit the possibility to reconstruct a phylogenetic tree of this system. However, the convergent piece of information, collected at different level of description (genetic, physiological, behavioral), might be sufficient to justify claims about phylogenetic affinities between unrelated species that possess brain areas with mirrorlike properties. Recent studies in mice (Jeon et al. 2010), for example, found that observation of conspecific fear involves the anterior cingulate cortex. The preservation of this area is crucial in order to recognize fear and to learn through observation in a fear-conditioning task. This study suggests that, in mice, observing others in pain and experiencing pain involve in part the same structures. Interestingly, similar results have been obtained also in humans by means of fMRI and with a different paradigm (Singer, Seymour, O'Doherty, Kaube, et al. 2004). In humans it has been proposed that an MS is present and activates in deeper brain structures, such as the anterior insula and the anterior cingulate cortex, for affective pain responses. Very likely this MS is retained from brain structures that have been preserved in evolution and that accomplish similar functions among several vertebrate species (not necessarily closely related to each other), such as mice and humans. This is one of the many instances showing how evolution has shaped not only anatomical, genetic, and developmental traits but also complex cognitive capacities and the underlying mechanisms.

Acknowledgments

This work was supported by MIUR (Cofin) 2004057380, the University of Parma, and NIH-P01HD064653–01 grants.

CHAPTER THREE

The Human Mirror Neuron System and Its Role in Imitation and Empathy

Marco Iacoboni

Introduction

Imitation and empathy are two building blocks of human social behavior. Through imitation, we learn from others, we create cultures, and we establish a basic form of bodily rapport that increases liking (Dijksterhuis 2005) and facilitates empathy (Chartrand and Bargh 1999). With empathy, we get attuned to the mental states, feelings, and emotions of other people, a prerequisite of prosocial behavior.

Many authors, over the centuries, have provided vivid descriptions of the relationship between imitation and empathy. Michel de Montaigne (1575), for instance, writes:

> Everyone feels its impact, but some are knocked over by it. On me it makes such an intense impression, my practice is rather to avoid it than to resist it . . . the sight of another's anguish gives me real pain, and my body has often taken over the sensations of some person I am with. A persistent cougher tickles my lungs and my throat.

And here is Nietzsche (1881):

> To understand another person, that is to imitate his feelings in ourselves, we . . . produce the feeling after the effects it exerts and displays on the other person by imitating with our own body the expression of his eyes, his voice, his bearing. Then a similar feeling arises in us in consequence of an ancient association. We have brought our skill in understanding the feelings of others to a high state of perfection and in the presence of another person we are always almost involuntarily practising this skill.

Another example comes from Edgar Allan Poe. In one of his most famous short stories (Poe 1982), the main character says:

> When I wish to find out how wise, or how stupid, or how good, or how wicked is any one, or what are his thoughts at the moment, I fashion the expression of my face, as accurately as possible, in accordance with the expression of his, and then wait to see what thoughts or sentiments arise in my mind or heart, as if to match or correspond with the expression.

Possible neuronal mechanisms for imitation and empathy have been recently discovered in the monkey brain. Chapter 2, by Pier Francesco Ferrari and Leonardo Fogassi, describes these discoveries in detail. In humans, there has been recently a wave of studies on the neural correlates of imitation and empathy, undoubtedly inspired by the neurophysiological discoveries in the monkey brain. This chapter is divided into three main parts. The first one discusses brain imaging and transcranial magnetic stimulation (TMS) studies of imitation in human subjects. The broader picture that emerges from these studies is a circuitry for imitation in which some areas seem to have properties reminiscent of the "mirror" properties observed in single neurons in monkeys (see Chapter 2 in this book). This imitation circuitry overlaps extensively with other neural circuitries in the human brain that are important for higher cognitive functions. This suggests that imitation is a building block of human cognition and sociality.

The second part of the chapter discusses the relationship between imitation and empathy. Many psychological studies have demonstrated correlations between the tendency to imitate automatically others and the tendency to empathize with them. Recent brain-imaging studies have explored the association between imitation and empathy in terms of neural systems. This line of work suggests a view of empathy—or at least some relatively simple forms of empathy—as embodied and based on mechanisms of simulation of the mental states of other people.

Finally, the third part of the chapter describes recent depth electrode recordings in human neurological patients. These recordings demonstrate the existence of "mirror" properties in individual neurons of the human brain. The remarkable aspect of these findings is that mirror neurons were found in many brain regions, suggesting that this system is pervasive and likely supports a flexible integration of many aspects of the actions of the self and of other people.

Neural Systems for Imitation in the Human Brain

The properties of mirror neurons in the human brain—that is, the firing during the execution of a goal-directed action and also during its observation when the action is performed by somebody else—make these neurons well suited for imitation. Thus, it made sense to posit that human brain areas important for imitation may also have "mirror" properties. To test this hypothesis, we performed a functional magnetic resonance imaging (fMRI) activation study (Iacoboni et al. 1999). Subjects were required to either imitate or observe simple finger movements, or to perform motor and visual control tasks. The general idea behind the experimental design was to model the activity of mirror neurons recorded in the monkey brain to test for mirror properties in areas activated during imitation. In monkeys, mirror neurons tend to discharge typically more during action execution than during action observation (Gallese et al. 1996). Thus, we similarly modeled a stronger response during action execution than action observation. During imitation, subjects were both observing and executing the action. Thus, we posited an additive response, approximately the sum of the activity during action execution and action observation. An area with mirror properties had to fit all these criteria, and its activity has also obviously to be higher during these experimental tasks than during the control tasks. We found two regions that reliably fit these criteria. One was located in the pars opercularis of the inferior frontal gyrus, and the other one was located in the anterior sector of the medial wall of the intraparietal sulcus, and the adjacent convexity in the superior parietal lobule. Figure 3.1 shows the normalized response of the inferior frontal area.

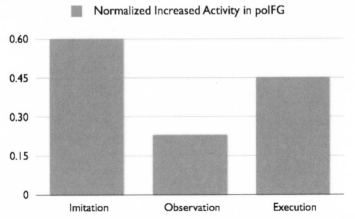

Figure 3.1. Normalized fMRI activity (with respect to baseline activity) in poIFG during imitation, action observation, and action execution. Data from Iacoboni et al. (1999).

Thus, this first study demonstrated that at least two human brain areas active during the experimental tasks had a profile of activity that suggested mirroring properties. Other brain areas in the frontal lobe, including the supplementary motor area (SMA), however, had a similar activity profile, although the level of activation did not reliably differentiate between tasks, and were not reported in the original paper (Iacoboni et al. 1999). We will see in the last section of the chapter that in light of the recent depth electrode recordings, those frontal lobe areas may also be interpreted as areas with mirroring properties.

After the publication of our first study on imitation, some raised the objection that the activity depicted in Figure 3.1 had nothing to do with imitation, but rather with covert speech. Indeed, poIFG (pars opercularis of the inferior frontal gyrus) overlaps with a major language area. This is one of the limitations of brain imaging. It gives you information that is only correlational, not causal. Luckily, the cognitive neurosciences have now many tools that can be used to investigate the relationships between brain activity and behavior. One of these tools is TMS. With TMS it is possible to interfere transiently with brain activity in a given brain region and test whether or not a given behavior is disrupted. By doing so, it is possible to test whether the activity of a given brain region is essential to imitation. The covert-speech hypothesis of the activity in poIFG predicts that interfering with poIFG activity should have no effect on imitative behavior. However, this is not the case. We performed a TMS study in which subjects performed an imitative task and a visuomotor control task while TMS was applied to poIFG and to a control brain region. We observed an effect of TMS only on imitation and only when poIFG was stimulated (Heiser et al. 2003). These results demonstrate that the activity in poIFG is not epiphenomenal but rather causally linked to imitation.

In real life, imitation is a very complex behavior, and some of its complexity is certainly not easy to reduce to tractable experimental paradigms. However, after that first study, we investigated some of the complex aspects of imitation using fMRI. For instance, when imitating hand movements facing the model, we can imitate with the anatomically correct hand (that is, if the model is using the right hand, the imitator will also use the right hand) or as if in front of a mirror (that is, if the model is using the right hand, the imitator will use the left hand). Early in life, there is a prepotent tendency to imitate as if in a mirror (Wapner and Cirillo 1968). This propensity can still be documented later on in adults with chronometric investigations (Wohlschläger and Bekkering 2002).

Using again a finger imitation paradigm, we investigated these two forms of imitation (mirror versus anatomical) with fMRI (L. Koski et al.

2003). The pars opercularis of the inferior frontal gyrus demonstrated a fourfold increase in signal during mirror imitation compared to anatomical imitation. Note that in this experiment, subjects were always using the right hand. The only difference between the mirror imitation compared with the anatomical imitation was the visual stimulus. In one case it was a left hand (mirror imitation), and in the other case it was a right hand (anatomical imitation). Still, such a subtle change in the visual stimulus produced quite a dramatic effect on the activity of the pars opercularis of the inferior frontal gyrus. The question is why.

Before speculating on why this happened, it is useful to make two other considerations. First, the experiment on mirror versus anatomical imitation reinforced the hypothesis that other cortical human areas may contain mirror neurons. The profile of activity of the SMA resembled closely the profile of activity expected for a region containing mirror neurons, as we had also previously observed but not reported due to lack of statistical significance. However, mirror neurons have not been documented yet in the monkey homologue of SMA, area F3. While the brain imaging data were suggestive, in absence of more compelling direct recordings from individual neurons demonstrating mirror properties in SMA, we downplayed (or rather, we said it sotto voce) the potential interpretation of mirroring activity in SMA in the original paper (L. Koski et al. 2003).

The second consideration is related to findings from a separate brain imaging study on imitation that we performed in our lab. The main objective of this experiment was to investigate the neural correlates of imitating the goal of an action. The study was inspired by developmental psychology experiments that suggested that children tend to prioritize different aspects of imitation, such that goals have higher priority compared with means (Bekkering et al. 2000). A simple way to investigate these issues at the behavioral level can be achieved with "Do as I Do" games between the experimenters and the children. Using a variety of settings and control conditions, it has been shown that when means and goals do not match, children tend to imitate the goal of the action.

We performed a brain imaging experiment that adopted very similar imitative tasks. Subjects performed imitation of finger movements. In some conditions the finger movements ended on red dots, suggesting implicitly that the goal of the movement was to reach the dot. The pars opercularis of the inferior frontal gyrus had much higher activity when subjects imitated finger movements that ended on the dots (L. Koski et al. 2002). Note, again, that the movement executed by the subject is the same. Furthermore, the finger movement observed by the subject is also the same.

The only difference between the two conditions is whether or not the finger movement ended on the dot.

Thus, the same frontal mirror neuron area that had much higher activity during mirror imitation had also much higher activity, in a separate experiment, during imitation with visible goals. On one hand, the findings on goal imitation are not surprising. The single cell recordings in monkeys on mirror neurons, reviewed elsewhere in this book, had suggested that mirror neurons may implement a coding of the goal of the observed action. On the other hand, the findings on goal imitation and on mirror imitation, taken together, suggest a speculation about the nature of the social relations facilitated by mirror neurons. Indeed, mirror imitation substantially means that model and imitator act in the same sector of space, they share that space, they get literally closer to each other. Maybe one of the primary goals of imitation is to facilitate an embodied intimacy between the self and other. The mirror neuron system seems the neural vehicle of such embodied intimacy. This basic, bodily based form of rapport may be the basis for an emotional understanding between people, for what we call empathy, which is the topic of the second section of this chapter.

Before moving to the relations between imitation and empathy, however, it is useful to summarize the findings from many brain imaging studies on imitation in a comprehensive model of neural systems implementing imitative behavior.

The first study on imitation of finger movements (Iacoboni et al. 1999) suggested that two cortical areas, one in the frontal lobe (pars opercularis of the inferior frontal gyrus, or poIFG) and one in the parietal lobe (rostral sector of the anterior intraparietal sulcus, or rAIP), had mirror properties. A third area, located in the parietal operculum where one would expect to find area SII, a major somatosensory area, demonstrated also increased activity for imitation compared with action execution. This region, however, had no response at all during action observation and thus cannot be considered a mirror neuron area. Indeed, the fact that this region was active during the motor tasks but not during the visual ones fits the known properties of this area. Its activity in our experiment most likely reflects re-afferent signals associated with the action of moving the fingers. The increased activity in this area during imitation compared with action execution may be a computationally simple way of distinguishing the subject's finger movement from the observed finger movement, a distinction that is most likely more complicated for areas, like the mirror frontal and parietal areas, that respond to both execution and observation of action.

The experiment on mirror versus anatomical imitation (L. Koski et al. 2003) revealed interesting responses from a fourth cortical area. This area is located in the posterior part of the superior temporal sulcus (pSTS), a cortical region well known for its complex visual properties and for responding to biological motion. During the imitation tasks, pSTS had much-higher activity for mirror imitation than anatomical imitation. As the reader may recall, this is the same pattern of activity observed in poIFG. Two possible explanations for this finding are as follows: It is possible that poIFG sends a corollary discharge—or efference copy—of the motor signal back to pSTS. According to this hypothesis, the higher activity of pSTS during mirror imitation is only secondary to the higher activity observed in poIFG during mirror imitation, compared with anatomical imitation. Alternatively, it is possible that the higher activity in pSTS during mirror imitation is due to increased attention. It is well known that attention modulates activity in higher-order visual areas. These two alternative hypotheses can be easily tested. Indeed, the corollary discharge hypothesis predicts that the higher pSTS activity during mirror imitation (that is, when subjects are watching a left-hand stimulus) compared with anatomical imitation (that is, when subjects are watching a right-hand stimulus), should disappear during the action observation tasks (because in absence of movement, there can't be any corollary discharge). In contrast, there is no obvious reason as to why the attentional hypothesis should not predict a similar difference in pSTS activity during observation of left- and right-hand finger movements. Thus, while the corollary discharge hypothesis predicts higher activity for the left-hand stimulus in pSTS only during the motor tasks but not during the visual tasks, the attentional hypothesis predicts similarly higher activity for the left-hand stimulus during both the motor and the visual task. The analysis of the activity in pSTS during the action observation tasks did not reveal higher activity for the left-hand stimulus, compared with the right-hand stimulus. This result supports the corollary discharge hypothesis but not the attentional hypothesis.

Thus, as shown in Figure 3.2, the information processing flow between these four cortical areas during imitation is most likely as follows: The higher-order visual analysis of the action to be imitated occurs mainly (but not exclusively, of course) in pSTS. This region sends the visual input to the mirroring areas in the parietal and frontal lobe. A corollary discharge of the imitative motor plan is sent back to pSTS. This would allow the pSTS to match the visual analysis of the action to be imitated with the imitative motor plan. When the imitative action is performed, re-afferent action-related signals are received by SII, which most likely

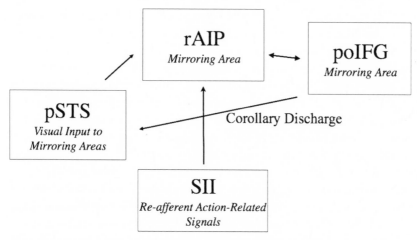

Figure 3.2. Network of brain areas involved in imitation and their presumed interactions.

feeds them back to the mirroring area (here, the parietal area most likely receives this information, given the anatomical contiguity between SII and rAIP).

From a functional standpoint, these visual and motor interactions that occur during imitation between mirroring areas and visual areas establish a correspondence between sensory and motor codes that has been posited to be extremely important in communication. This concept had been also proposed for human speech by the motor theory of speech perception (Liberman and Whalen 2000; Liberman and Mattingly 1985; Liberman et al. 1967). Inspired by the work reviewed above, recent studies have demonstrated the existence of mirroring speech responses in humans, showing increased activity in speech motor areas during speech perception (Fadiga et al. 2002; S. Wilson et al. 2008; S. Wilson and Iacoboni 2006; S. Wilson et al. 2004). What those studies could not demonstrate, however, was that the activity in speech motor areas during speech perception is essential to the perceptual process. TMS can provide a causal link between a brain area and a given behavior with the "virtual lesion" approach. With this approach, it is possible to interfere with activity in a brain region of interest and study the effects on behavior. A recent TMS study perturbed activity in a speech motor area typically activated during speech perception while subjects performed a perceptual speech task. Speech perception was indeed impaired, thus providing for the first time evidence that disrupting activity in a motor area reduces perception (Meister et al. 2007).

These new results do not obviously undermine the role of classical auditory cortices in speech perception. However, they beg the question: what is the interplay between speech motor areas and more classical auditory cortices during speech perception? A recent model proposes that auditory neurons in the superior temporal cortex would provide acoustic analysis of speech sounds, whereas motor speech areas would provide a "simulation" or "inner imitation" of phoneme production. This simulative process would generate a predictive model of the acoustic consequences of phoneme production. This prediction would be compared in the superior temporal cortex with the acoustic analysis of the heard speech sounds. If necessary, an error signal would be generated to allow correction of the simulated phoneme production used for phoneme categorization (Iacoboni 2008; S. Wilson and Iacoboni 2006). From a functional standpoint, this information-processing flow is very similar to the one proposed in Figure 3.2 for imitation of observed actions.

The Role of Mirroring in Empathy

Empathy is a complex phenomenon that takes many forms. The study of the neural underpinnings of empathy is relatively recent in neuroscience and neuroimaging. Obviously, given its complexity, the neural correlates of empathy can be studied in many different ways. Others have published extensively on this topic, adopting specific vantage points (Avenanti et al. 2010; Bufalari et al. 2007; Avenanti et al. 2005; Singer et al. 2006; Singer et al. 2004). One of these vantage points is nicely summarized in Chapter 10 in this book, coauthored by Tania Singer and Grit Hein. Here, this section of the chapter discusses the role of neural mirroring on empathy as studied mostly with observation and imitation of facial emotional expressions during fMRI, and with correlations between behavioral variables and brain activity.

Theodor Lipps proposed more than a century ago a concept of empathy, or *Einfühlung*—a term that means roughly "in-feeling" or "feeling into"— according to which we share and understand the emotions and feelings of others by using some sort of projection of the self into the other. He suggested that "when I observe a circus performer on a hanging wire, I feel I am inside him," as cited by (Gallese 2001). Lipps proposed that at the basis of our ability to empathize there is a process of inner imitation. The neural mechanisms of mirroring reviewed above seem ideal also for this process of inner imitation. Indeed, well-controlled studies show that being imitated increases liking and that more-empathic individuals tend to imitate automatically other people more than less-empathic individuals do (Chartrand and Bargh 1999). These psychological studies suggest functional links between mirroring areas and neural systems specialized in emotional processing. The anatomical connectivity in the primate brain suggests that the superior temporal cortex (where pSTS is located), the posterior parietal cortex (where rAIP is located), and the inferior frontal cortex (where poIFG is located) are connected to areas traditionally associated with emotional processing—such as the amygdala—through the insula (Augustine 1996, 1985).

Thus, a conceivable mechanism of empathy as proposed by Lipps may require the interactions between all these neural systems, such that mirror neuron areas would provide the simulation (or inner imitation) of the facial and bodily emotional expressions of other people and subsequently send information to the limbic system via the insula, such that the evoked activity in the limbic system makes the observer feel what others feel (Figure 3.3). This model makes two predictions: first, a brain imaging experiment of imitation and observation of, say, facial emotional expressions should demonstrate activation of mirror neuron areas, with also activation of insula and amygdala, during both observation and imitation of facial emotional expressions; second, throughout the whole network of areas the activity during imitation should be higher than during observation, as the previous study on imitation of finger movements had shown (see Figure 3.1). Importantly, the higher activity during imitation should not be restricted to mirror neuron areas only. Indeed, if empathy is produced by the simulation of others' actions and by activity evoked from mirror neuron areas (or the whole imitation circuitry) to limbic areas, one would expect that the higher activity during imitation should also spread to the insula and limbic areas. Empirical data from an fMRI study confirmed both predictions (Carr et al. 2003).

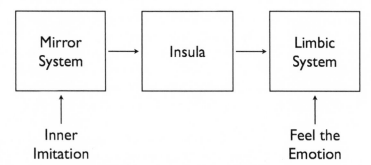

Figure 3.3. Model of simulation-based empathy based on interactions between the mirror neuron system, the insula, and the limbic system.

This initial study suggested a large-scale neural network supporting empathy via a simulative process implemented by mirror neurons. However, the study did not provide any direct evidence linking the activity in this neural network and individual differences in empathy, because behavioral data were not collected in that study. Three fMRI studies have recently correlated brain activity with empathy scores. In one study, subjects listened to action sounds, which trigger a discharge in mirror neurons. The prediction here is that human brain areas with mirror neurons should also be activated while listening to action sounds. Indeed, the study demonstrated that action sounds increased the activity of the frontal mirror neuron system (Gazzola et al. 2006). Furthermore, subjects with high empathy scores had higher activity in the frontal mirror neuron system than subjects with low empathy scores.

Another fMRI study measured brain activity while healthy subjects observed different kinds of grasping actions (Kaplan and Iacoboni 2006). Grasping observation activated the frontal mirror neuron area, and activity in this area was also correlated with empathy. It is important to note that this study on grasping observation and the previous one on action sounds used stimuli that were not associated with emotions, but rather with everyday actions. Even with these relatively simple stimuli, it was possible to observe a correlation between the activity in mirror neuron areas and the tendency to empathize of subjects. A slight difference between the two studies was related to the type of empathy that correlated with mirror neuron activity. While in the study using action sounds the activity in mirror neuron areas correlated with cognitive empathy scores, in the grasping observation study mirror neuron activity correlated more with emotional empathy. This makes sense, if we assume that listening to action sounds without seeing the action is a less familiar and less immediate form of perception of the actions of other people.

A more recent study investigated the relationships between activity in mirror neuron areas and empathy in children (Pfeifer et al. 2008). The children imitated and observed facial emotional expressions displaying basic emotions. As in the previous study on adults (Carr et al. 2003), mirror neuron areas, the insula, and the amygdala were activated for both observation and imitation of facial emotional expressions, with higher activity during imitation. Correlation analyses were performed between brain activity and two types of scores: empathy scores and interpersonal competence scores. Emotional empathy scores correlated with activity in mirror neuron areas during observation of facial emotional expressions. Furthermore, mirror neuron activity during imitation of facial emotional expressions correlated with interpersonal competence scores. As overtly

mirroring the emotions of others plays an important role in social interactions, because it is through this mirroring that we communicate to other
people that we understand what they are feeling, these results are also
not entirely surprising. However, the fact that activity in mirror neuron
areas correlates with interpersonal competence during emotion imitation
suggests that the mirror neuron system is a fairly nuanced biomarker of
sociality.

The evidence that associates mirror neurons with imitation, empathy,
and sociality supports an embodied view of social cognition. In psychology, cognitive science, and philosophy, the embodiment movement emphasizes the role of the body in shaping cognitive processes and the mind. That
is, mental processes are constrained by how the body is built and interacts
with the external world. If social cognition is embodied, then it is conceivable that interventions on disorders of social cognition can be based
on bodily practices. Indeed, the case of autism spectrum disorders (ASD)
seems to support this view.

Autism is a severe developmental disorder characterized by a triad of
symptoms, including impairments in reciprocal social interactions, language and communication, as well as a range of behavioral deficits such
as restricted interests, sensory sensitivities, and repetitive behaviors. Although the symptoms of autism have been well characterized, and there
is general consensus that both brain-based and genetic mechanisms are
implicated in the etiology of this disorder (Volkmar et al. 1994), identifying its underlying neurobiological underpinnings has been challenging. Both structural (see Bauman and Kemper 2005 for a review) and
functional brain abnormalities have been reported in autism, often in
structures and functional networks relevant to social behavior (see Pelphrey et al. 2004 for a review).

It has been proposed that a dysfunction of brain systems supporting
overt and covert mimicry early in development may give rise to the cas-
cade of impairments that are characteristic of autism (J. H. G. Williams
et al. 2001), including deficits in imitation and social communication. In
the past years, several independent studies using different electrophysiological and neuroimaging techniques (i.e., electroencephalography, or
EEG; magnetoencephalography, or MEG; TMS; and fMRI) have consistently reported abnormal neural responses in adults with ASD during imitation and action observation (reviewed in Iacoboni and Dapretto 2006;
see also the recent Minio-Paluello et al. 2009). But perhaps the strongest
empirical support for this theory of autism is provided by our study (Dapretto et al. 2006) where we examined brain activity in the context of a
socio-emotional task and in a developmental sample. Here, high-functioning

children with autism and matched controls underwent fMRI while they imitated and observed facial emotional expressions. The results of this study suggest that while both groups performed the tasks as required, the neural strategies adopted by typically developing and ASD children were quite different. Typically developing children could rely upon the large-scale neural network previously described in this section of the chapter and composed of mirror neuron areas interfacing with the limbic system via the insula—whereby the meaning of the imitated (or observed) emotion was directly felt and hence effortlessly understood. In contrast, this mechanism was seemingly not engaged in ASD children, who likely adopted an alternative strategy of increased visual and motor attention, whereby the internally felt emotional significance of the imitated facial expression was not likely experienced. Importantly, the level of neural activity in mirror neuron areas for imitation and action observation measured in individual children with ASD was negatively correlated with symptom severity in the social domain.

These findings on ASD subjects suggest that reduced activity in mirror neuron areas may represent a core deficit in autism. They also suggest that the use of interventions based on imitation in these patients may be able to retrain mirror neuron activity in children with autism, thus being highly beneficial for these children. Data on naturalistic interventions based on imitation support this idea (Ingersoll 2010, Ingersoll et al. 2007, Ingersoll and Gergans 2007, Ingersoll and Schreibman 2006). While the benefits of these interventions based on imitation cannot be conclusively attributed to increased mirror neuron activity, given the properties of these cells it is reasonable to assume so.

Individual Mirror Neurons in Humans

The study of the human mirror neuron system is obviously complicated by the fact that the activity of individual neurons cannot be easily monitored. Even when rare clinical conditions allow neuroscientists to probe individual cells in humans, the constraints of the clinical settings preclude the extensive studies that can be performed in monkeys. However, the progress in brain imaging and of our understanding of the relations between spiking activity in individual neurons and the activity that can be recorded by brain imaging techniques in humans will help in filling the gap from the monkey studies to the investigation of the human mirror neuron system.

We have recently published a study that represents an initial step at filling this gap. We took advantage of a rare clinical opportunity. Some patients with epilepsy undergo surgery to remove the epileptic focus. In order to localize the focus of epilepsy, the neurosurgical team implants depth electrodes in the brain of these patients. Subsequently, the anti-epileptic medications are no longer administered to the patients. Eventually, the patient seizes and the neurosurgical team can evaluate with precision, through the recordings obtained with the depth electrodes, the location of the epileptic focus and remove it. It typically takes one or two weeks to obtain this valuable medical information. During those days, the patients have implanted electrodes that can be used not only to monitor epileptic activity, but also to record brain activity for research purposes. Typically, the depth electrodes used for these procedures can only measure EEG activity. However, at UCLA, Dr. Itzhak Fried has modified those depth electrodes so that his team can also record spiking activity from individual neurons in the human brain.

We recorded the spiking activity of individual neurons from 21 patients with epilepsy undergoing the procedure described above. Obviously, given that this procedure is highly invasive, the location of the electrodes was determined exclusively on the basis of medical considerations. Nevertheless, we had the opportunity to test mirroring properties of individual neurons while patients performed and observed simple grasping actions and facial emotional expressions. The patients also performed control conditions to test the specificity of the neuronal activity recorded (Mukamel et al. 2010).

We recorded the activity of 1,177 neurons in our patients. We found mirror neurons in four regions with implanted electrodes. One of these regions is a frontal area, the supplementary motor area (SMA). As you may recall from the section on imitation in this chapter, our early fMRI studies of imitation had indeed suggested that SMA may contain mirror neurons (L. Koski et al. 2003). However, at that time there was no evidence of mirror neurons in the macaque homologue of SMA (an area of the macaque brain called F3). The ambiguous nature of the fMRI signal made it difficult to conclude with certainty that there were mirror neurons in SMA. However, the new data obtained in the series of neurosurgical patients are now conclusive: SMA is another cortical region of the primate brain (or at least of humans) that contains mirror neurons.

SMA is a premotor region associated with both action sequencing and action initiation. Indeed, inspection of the firing activity of the mirror neurons recorded in SMA in our study suggests that these neurons code the initiation of the action. This is at variance with the mirror neurons typically recorded in the lateral wall of the frontal and parietal cortex of the macaque, which seem to be coding the goal of the action. This suggests that neural mechanisms of mirroring may code many different aspects of the action of the self and of other people. This is a new concept for neural mirroring. The initial observations all pointed to the role of mirror neurons in coding the goal of the action, and the theoretical discourse has been mostly centered on action goals. It makes sense, however, to hypothesize that mirror neurons, as a whole, may implement the mirroring of many different aspects of actions of others. This neural mechanism may provide a much richer coding of the actions of others that can be used for imitation and social cognition. Furthermore, observations on human behavior fit these new findings. When people interact socially, they often mimic only some aspects of the actions of other people. For instance, when people are dining together, they often initiate at the same time different kinds of actions. One may reach for the glass while the other goes for the napkin. In this particular scenario, the initiation of the movement is mirrored, but the action itself and the goal of the action are not.

We also found mirror neurons in three areas of the medial temporal lobe: the parahippocampal gyrus, the hippocampus, and the entorhinal cortex. This was unexpected. These regions of the brain are known for both higher-order visual processing and memory processing. They are certainly not known for motor properties. Mirror neurons in the medial temporal lobe were found for both grasping actions and facial emotional expressions (this was the case also for mirror neurons in SMA). While the mirror responses for grasping actions may in principle be simply explained in terms of visual processing (because of the sight of the object to be grasped and the grasping hand of the patient during the motor task), this was not the case for facial emotional expressions. Patients could not see their own face when executing facial emotional expressions. Thus, the deflationary "visual" explanation cannot account for the responses of these cells. Our hypothesis is that these mirror neurons in the medial temporal lobe implement some form of memory mirroring. When the patient performs the action, these cells may encode the memory of performing the action. When the patient observes the action performed by somebody else, the memory trace of the same action performed by the observer is reactivated. While this is a post hoc explanation of an unexpected finding, if proven correct, it would support the idea that mirror neurons, as a whole population of cells, can provide a very rich mirroring of many aspects of the actions of self and other.

Approximately 20 percent of recorded human mirror neurons in our study had opposite firing-rate changes in action execution and action observation. The majority of these cells demonstrated increased firing rate for action execution and decreased firing rate for action observation.

This pattern of firing-rate changes may be useful for at least two reasons. On one hand, it may help in preventing unwanted imitation. On the other hand, it may help differentiating between the actions of the self and of other people.

To summarize, this study sheds new light on the properties of mirror neurons, while also beginning to fill the gap between the recordings of individual cells in macaques and the measurements of large neuronal ensembles that are typically obtained with brain imaging techniques in humans. Obviously, many more studies will be needed to better understand and compare the properties of mirror neurons in humans and nonhuman primates.

Conclusions

The human mirror neuron system and more generally neural mirroring mechanisms seem essential for imitative behavior, a cornerstone of social cognition. They also seem to play an important role in empathy, by making it possible to simulate what other people are feeling. This is probably why mirror neurons were selected by the evolutionary process. They allow us to connect deeply with other individuals. Mirror neurons solve gracefully the problem of other minds, which is fundamentally a problem of having access to the mind of other people. Mirror neurons seem to let us have that access in an effortless, automatic way.

The research on the mirror neuron system and its role in imitation and empathy is only at its beginnings. Many more studies will be required for a complete understanding of this neuronal system and its role in social cognition. This research is by necessity interdisciplinary, requiring the collaborative effort of neuroscientists, cognitive psychologists, developmental psychologists, linguists, and so on. This can be at times a challenge, because different disciplines tend to use different jargons and mentality. However, the questions to be investigated are exciting, making the effort to integrate knowledge between disciplines worthwhile.

Acknowledgments

For generous support the author thanks the Brain Mapping Medical Research Organization, the Brain Mapping Support Foundation, the Pierson-Lovelace Foundation, the Ahmanson Foundation, the William M. and Linda R. Dietel Philanthropic Fund at the Northern Piedmont Community Foundation, the Tamkin Foundation, the Jennifer Jones-Simon Foundation, the Capital Group Companies Charitable Foundation, and the Robson Family and Northstar Fund.

CHAPTER FOUR

Social Rules and Body Scheme

Naotaka Fujii and Atsushi Iriki

Introduction

Humans are the paragon of social animals (Adolphs 1999, 2003; Beer et al. 2006). The complexity of our social systems beggars that of any other species on earth. There are several neurocognitive reasons for this, one of which is our ability to follow rules that constrain and channel our behavior. Of course, while rule following is necessary, it is not sufficient; after all, ant colonies and beehives also have complex societies that emerge from the rule-governed behavior of their members, yet they are not nearly as complex as we are. Human social systems are complex not just because we follow rules, but because our rule-generating and rule-following faculties are highly flexible, context sensitive, and multifarious. Rule systems vary widely between cultures, families, friendships, workplaces, legal systems, and creeds, and rule expression can vary based on diverse contextual factors including time, place, and milieu. All human social environments are pervaded by both explicit and unstated standards of behavior. Rules for every kind of occasion are all mingled together in our brains and are flexibly and (usually) adaptively expressed as conditions fluctuate and evolve. Understanding how individuals and groups create, acquire, follow, alternate, and adjust social rules is one of the most important questions in social neuroscience.

The neural mechanisms of rule-governed behavior have been much studied (Matsumoto and Tanaka 2004; Donohue et al. 2005; Bunge et al. 2005; Bengtsson et al. 2009; Strange et al. 2001; Wise et al. 1996). One of the most famous tasks used in such studies is the Wisconsin Card Sorting

Test (WCST) (Milner 1963; Drewe 1974; Mansouri et al. 2006). In this task, a subject is given a deck of stimulus cards that are printed with varying numbers of colored, abstract shapes and is presented with a row of face-up sample cards on a table. The subject is asked to draw cards from the deck one by one and match each new card to one of the sample cards using one of three possible rules: matching by shape, by color, or by number. The subject is never told which rule is in effect, but receives feedback (right or wrong) after matching each card. At several points during the task the experimenter switches rules without notice, and the subject has to use trial and error to figure out which rule is now in effect. There are two important factors in performing the task correctly: rule-following behavior, and rule-switching behavior (Zanolie et al. 2008). Patients who have suffered injury to the prefrontal cortex (PFC) often have deficits in both of these capacities. They tend to "perseverate"—that is, to keep applying the same rule over and over despite ample feedback that the rule is incorrect. PFC damage can also disrupt rule following, which is often made conspicuous by wildly inappropriate social behavior (H. Damasio et al. 1994; Sanfey et al. 2003). Such inappropriate social behavior could be caused by any malfunction of social cognitive functions that PFC is deeply involved in. A key feature of the social cognitive function is integration of multiple social factors. The integration of different modalities of information is processed in association cortices, especially in PFC where concerning decision making (Kable and Glimcher 2009; C. Frith and Singer 2008). PFC has been thought to support rule-governed behavior through two important functional domains: the maintenance and the manipulation of temporal information (Fuster 2000). PFC represents rule, goal, motivation, and contextual information in temporal work space (Watanabe and Sakagami 2007; Levy and Goldman-Rakic 2000; Tanji and Hoshi 2001; Saito et al. 2005) and guides us to socially correct behavior by suppressing inappropriate impulsive behaviors (Kable and Glimcher 2009; Funahashi 2008).

Despite much progress, a great deal of research remains to be done to elucidate the neural mechanisms of rule-governed social behavior. One limitation of studies like the WCST is that such tasks employ highly artificial, abstract rule sets and thus offer limited insight into *social* rule-set following and switching. Few neuroscientific studies have looked at the rules that emerge and hold sway naturally during real-life social interaction. One factor that has limited such studies is the fact that unlike the abstract, rationally designed rule sets of the WCST and similar cognitive-psychological probes, novel social rule sets cannot be invented by experimenters and internalized by subjects. Another limiting factor has been the extreme technical difficulty of measuring the neural correlates of social adaptive behavior in humans in authentic social situations. In the real world, where real, authentic social intercourse occurs, modern brain imaging devices cannot be deployed. MEG and fMRI are unusable because subjects have to be isolated from real social environments and must remain unnaturally still and calm in the scanner. EEG is inadequate because— its inherently poor spatial resolution aside—real-life social conditions radically diminish its signal-to-noise ratio. The ability to study, under realistic conditions, human social brain function, which allows us to make social adaptive behavior, probably awaits some future breakthrough in brain-imaging technology. In the meantime we can still advance the field by studying social brain function in other social species—namely in monkeys, who share many social faculties with us and whose brains possess the neural precursors of all our more highly developed social-adaptive and cognitive abilities.

In the experiments we are about to describe (Fujii et al. 2007, 2008; Fujii et al. 2009), we created a setup that allowed us to observe monkeys' social behavior while recording simultaneous, real-time neural activity data from their brains. The setup created a social environment that, while not fully naturalistic, allowed the monkeys to interact and compete in a free and authentic fashion, on their own terms, in an unstructured task.

How Monkeys Develop Social Rules

Monkeys are highly socially adept animals. The most commonly studied monkey in behavioral neuroscience is the Japanese macaque (Matsuzawa 2001). Their troops tend to consist of around 30–40 individuals and are governed by a strict social hierarchy based on dominance. It has been well described how an alpha male dominates each group and how the dominance is maintained and succeeded over decades (Mito 1979). This hierarchy emerges between each pair of monkeys through their competitive interactions. Once a dominance relationship is established between two monkeys, the submissive monkey accepts the relationship until the dominant monkey loses his position (Mori et al. 1989). The hierarchy is robust, meaning it endures for long periods. However, when there is no competitive situation—that is, when no food or ovulating females are available—the hierarchy becomes relatively vague and relaxed, and the monkeys experience little social stress. This suggests that their hierarchical social structure evolved to provide social solutions under competitive circumstances. Grooming between individuals is another form of adaptive social behavior that reduces social stress due to complex social hierarchy. It is reported that grooming seems to work as a currency circulating within the group (Muroyama 1991). These characteristics of monkey social organization have been studied ethologically for decades (Matsu-zawa 2001), but so far very few studies have been designed to investigate their neurocognitive underpinnings.

For our first experiment, we brought into the lab room two male monkeys, M1 and M2, who had never met each other. Each monkey was seated in a primate chair and could move his hands, arms, head, and torso freely. We seated the monkeys facing each other across a small table. They could not reach each other's face or torso, but both could reach the center of the table easily with their hands. The experimental task was extremely simple: in each trial we placed a piece of food at the center of the table and let the monkeys do whatever they wanted. We did not prompt, encourage, or force them to do anything during the task. If they wanted to take the food they were free to reach for it, but they were equally free to refrain. Both monkeys were clearly eager to take the food. Since there was no pre-established dominance hierarchy between them, during the first several trials both monkeys tried to snatch up each food reward as fast as possible. During intertrial intervals, when no food was visible, the monkeys tended to ignore each other, as if the other monkey did not exist. After about 20–30 minutes of repeating this food-grab task, a hierarchical relationship began to emerge. At one point M2 started to show signs of hesitation to take the reward. This tendency strengthened rapidly, and within just a few minutes, M2 had completely stopped reaching for the rewards. From then on, M1 took all the rewards unchallenged. Evidently some social suppressive mechanism had been engaged in M2. We confirmed that M2 had not simply lost interest in food because of feeling full by offering him food immediately after he and M1 had been separated. Removed from that social context, M2 resumed grabbing and eating without hesitation.

We went on to observe this same dynamic with several other monkey pairings. The same social suppressive mechanism manifested in one monkey in nearly all pairings we observed. The only exceptions arose in the case of two juvenile monkeys, each of whom was under three years old. These young monkeys developed social suppression similar to that seen in mature monkeys, but they often seemed to forget to follow the rules of dominance and reached for the food anyway. When this happened the dominant monkey would often threaten the younger monkey. This suggested that the younger monkeys had a weaker behavioral-suppressive function, similar to human children.

Thus, it may be safe to say that in this first round of experiments we were able to observe the emergence of one of the most elemental simian social rules—the dominance hierarchy—within pairs of monkeys. Each rule was developed organically through the monkeys' social interaction, and the competitive situation was eventually settled by the engagement of behavioral suppression in one of them. Notably, none of these rules were general. Each rule was effective only between a specific pair of monkeys, and a given monkey might be dominant in some pairings and submissive in others. Wild monkey troops can consist of dozens of individuals, so monkeys are capable of learning and adaptively applying dozens of social rules. This means that monkeys, not unlike humans, can learn large sets of rules and retrieve the one that fits the current social context.

Adaptive Behavior during the Food-Grab Task

We decided to try using a modified version of the food-grab task as a behavioral paradigm for probing the neural mechanisms of social adaptive behavior. We initially employed two monkeys, M1 and M2, and later added two more monkeys, M3 and M4. All the monkeys were mature males; their weight ranged from 4.5 to 8 kilograms.

In the new food-grab task, M1 and M2 were seated around a 20-inch-square table in one of three different relative positions: A, B, or C (Figure 4.1). Position A was conflict-free: the monkeys sat at opposite ends of the table, and their reachable spaces did not overlap. Positions B

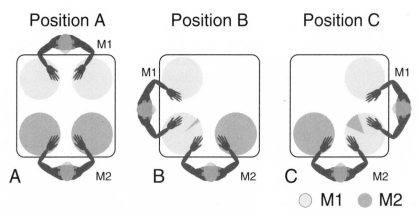

Figure 4.1. Schematic overhead view of the task environment for positions A, B, C. During the task, two monkeys were seated in one of three relative positions (A, B, C) around a table. Each monkey is marked by text: M1 and M2. In each trial, a food item was placed at one of the four reachable locations in Position A or at one of the three reachable locations in positions B and C, indicated here by circles on the table. Circles were not present on the actual table. In positions B and C, one corner was contested space, since both monkeys were equally capable of taking food items from that location. Each circle on the table is a pie chart that depicts each monkey's success ratio for food retrieval from that location in that position.

and C were conflict conditions: The monkeys shared a common, contested corner that both of them could reach easily.

In each trial, we placed a piece of food on the table where at least one monkey could reach it. In Position A, each monkey successfully took the food in every trial when it was placed within his reach. In Positions B and C, their behavior was markedly different. If food was placed in a location where only one monkey could reach it, that monkey unfailingly took it without hesitation. But if food was placed at a contested corner, M2 showed almost no inclination to reach for it, and M1 took it in almost every trial. However, the monkeys' behavior was not identical between Positions B and C. In Position C, M2 showed a significantly higher success rate (13 percent, t-test, $p < 0.05$) in taking food from the contested area than he did in Position B. Also in Position C, M1 looked at M2 more frequently than he did in Position B and often threatened M2, especially when M1 saw that he had lost the trial. We are not certain why this asymmetry existed, but we suspect it was because both monkeys were right-handed, giving the advantage to M1 in Position B and to M2 in Position C. However, in most of the trials where M1 lost, it happened because M1's attention had momentarily wandered away from the table and M2, who was vigilantly observing the social situation, seized this brief window of opportunity to snatch up the reward for himself.

M2 also performed the food-grab task with M3 and M4, separately. The monkeys' aggregate social hierarchy was M1 > M3 > M2 > M4. In other words, M2 was dominant when paired with M4 but submissive when paired with M3. These behavioral patterns were consistent for each pairing over months. Thus, we believe the monkeys were retrieving and switching between behavioral rules based on immediate social context.

The Relevance of Monkeys' Adaptive Social Behavior to Humans

Let us take stock of what we have seen so far. It seems that monkeys' basic social strategy is characterized by two distinct behavioral modes: dominant and submissive. Under dominant mode, a monkey will eagerly take food rewards without hesitation, regardless of the presence of a competitor. Under submissive mode, a monkey will continually suppress his impulse to take contested food. He will also be cautious to signs of conflict with respect to the dominant monkey and will reach for food only if the dominant monkey does not appear intent on taking it. Each monkey employed one of these modes depending on immediate social context. It appears that every monkey's default mode is dominant, since this is how they always behave when they first meet someone new (including humans).

But once the hierarchical relationship is settled, their behavioral mode always assumes the same pattern whenever the two are placed together in a competitive situation.

From these observations, we concluded that monkey social behavior could be understood as a function of modulation of behavioral suppression. Under conflict conditions, balancing forwardness and deference is the most important factor for managing social risk. This appears to be why monkeys have evolved to carefully manage their behavioral mode depending on social context. If one monkey suppresses his desires in a competitive situation, the other can satisfy his desires instead, and the physical and social risks that overt fighting could entail, both to the individual competitors and to the troop, are greatly mitigated. This behavioral strategy provides a simple, general, and powerful method for managing just about any sort of conflict. Our hope is that by elucidating the underlying mechanisms behind this dynamic, we will also shed light on human brains and human social behavior.

Humans are clearly a great deal more cognitively and socially complex than monkeys, but nonetheless, the monkey brain provides the neural and social-cognitive ground plan for our own brain. It can be tempting to think that the simple, rigid dominance hierarchies of monkeys have been relegated to a distant "backseat" role in the more highly evolved human brain, but we should not flatter ourselves too much by overstating the case. Consider the fact that the dramatically "flattened" and egalitarian social structure of most modern industrialized societies is a historical aberration of extreme recency. Throughout the vast majority of recorded history, in every region of the world, strict, pyramid-shaped social hierarchies have been civilization's chief organizing principle. Modern material advances, various social forces, neuroplasticity, and a handful of other factors have apparently led to a significant, culturally induced attenuation of this basic social-organizational mechanism; yet underlying human nature has not been changed at all. Thus, the more insight we can gain into the neural mechanisms of monkeys' suppressive social behavior, the better we will be able to understand our own social-cognitive foundations.

Our next set of experiments involved recording neural activity, at first, in monkeys' prefrontal cortex during the food-grab task in order to search for the characteristic neural modulation patterns reflecting social context and behavioral mode, since PFC is known to be involved in rule-dependent behavioral control (Mansouri et al. 2006; Buckley et al. 2009; Wallis and Miller 2003) and disruption of PFC causes abnormal social behavior (J. Wood 2003; Mah et al. 2004; A. Damasio 1995; Brunet-Gouet and Decety 2006).

Context-Dependent Modulation in PFC

We recorded from dorsolateral PFC simultaneously in M1 and M2 while they performed the food-grab task (Fujii et al. 2009). The monkeys' behavior was recorded using a motion-capture system; limb trajectory data were analyzed later off-line. We isolated 196 single neurons (M1: n=95, M2: n=101) in their left hemispheres around the principal sulcus throughout 10 recording sessions. We found many different types of neural pattern that responded to task events or conditions. Figures 4.2A and 4.2B show typical PFC neurons whose firing rates modulated with relative position. Some neurons showed less activity in Position A but increased activity in Positions B and C (Figure 4.2A). Another set of neurons showed the reverse pattern: higher frequency in Position A, lower frequency in Positions B and C (Figure 4.2B).

However, the position-dependent modulation shown in Figures 4.2A and 4.2B could also be explained as responses related to the monkeys' volitional movements, rather than to social-contextual manipulation. To exclude this motor interpretation, we extracted neural activity from just the no-motion periods when neither monkey moved an arm and statistically compared it between Positions A, B, and C (Wilcoxon test, $p < 0.05$). This analysis revealed the same pattern as before, thus confirming that the modulation was indeed related to positional switches and not to motor activity.

Next, we defined a neuron as social-state dependent (SSD) if it showed significant modulation in its firing rate during no-motion periods after positional switches. Seventy-six PFC neurons were identified as SSD. During recording, position was sequentially alternated from A to B to C. At each positional switch, a neuron could show one of three possible modulation patterns: increase, decrease, or no change. Thus, there were nine (3x3) possible modulation patterns through the A-B-C sequence of switches. Figure 4.2C schematically depicts eight of these patterns (it excludes the one that corresponds to no change across all three position blocks). Figures 4.2D and 4.2E indicate the proportion of SSD neurons from each monkey assigned to each of these eight categories. There was a significant bias in the distribution of modulation patterns between the dominant and the submissive monkey (Fisher's exact test, $p < 0.05$). Their PFCs showed SSD modulation in opposite directions. Among M1's SSD neurons, the proportion of increasing neurons (modulation categories 1, 2, and 3) was significantly larger than that of decreasing neurons (modulation categories 6, 7, and 8). In contrast, among M2's SSD neurons the proportion of decreasing neurons was significantly larger than that of increasing neurons.

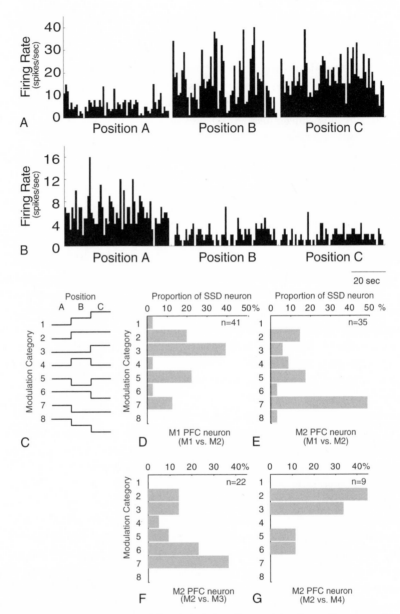

Figure 4.2. PFC neurons modulated their baseline activity depending on social context. (A, B) Histograms of two SSD (social state dependent) neurons recorded from (A) M1 and (B) M2 in three positions. (C) Schematic depiction of eight categories of modulation pattern. (D, E) Proportion of SSD neurons for all eight modulation categories recorded from (D) M1 and (E) M2 when M1 and M2 were paired. (F, G) Proportion of M2's SSD neurons for all eight modulation categories when M2 was paired with (F) M3 and (G) M4.

When we compared the proportion of increasing and decreasing SSD neurons between M1 and M2, M1 showed significantly more increasing SSD neurons and fewer decreasing SSD neurons. These findings suggested that these SSD neurons were discriminating social context. And because they were located in PFC, the theoretical seat of executive function, they might also have been participating in social behavioral management.

However, several questions had to be addressed before we could settle on these conclusions. First, would we see the same modulation patterns if we paired M2 with another dominant monkey? Second, would M2's modulation pattern be similar to M1's pattern when M2 was behaving in dominant mode in a different pairing? Third, would the same pattern emerge from a different sequence of positional switches? To address these questions, we brought back M3 and M4 and again paired them separately with M2 while continuing to record M2's PFC-neuron activity. In these sessions, relative position was sequentially altered from Position B to A to C. Recall that M2 was submissive when paired with M3 but dominant with M4. The PFC-neuron activity data from the M2-M3 and M2-M4 pairings were analyzed using exactly the same methods used on the data from the M1-M2 pairing. All our previous results were confirmed again in these new pairings. Thirty-eight percent of M2's PFC neurons showed SSD response. When M2 was paired with M3, his SSD neurons decreased their firing rates (Figure 4.2F). When M2 was paired with M4, his SSD neurons' firing rates tended to increase (Figure 4.2G). This tendency was confirmed by statistical testing.

These results showed that PFC neurons could discriminate multiple social-*behavioral states* through binary response. SSD neurons entered an "up" state in dominant mode and a "down" state in submissive mode. Thus, PFC might have been using these patterns of up and down states among SSD neurons to broadcast the current social state to other brain areas, thereby providing an internal reference for adaptive social behavior. However, such information would *not* have been suitable for detecting fast or fleeting changes in social context: The time constant of these neurons' baseline modulation was on the order of tens of seconds, not on the order of seconds or fractions of seconds, which is what would be required for a submissive monkey to detect those brief chance windows during the food-grab task. Furthermore, to be able to detect abrupt changes in context, the brain has to know how each agent (including the self) is behaving on a real-time basis. There was no indication in our data that SSD neurons in PFC were representing this sort of information. Thus we suspect the observed SSD modulation was representing a general principle of behavior, a strategy, for adaptive behavioral control within the operative social context. If this is correct, the detection of fast changes in context and the representation of agency must have been performed elsewhere in other association cortices. Thus we expanded our search to include the parietal cortex, the hub of spatial and somatosensory cognition. We hoped to see much more dynamical information than PFC about the body and space (Obayashi et al. 2001; Fujii et al. 2007) because the parietal cortex is thought to represent body scheme by integration of multimodal information (Iriki et al. 2001; Maravita and Iriki 2004).

Social Cognition in Parietal Cortex

We recorded the activity of 176 parietal neurons from area 5 in the left hemispheres of M1 and M2 while they performed the food-grab task. We chose area 5 because neurons there tend to show somatosensory responses to the shoulder, forearm, and/or hand. The neurons in area 5 are also known to have multisensory properties and are theorized to contribute to the generation of an internal, real-time representation of the physical self—a construct known as body image or body scheme. It is also known that these neurons' receptive fields are extended by tool use: when a monkey (and presumably, a human) wields a hand-held rake that extends his reach, neurons in his parietal lobe start to fire as though the arm had actually been lengthened. In other words, external objects can literally be incorporated into the body scheme. Given such representational flexibility, it was conceivable that the neurons of area 5 might also modulate their activity in response to social factors—for example, to represent the identities, body parts, locations, and purposeful movements of individuals (including the self) in the local spatial environment.

Our next analysis was designed to see whether and how parietal neurons responded to the actions of self and other. We analyzed the motion-capture data and extracted five categories of "motion episode," defined by arm movements: control, M1 right arm, M1 left arm, M2 right arm, and M2 left arm. Control periods were the same as the no-motion periods defined earlier, and the other four categories were defined as periods when one monkey moved one arm. We defined neurons as motion-related (MR) if their activity during one motion episode was significantly greater than during the control period ($p < 0.05$, Wilcoxon test). We analyzed the MR response of each neuron at each position separately. The four independent motion factors characterized the response of each neuron. We found 91/176 parietal neurons that showed MR response in at least one motion factor in one position, so that we could characterize MR parietal neurons using four independent motion factors. However, this analysis

Rule Emergence and Retrieval in the Cortex

Our results showed that PFC assumed different tonic baseline states depending on social context. It based this discrimination on one's hierarchical position and the spatial relations between individuals. It is well established that PFC plays an essential role in rule-dependent behavior (Bengtsson et al. 2009; Brunet-Gouet and Decety 2006). However, this does not mean the rules themselves are represented in PFC; in fact, most rules almost certainly reside elsewhere in the brain. This is indicated by, among other evidence, the fact that the neural response patterns in PFC are often tonic and abstract in form (Funahashi et al. 1993; Wallis et al. 2001). PFC's role seems to be to discriminate context, to retrieve and activate context-appropriate rules, suppressing context-inappropriate rules or behavioral impulses or goal states, and to keep them active for as long as they remain appropriate (Donohue et al. 2005). This sort of stable representation of context is not suitable for orchestrating specific, rapid cognition and action but rather provides a global reference frame for decision making by other brain areas (Fujii et al. 2009).

Our results also showed parietal cortex modulating its response patterns based on social context. This activity apparently reflected modifications to the body scheme. In other words, the parietal representation of the bodily self was altered to some extent to incorporate social conflict as well as the actions of a nearby, nonself social agent. This may come as a surprise, since the body scheme was originally conceived of as being a more or less internal function, integrating visual, auditory, and interoceptive inputs to maintain a real-time working model of the body in space (Iwamura 1998). Body scheme was thought to adaptively reconfigure itself to reflect changes in the body's configuration (e.g., moving limbs and joints) and in the body's relation to its environs (e.g., whether a nearby object is within reachable distance). But eventually evidence started to accrue that the body scheme can be a great deal more inclusive than this. For example, it has been shown that body scheme can be reconfigured by tool use (Iriki 2006) and by a variety of multimodal illusions (e.g., the rubber hand illusion [Makin et al. 2008]). Our results extend this line of findings even further, showing that our monkeys' social rules were represented at least partly through a modified body scheme.

Taken together, our results demonstrate that PFC and parietal cortex support social adaptive behavior by modulating their neural response patterns based on the existence of social conflict. PFC maintains a tonic state that reflects this context and signals other brain areas to activate or stand ready to activate context-appropriate behaviors while keeping context-inappropriate behaviors actively suppressed. Parietal cortex adapts the body scheme in ways that appropriately constrain the monkey's behavior in the given context. This modified body scheme may in turn serve as a reference for PFC to use to decide if its representation of social context is still correct.

Conclusion

We are proposing body-scheme-based social-rule representation in monkeys. Our findings so far are limited to just a few areas of the monkey brain. Moving forward, it will be necessary to learn many more details about the cortical, subcortical, and network connectivity and dynamics that support social brain function. Our findings so far are also limited to a single kind of social context (competition), and involving only two monkeys at a time. We hope that with more research, body-scheme-based social-rule representation can be shown to be a widely instantiated—and perhaps even the universal—mechanism of social-rule representation in the monkey brain. Other kinds of social rules that could be investigated include more-complex situations involving more than two monkeys, and the rules that govern mutual grooming behavior, which monkeys use to cement social ties and ease social stress.

For now, body-scheme-based rule representation is just a hypothesis, and the evidence for it, being so-far limited to monkeys, cannot readily be expanded to human rule representation and social brain function. Even if our hypothesis is borne out, it may prove very difficult to retrace the path between monkey and human social brain function, because some of these networks and mechanisms may have evolved almost beyond recognition as we became a tool- and language-using species. But it is not a hopeless task. For example, we proposed that tool use, which modulates body scheme, may have acted as a catalyst for developing a more robust theory of mind via a more flexible way of representing the relationship between the self, external objects, and fellow social agents.

CHAPTER FIVE

What, Whom, and How: Selectivity in Social Learning

Ludwig Huber

N THIS chapter I will discuss social learning phenomena from the point of view of selectivity. The common picture emerging from the many studies conducted with human and nonhuman animals on their ability to learn from others shows that observers are selective in many respects. They are not copy machines that blindly copy what they see, but rather, they reproduce only specific parts of the demonstration. Call and Carpenter (2002) presented a framework for investigating social learning based on the different types of information that observers are able to extract from models. The main idea behind their multidimensional framework is that the model produces several sources of information simultaneously and observers attend selectively to some of these sources but not to others. The authors distinguish between goals, actions, and their results. In this chapter I will extend this framework by showing that observers are selective in several additional respects. Observers can differ with respect to the what, when, and from whom to learn. This selective nature of social learning is confirmed by theoretical models of the adaptive advantages of social learning, suggesting that individuals should adopt strategies that dictate the circumstances under which they copy others (Galef and Laland 2005) and from whom they do so. The theoretical analyses predict that, among other strategies, organisms will copy when uncertain, copy the majority, copy if better (Laland 2004), copy the older, higher-ranking individual (Horner et al. 2010). Here I will review experiments that show a) the selectivity of an observer's distribution of attention in a social learning situation and b) the influence that knowledge about the social and physical environment has on selectivity.

Selectivity in What and Whom to Observe

One crucial aspect of living in a socially complex society is paying attention to other group members, their interactions and behaviors. Social learning is expected to occur in circumstances in which the observer can acquire some new knowledge, for example about predators, spatial utilization, or what food is palatable. Furthermore, only if an individual monitors the behavior of others with regard to affiliation, dominance, and tolerance will it be able to make the correct decisions of whom to solicit in agonistic conflicts, whom to groom, and whom to avoid. Monitoring the behavior of others is therefore a prerequisite for any form of behavioral adjustment during co-operation, competition, and communication. But it may also be of crucial importance when it comes to the acquisition and spread of social information. However, time and/or habitat constraints might restrict the opportunity to observe every other animal within the group or every action performed, forcing animals to be selective.

Which aspects might determine the decision as to what and whom to observe? First, individuals might vary in dominance rank. Monitoring an old high-ranking female might be more important than monitoring a young low-ranking female, because the behavior of the latter will have fewer consequences for the social dynamics within the group. Over the years, evidence accumulated that these features affect how much individuals visually attend to others or even socially learn from others (Nicol and Pope 1994; Valsecchi et al. 1996).

A further important aspect influencing social attentiveness is tolerance. An important determinant of tolerance is the social structure—in particular, social dynamics. Coussi-Korbel and Fragaszy (1995) proposed a model relating social learning to social dynamics among members of a group, which predicted that more-extensive and more-frequent behavioral coordination in time and space will be achieved among groups exhibiting an egalitarian or tolerant style of social dynamics than those exhibiting a despotic style. Individuals in tolerant social groups experience many opportunities for close behavioral coordination in space and time with other group members.

Finally, the frequency and duration of monitoring bouts may also depend on the specific context (e.g., food versus object) and on the type of information that is acquired. If monitoring is used to update old, status quo information, short but evenly distributed looks might be sufficient. On the other hand, if new information is being gathered—for example, learning how to extract food items from small crevices with a stick—

long observation bouts would be expected during the demonstrator's manipulation of the stick.

Comparing Attention Getting and Attention Holding in Six Species

In three related studies we investigated the attention pattern of human and nonhuman animals in a social learning paradigm: in common marmosets *(Callithrix jacchus)* (Range and Huber 2007); ravens *(Corvus corax)* and jackdaws *(Corvus monedula)* (Scheid et al. 2007); and dogs *(Canis familiaris)*, keas *(Nestor notabilis)*, and three-to-six-year-old human children *(Homo sapiens)* (Range et al. 2009). We were interested how much and for how long individuals would observe a model while searching, manipulating, or feeding (eating), as well as how age, sex, and dyadic relationships influenced processes of attention holding and attention getting in individual animals.

From the avian class, we chose one parrot and two corvid species. Keas are exceptional parrots showing a number of strange behavioral aspects. For instance, they have been shown to selectively manipulate conspecifics for gaining access to enclosed food (Huber et al. 2008; Tebbich et al. 1996). The apelike combination of extractive foraging, high sociality, extreme behavioral flexibility, and delayed maturation and lenience by adults toward the young seemed to us ideal prerequisites for the study of social and physical cognition in nonhuman animals (Huber and Gajdon 2007).

From the corvid family, we chose ravens and jackdaws, because they show advanced sociality but different grouping behavior. In particular, they differ in their social dynamics (stable versus fluctuating), as well as in foraging style and diet. Ravens are facultative social animals and assemble particularly as nonbreeders for foraging, roosting, and playing (Heinrich 1989; Ratcliffe 1997). Competition for food is fierce, with birds showing elaborate tactics to gain access to and protect food from others (Bugnyar and Kotrschal 2002). Jackdaws, on the other hand, are highly social, breed in colonies, and forage in groups (Roell 1978). As compared with ravens, they show moderate levels of scrounging and rarely cache food. Both species, however, form monogamous pairs and also establish valuable relationships ("friendship") with same and/or different sex partners (de Kort et al. 2006).

Over the past decade, behavioral scientists have uncovered a surprising set of social-cognitive abilities in the domestic dog (Míklósi 2007). These abilities seem to be exceptional, not only in comparison to their closest relatives, wolves, but also in comparison to great apes. In direct comparisons, dogs are more skilled than chimpanzees at using human communicative cues when searching for food (B. Hare and Tomasello 2005; Míklósi 2007).

From the nonhuman primates we have chosen a New World monkey, the common marmoset. Like all callitrichids, they breed cooperatively with biparental and sibling care and show an egalitarian social organization characterized by high levels of within-group cohesiveness and coordination and low levels of overt aggression (Huber and Voelkl 2009). Adult marmosets and tamarins can effectively learn by either jointly interacting with other social group members, or attending to visual and/or vocal cues provided by others or to the actions of others (Bugnyar and Huber 1997; Caldwell and Whiten 2004).

Finally, human children are well known for their ability to imitate, share attention, and cooperate (Tomasello 1999; Tomasello et al. 2005). Humans do not stand outside the evolutionary trends found in primates but appear to have only selectively strengthened some important cognitive abilities in the social domain, such as joint intention and attention, ostensive-communicative processes, imitation, teaching, and perspective taking. Thus, human cognition is not unique, but a specific (in the literal meaning of the word) instance of primate cognition.

The Peep-Hole Experiments

In order to make the task comparable across species, and observation patterns measurable, we thought to study the behaviors of observer and the demonstrator in two compartments separated by an opaque partition. Direct attention to the model was measured by providing only restricted visual access through two holes in the partition, which allowed for measurements of attention-getting and attention-holding processes. The rationale for using a two-hole procedure is that if animals are interested in the actions of the model, they will look through one of the two holes during the demonstration. Two holes were provided to allow some flexibility for the observer with regard to the angle of vision as well as the position within the animal's own compartment. This method can easily be applied to mammals and birds and allow measurement of attention/interest toward a particular conspecific and its actions.

The models of all six species (in all three studies) were required to show the same sequence of behaviors: searching, manipulating, and feeding (eating), while the observer was allowed to watch through the holes in the partition wall (Figure 5.1). The food or sweet (children) was enclosed in a film canister (marmosets, keas, children), a tetra pack box (corvids), or tightly crumbled into a piece of paper (dogs). These containers were hidden by either being buried in sand (corvids) or wood chips (marmosets, keas, corvids) on the ground or placed in a box (children). Trials ended once the food was retrieved and eaten.

The results of all three studies converged to the same overall effect, namely that observing behavior is dependent on several factors, such as the behavior and age/sex/identity of the model. In addition, we found striking differences between species, with some but not all species being selective to when, what, and whom to watch. Overall, most differences

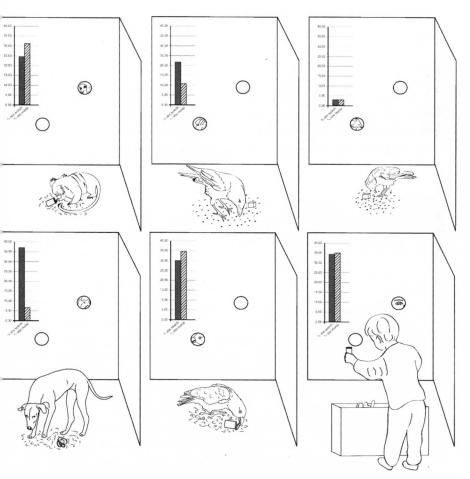

gure 5.1. Sketch of the experimental setup for marmosets, ravens, jackdaws, dogs, keas, d children in the peephole paradigm. Inserts depict the relative amount of observation ne during the model's searching (solid bars) and manipulation activity (cross-hatched rs). After Range and Huber 2007; Scheid et al. 2007; Range et al. 2009.

within and between species were found in the attention-holding rather than the attention-getting processes, which is more important for learning about the sequence and coordination of actions and their consequences. This suggests that the tested species differed less in attentiveness in the service of updating old, status quo information, than in the attentiveness in the service of gathering information about new foraging possibilities.

One of the most surprising result was the low proportion of observation from the total time of demonstrated foraging behavior (around 30 percent) and the especially small duration of looks in all nonhuman animals (with the exception only of juvenile keas). Most of the looking bouts were only a few (< 10) seconds long, even if the observer was watching the demonstrator's manipulation of the food container. This is little time for an observer to perceive an action that is composed of a sequence of novel action elements (as in complex tool-using actions). However, in experiments that demonstrated imitation in marmosets (Bugnyar and Huber 1997; Voelkl and Huber 2000, 2007) or ravens (Fritz and Kotrschal 1999) the demonstrated action consisted of a single movement. Furthermore, these movements were shorter (< 5 s) than the attention-holding capacity of the species under investigation and were demonstrated many times. Insofar, the sometimes reported failure of marmosets to copy the actions necessary to open an artificial fruit (Caldwell and Whiten 2004) might have been a problem not of their imitative ability, but rather of their short attention span in relation to the length of the demonstrated action (sequence). In those cases, in which the subjects learned action sequences or hierarchically organized foraging tasks, human trainers or caregivers were used as demonstrators (Horner and Whiten 2005; Whiten 1998). A human demonstrator usually adjusts his/her demonstration of the relevant motor actions according to the attention of the observer.

Differences in attention levels on the individual level (range: 10–50 percent) were due to the age and the sex of the observers. For instance, juvenile keas were more interested in the demonstration than adults and not selective in regard to whom they watched—peer or adult—in contrast to the adults. Concerning the effect of the demonstrator's age, three of four juvenile marmosets were much more attentive to their siblings than to their parents, which confirms reports of a trend toward interacting with other group members besides the parents (Schiel and Huber 2006).

In children, there was only an interesting sex effect of observers, with boys watching significantly longer than girls. Concerning the effect of the sex of the demonstrator, we found only two significant trends. Ravens were more interested in the demonstration if the model was of opposite sex, which might reflect their interest in potential mates. Marmoset observers were more attentive toward demonstrators of the opposite sex than toward animals of the same sex in the relative proportion of observation time and the mean duration of looks during food-directed behavior.

In the remaining species, no such age or sex effects appeared, but this may be due to the small sample size. However, several other interesting observation differences emerged in these studies, which could be accounted for by differences in the behavior and identity of the demonstrator.

Selectivity about What to Look At

Only children and jackdaws exhibited about the same amount of interest in the demonstrator's searching and manipulating behavior. For the other species we found two main trends (Figure 5.1).

Marmosets and keas looked longer when the model was engaged in manipulating the food container rather than in searching, suggesting that they were mainly interested in how to obtain the food (Range and Huber 2007; Range et al. 2009). Dogs and ravens showed the contrary pattern; they looked longer when the model was searching than when it was manipulating the encapsulated food (Scheid et al. 2007; Range et al. 2009). While this makes sense for dogs, as they usually do not need to process encapsulated food, it remains an open question for ravens.

In the corvids the results confirmed the predictions of the social dynamics model of Coussi-Korbel and Fragaszy (1995). As predicted, the distribution of attention toward different models was more asymmetric in ravens than in jackdaws. Ravens engage in various types of cooperation, such as recruitment at roosts and food calling, joint object exploration and alliance formation, some of which appear to work on the basis of affiliate social relationships. Accordingly, they showed more attention toward friends than toward nonfriends in the experiment.

Selectivity in Learning about the Environment

Selectivity is also apparent in terms of the aspects of the physical environment that are assimilated into the observer's knowledge about the world. Sometimes the observer only "learns" that somewhere in the environment something is interesting. Insofar, the observer has learned nothing specific about the environment but is simply attracted by a conspecific to have a closer look. A nice example is the juvenile marmoset when exploring the environment to forage. Already when starting to move independently (in the second month of life), marmosets follow their group mates to places where these are foraging and sometimes manipulate the same objects thereafter (Schiel and Huber 2006). To some extent, these objects have been signified by the group mates, mostly the parents. However, when the marmosets actually learn something useful, for instance that insects are often below leaves, then this is not a direct result of local or stimulus enhancement. Rather, through repeated encounters, the animals form an association between the two objects (the leaf and the insect) by classical (Pavlovian) conditioning. Being attracted by conspecifics is only a prerequisite of that learning. It increases the probability of its occurrence. One should, therefore, speak of attentional or perceptual effects, rather than learning (Zentall 2006).

Of course, animals may learn much about the functional significance of objects in nature by first having their attention drawn to those objects when manipulated by conspecifics. Without such opportunities, it would perhaps be never noticed. For example, Lorenz (1935) noted that ducks enclosed in a pen have difficulties finding the way out through a hole, unless they happen to see another duck as it is escaping from the pen. The observer may actually not have learned how to pass through or that the hole is large enough to escape through. It may be sufficient that the escaping duck simply called the attention of the observer to the hole.

Similarly, a *detour* (barrier circumvention) problem may be learned only by first seeing another individual solving it. Attention may have been drawn to the path that led around the barrier, and local enhancement is sufficient to facilitate acquisition. In fact, without demonstration, a detour proved to be difficult for dogs. While trial-and-error learning improved the performance slowly, observation of a human demonstrator led to immediate improvement in detouring behavior of dogs (Pongrácz et al. 2001). These results are fully compatible with converging evidence that dogs are exceptional in paying attention to human actions and in using even subtle cues to solve problems. But would other, less socially gifted animals also profit from the demonstration of skilled models in such problems?

Can a solitary species make use of the behavior of a conspecific to solve a navigation problem? We presented eight subjects from a solitary reptile species, the red-footed tortoise *(Geochelone carbonaria)*, with a detour task: the animal has to make a detour that involves a temporary distancing from the target (Wilkinson et al. 2010). The tortoises were divided into two groups: a nonobserver group and an observer group, of four individuals each. In each trial a tortoise was placed on the outside of a V-shaped fence where a preferred food was available in the center. To successfully reach the goal the tortoise had to move away from the food, circumvent the fence, and return to the food on the other side. Only the four animals of the observer group watched a conspecific demonstrating the task before being allowed to solve the task alone.

Interestingly, none of the four nonobserver tortoises reached the goal in any of their 12 trials. They readily approached the reward but were unable to successfully navigate the detour. In sharp contrast, two of the observer group completed the detour on their first attempt, and all managed to complete it at least twice. Overall, the observer group was significantly better at reaching the goal than the nonobserver group (Wilkinson et al. 2010).

What had these animals learned? Had the tortoises simply followed the scent cues of the demonstrator to complete the detour, or did they learn something more general from observing its behavior? The fact that the observer tortoises completed the detour in a leftward as well as the demonstrated rightward direction supports the second possibility. It also excludes a simple local enhancement explanation. Even a generalized stimulus enhancement is implausible, because some control animals also approached the distal end of the fence but were unable to circumvent the fence there and complete the track. Therefore the tortoises may have learned either the detour principle (first move away and then return on the other side) or the principle of circumvention. In both cases they would grasp something about the spatial relationships between the food bowl, the fence, and themselves. Moses, a member of the observer group, solved the task in 11 of 12 trials. In the only trial in which he failed, he stopped at the distal end of one branch of the fence, then turned around once, but in the direction away from the fence, and then returned to the start point on the wrong side of the fence. It seems, therefore, that he executed the turn with an incomplete understanding of the spatial relationship of the situation.

Learning about Actions, Objects, and Effects

In all cases of social influence or social learning described so far the observer had been confronted with a more or less static feature of the environment. It approached a locus or an object because of interest, and showed approach or avoidance responses because of positive or negative valence formation. But social learning may also involve the delivering of a new response to a former neutral stimulus, such as a manipulative action with the object. Even if there is no immediate reward for executing the response, the observer seems to attempt to reproduce the result or change in the environment it has witnessed before. But how is this accomplished? What has the observer learned here or understood about the observed change of state in the environment?

To understand results of what others do one needs to relate the action and the effect. If the action itself is not copied, which is imitation per definition, the environmental change must be understood in physical or causal terms. It is, however, questionable whether this alternative to imitation is in fact more common than attempts to copy the action. Reproducing behavior, especially if it consists of simple actions, could be an automatic act (response facilitation). But the enormous difficulties of chimpanzees to understand weight (Penn and Povinelli 2007a), the long developmental period of chimpanzees to learn nut cracking (Matsuzawa 2001), and the often reported failure to learn complex tool-using behavior solely by observation (Visalberghi and Fragaszy 2002) indicates that learning tool use by observation is everything but evident.

Emulation

Since the first study in which learning of tool use by observation was claimed (Tomasello et al. 1987), this type of learning has undergone both methodological and theoretical developments in its usage (e.g., Byrne 1998a; Tomasello 1996; Whiten et al. 2004; Zentall 2006). By observing a demonstrator successfully obtaining food, the observers don't just learn to manipulate the tool but rather to use it functionally. More generally, emulation has been considered as learning about the operating mechanisms of objects or environment (Whiten et al. 2004), properties of objects, relations between objects and functions (Byrne 1998a), and the causal structure of the task (Tomasello 1996; Byrne 1998a).

An understanding of either the causal relationships or the intention of the conspecific model's behavior is likely to reduce what is learned about the finer details of an action. For example, the imitator may attend to some functional properties of an object, the goal of the behavior, the serial order of the actions involved, or the hierarchical structure of an action sequence (all reviewed in Whiten et al. 2004). If, however, an understanding of these aspects of the task is not available, or is beyond the cognitive capacity of an observer, then faithful (slavish) copying of the behavior of a model, called mimicking (Tomasello 1996), is a valuable alternative (Huber 1998). In contrast, emulation may represent the cognitively advanced option (the "cognitivist's answer to local enhancement," Tomasello 1996, 321), the ability to select intelligently just those pieces of information that are useful, neglecting details of behavioral form judged to be redundant. Here again, I would suggest to have a fresh look at social learning phenomena from the viewpoint of selectivity.

In a study with keas, we attempted to disentangle effects of imitation and emulation (Huber et al. 2001). Five individuals were allowed to observe a trained conspecific that iteratively demonstrated several techniques to open a large steel box. The lid of the box could be opened only after several locking devices had been dismantled: a bolt had to be poked out, a split pin had to be pulled, and a screw had to be twisted out. The observers' initial manipulative actions were compared with those of five naive control subjects (nonobservers). The observers explored more, approached the locking devices sooner, and were more successful at opening them. Most important, their initial attempts did not match the response topography or the sequence of the model's actions. For instance, for testing the observers we used a "fake screw" instead of the real screw that was most saliently twisted out by the demonstrators. This "nonfunctional screw" was identical to the real screws except that it was not threaded, so that it could be poked out of the rings without twisting it. Interestingly, despite manipulating this object effectively, the observers never copied the actions they witnessed before. We therefore suggested that their improved efficiency at unlocking the devices seemed to reflect the acquisition of some functional understanding of the task through observation.

According to the emulation hypothesis, observers might have acquired specific information about the function or the potential use of the locking devices. Rather than learning that the split pin, for instance, is a signal for food, or is attractive as a manipulandum, the observer may have learned that the split pin "can be removed" or "can be pulled out." However, this would imply that the observer formed an association between an object and the particular action the model used to produce a specific change of state in the environment. At least at a detailed level, such response matching (e.g., twisting the screw) was not found. An alternative formulation in the sense of what *can* be done (declarative, e.g., removing) with the object rather than what *must* be done (procedural, e.g., twisting) is implied if we think of the notion of "affordance" by Gibson (1979). From detailed observation in the wild, Diamond and Bond (1999) concluded that the keas' interest in objects for demolition is related more to an object's affordances than to its potential for providing food resources.

Selectively Learning about Spatial-Temporal-Causal Relationships

Although the concept of affordance was originally developed as the main feature of an ecological theory of perception by Gibson (1979), it was developed further into a means to explain the organism's interactions with objects. More specifically, it became a conceptual tool to explain how inherent "values" and "meanings" of things in the environment can be directly perceived, and how this information can be linked to the action possibilities offered to the organism by the environment. For most ecological psychologists in the tradition of Gibson, affordances are opportunities for behavior; they are properties of the environment, but taken relative to an animal. Other Gibsonians, however, argue that affordances are not properties of the environment, and not even properties. Affordances are relations between the animal and its environment that have consequences for behavior (Chemero 2003).

But what in the environment would qualify to contribute to an affordance relation? Gibsonians argue that people must often cross gaps of one kind or another; they walk across a river on a railroad trestle; they hop from one bank of a stream to another; they jump from a dock onto a boat as it pulls away. These are parts of the natural environment, parts that have been constant for quite some time so that organisms could adapt to them. To get by in these situations, one must be able to perceive whether a particular gap affords crossing and, if so, whether it can be crossed by a casual step or by some other means (jumping, hopping, lunging, etc.). Experiments on stair climbing and gap crossing support the theory of affordances by showing that the perception of the environment depends on the perceiver's body or its abilities. For instance, Warren (1984), in attempting to quantify affordances for stair climbing ("climbability"), quantified them as an aspect of body scale, the ratio between leg length and riser height. And Chemero et al. (2003) provided preliminary evidence that gap-crossing affordances are perceived not only in terms of body scale, but of ability. Abilities are functional properties of animals, which depend on the individual animal's developmental history or the evolutionary history of the species.

The Gibsonian concept of affordances has recently been reformulated into "embodied cognition." It is a specific mechanism through which the brain/body system models its interactions with the world (Gallese 2001; but see Goldman and de Vignemont 2009). More specifically it is any form of cognitive processing that is performed in representational codes that are specific to the body (Bastiaansen et al. 2009).

Limitations in the capacity for understanding goals or forming particular associations between an object and the observed action are not entirely dependent on the perceptual system of a given species or on its learning skills, but also depend on its motor skills. Understanding object properties is tightly linked to the motor system. As soon as we see an object, its affordances selectively activate groups of anterior intraparietal area (AIP) neurons. These are connected with visuomotor neurons in the premotor cortex area F5 and so generate visuomotor transformations. This circuit, however, does not code the individual affordances, but the motor acts that are congruent to them (Rizzolatti and Sinigaglia 2006).

With this modern concept of affordances in mind it is worth reconsidering the question of affordance learning through observation. Researchers have claimed affordance learning through observation in tool use. Is there any developmental or evolutionary reason for objects to have affordances or to participate in environment–animal relations if they are completely unknown to the animal? Is learning about "how the environment works" really learning about affordances? If, for example, pigeons learned without a conspecific demonstrator (the so-called "ghost control" condition) to move a screen in the same direction as it was seen before (Klein and Zentall 2003), what kind of developmental or evolutionary history exists in the relation between the screen and the pigeon's pushing abilities?

It seems to me that the importance of affordance learning is overestimated. I suggest, instead, that animals need some understanding of possible effects of their actions on objects, which they acquired during ontogeny. This understanding of spatial-temporal-causal relationships might then be generalized to the model's demonstration and the observer's own later attempt. It might be sufficient to selectively choose the efficient solution to the task. This selectivity and goal-directedness is likely to lead to a divergence of actions between demonstrator and observer if the actions demonstrated by the model vary considerably in their effectiveness. Having learned about the affordances of various objects, an individual may be able to exploit this knowledge to select the most efficient way to the goal—that is, the desired end state of the manipulation (Want and Harris 2001, 2002). A consequence of affordance learning would be the acquisition of some kind of "technical' knowledge," which then could be applied to similar physical problems. Want and Harris (2001) showed that young children ignored the unproductive part of a model's actions (but see the recent reports of human overimitation: Lyons et al. 2007; Whiten et al. 2009).

The same selectivity was later found in young chimpanzees (Horner and Whiten 2005). The authors interpreted this astonishing result in terms of *contextual* selectivity, as chimpanzees seem to selectively opt to imitate or emulate according to one or several factors that make the choice adaptive. The human demonstrator showed several tool-using actions on a complex food container, using a mixture of effective and ineffective actions. This was done in two conditions that differed in the visibility of the box (and the effects of the actions). The young chimpanzees demonstrated context-sensitive, flexible social learning by copying reliably both types of demonstrated actions (causally relevant and irrelevant ones) only if they couldn't see the causally irrelevant actions demonstrated. If they could see them, they ignored the ineffective parts and predominantly tried the effective ones instead.

Inferential Selectivity

The final example of selectivity concerns efficient, goal-directed reproduction of a demonstrated food retrieval technique in the social context. Recent advances in child psychology demonstrate that already 3- to 12-month-old infants interpret others' behavior as goal directed (Tomasello 1999; Woodward 1998) and, as a result, predict the most efficient action to achieve a goal within the constraints of a given situation (Gergely and Csibra 2003). This inferential competence does not require the attribution of mental states to others but relies simply on the evaluation of the observable facts: the action, the goal state, and the situational constraints. One situation in which human infants are thought to manifest this nonmentalistic inferential process is their selective imitation of goal-directed actions. In an ingenious study, Gergely et al. (2002) showed that 14-month-old children copy a nonpreferred action only if the model has the opportunity to use the preferred action to solve the same problem, but do not copy it if constraints "force" the model to avoid the preferred action. A similar finding was recently reported from dogs (Range et al. 2007) and chimpanzees (Buttelmann et al. 2007).

Our aim was to simulate the Gergely et al. (2002) study as closely as possible, to make a comparison between "humans and their best friends" maximally valid. We designed a manipulative task that required dogs to pull down a wooden rod to open a food container connected to the rod in order to gain a food reward (Range et al. 2007). Manipulations of 14 dogs in the control group that had not seen a demonstration of the task showed that the dogs' preferred method to solve this problem is pushing the wooden rod down with their mouth. In two experimental groups we let other dogs observe an adult female dog that used her paw to obtain food from the apparatus. In one group the demonstrator dog's use of this nonpreferred method was explicable because she was carrying a ball in her mouth (mouth-occupied group, n = 21), while in the second group no constraints were present to explain the demonstrator's choice of the nonpreferred method (mouth-free group, n = 19). The dogs in the two experimental groups were first allowed to watch the model dog producing food from the test apparatus 10 times by using her paw. After each of the 10 demonstrations, the observer dogs were encouraged to manipulate the apparatus to get the food by themselves. Interestingly, while dogs in the control group and in the mouth-occupied group predominately used their mouth to manipulate the wooden rod, 83.33 percent of the dogs in the mouth-free group used their paw. We found the same significant differences when analyzing the percentage of cases where the paw was used in all manipulations, including unsuccessful attempts. Thus, the dogs' performance was comparable to that of the infants in the Gergely et al. (2002) study; they *selectively imitated* the model's action (Huber et al. 2009).

Although all the dogs watched the same number of demonstrations, it is still possible that the selective reenactment resulted from a different level or focus of attention in the two experimental groups rather than from an inferential process—for example, in the mouth-occupied group the presence of the ball may have attracted the observers' attention and thus led to a failure to notice the model's paw action. After the first trial, however, we continued testing the dogs in an additional seven trials, placing food in the container and letting the dogs operate the rod repeatedly. Significantly, by the second trial, the dogs in the mouth-occupied group also started using their paws, whereas the control dogs stuck to the mouth use (Range et al. 2007). This later performance shows that the dogs in the mouth-occupied group learned the same information from the model as the dogs in the mouth-free group but did not manifest it in their initial performance. Thus, attention alone cannot explain the different performance of the dogs; it must be assumed that other processes contributed to the decision-making processes.

As with the Horner and Whiten (2005) and Buttelmann et al. (2007) studies, our findings illustrate selective and functionally appropriate use of either imitation or emulation in relation to variations in both physical and social contexts (but see Tennie et al. 2009; Whiten et al. 2009). The functional significance of this inferential selectivity for "cultural learning" is obvious. The transmission of cultural knowledge requires learners to identify what relevant information to retain and selectively imitate when observing others' skills. Simply copying in a blind manner would be maladaptive. Human infants and some nonhuman species—without relying on language or theory of mind—already show evidence of this ability. So far, only dogs and chimpanzees have been tested in this respect. Further research is necessary to determine the taxonomic range of this ability.

Future research should also focus on the attempts to modify the observer's action repertoire by incorporating new actions demonstrated by a model, on the opaqueness of the demonstrated action (if the copied action can be evaluated by the observer), on the difference between object-oriented and gesture-like actions (as the former seem to be much easier to copy or to achieve a high matching degree), and finally the relation-ship between model and observer in rational imitation tasks (as this may strongly influence the ability to grasp the intention or goal of the model). In this respect the use of the do-as-I-do paradigm seems to be a fruitful strategy (Huber et al. 2009).

CHAPTER SIX

Learning How to Forage: Socially Biased Individual Learning and "Niche Construction" in Wild Capuchin Monkeys

Elisabetta Visalberghi and Dorothy Fragaszy

HIS CHAPTER aims to provide a way to think about how naïve monkeys become proficient foragers. In general, young primates (at the time of weaning and for some period thereafter) are less effective foragers than adults of their species. Primates have complex diets, live highly social lives, and spend months to years as juveniles. These characteristics, taken together, suggest that social partners may influence how young monkeys learn about food and feeding. Much research has addressed psychological processes occurring in the short term and within the learner that allow an individual to match another's behavior (such as imitation, emulation, or social facilitation; for review of these processes in relation to foraging, see Rapaport and Brown 2008, and also this volume). Here, we adopt an ethologically grounded approach to social learning, focusing on how young individuals acquire foraging skills in natural contexts. One of our major tasks is to explain why, in the case of difficult foraging tasks, young animals engage in patently ineffective foraging behaviors over some period of time while they are acquiring the skill. During most of this extended period, their efforts are not reinforced in the usual manner (by obtaining food). Thus, other conditions must support persistent practice. We argue that other monkeys, through their own foraging, construct a niche for young monkeys (*sensu* Odling-Smee, Laland, and Feldman 2003; see also Laland et al. 2000) by providing young monkeys with the opportunity and/or the motivation to practice foraging for those foods that are difficult to find or difficult to obtain.

Niche construction refers to the consequences of individuals' actions that in effect define or partially create their own living environments.

Termites provide a familiar example of niche construction: termites build mounds, which multiple successive generations inhabit. The mound provides a particular environment (temperature, humidity, etc.) that termites require. Through building and maintaining the mound, the behavior of individuals impacts the lives of others at that time and later. Niche construction modifies sources of natural selection and thus, in principle, affects the evolutionary process. For this reason, niche construction theory is one component of an extended theory of evolution (see http://www .nicheconstruction.com). Social influences on learning are a form of niche construction, and thus they carry evolutionary as well as developmental importance.

Wild capuchins (New World monkeys in the genus *Cebus*) provide cogent examples of how social learning about foraging can be examined in this ethological way. Capuchins are generalists, meaning they live in a wide variety of habitats and thus eat a wide variety of foods. The diet of capuchin monkeys in one area may have little in common with the diet of monkeys of the same species living in another area, and both populations can exploit hundreds of species of plants and animal foods (Fragaszy et al. 2004, appendix I). They are omnivorous, eating insects, mollusks, vertebrates, fruits, seeds, flowers, roots, leaf buds, fungi, tree gum, underground plant storage organs such as tubers, etc. These foods may contain toxic substances, be hidden from view, be encapsulated in a tough outer covering of some sort, and/or need specific processing techniques (for review see Fragaszy et al. 2004), and these features vary locally. The tufted species of capuchins (*C. apella, C. libidinosus, C. nigritus,* species living in the southern and eastern part of the range of the genus) in particular include tough foods in their diet (Wright et al. 2009). Learning about food and feeding from group members is thought to be particularly relevant in generalist species (Galef 1993).

Young monkeys have lesser physical resources for foraging than do adults, with respect to teeth, bite force, and manual strength. During and after weaning, young capuchin monkeys are much smaller than adults, and most of their permanent teeth have not yet erupted. They are also less experienced at foraging. Thus they seem to be ill-equipped to forage on the same foods as adults. The period after weaning is a dangerous one for young capuchins, and they grow slowly, suggesting that acquiring sufficient food is indeed challenging for them (Janson and van Schaik 1993). Thus young monkeys face strong challenges to find and to process many foods common in their diet.

Yet capuchins have certain physical, behavioral, and social resources that allow them to navigate this dangerous period (reviewed in Fragaszy et al. 2004). Capuchins exhibit well-developed manual dexterity (superior to other New World monkeys) and a strong propensity to explore objects and surfaces in diverse ways, which in some settings is expressed in spontaneous tool use. They have proportionally large brains for their body size, suggesting well-developed perceptual and cognitive processes supporting learning. They are weaned gradually over a period of many months and spend several years as juveniles, providing ample time for learning. Finally, and the feature that is most relevant to the issue of social supports for learning, capuchins live in cohesive social groups, and members of a group exhibit a high degree of tolerance toward each other, especially toward infant and juveniles, both in the wild (Izawa 1980; Janson 1996; Perry and Rose 1994) and in captivity, where food is sometimes transferred from one individual to another (de Waal et al. 1993; Fragaszy et al. 1997; Thierry et al. 1989). Young capuchins are highly motivated to watch others foraging or smell others' food, especially when the food is novel or difficult to acquire (Fragaszy et al. 1997, Ottoni et al. 2005; Perry and Ordoñez-Jiménez 2006; Drapier et al. 2003).

Overall, this suite of characteristics parallels those of humans in important ways. For example, humans also display extended juvenescence, well-developed learning abilities, curiosity about objects, tolerance toward young individuals, and a diverse, challenging, and locally variable diet. All these parallels make study of capuchin behavior particularly interesting for comparative purposes. Capuchins seem an ideal taxon through which to explore the issue of how youngsters learn to find and process foods, and how social context supports this learning. Our goal is to describe how specific foraging skills are acquired by young wild capuchins living and acting with their group members in natural settings. We focus on examples from three different species of capuchin monkeys: learning to find larvae hidden inside bamboo stalks (*Cebus apella,* the tufted capuchin), learning to use hammer-and-anvil tools to crack palm nuts (*Cebus libidinosus,* the bearded capuchin), and learning to access the encased seeds of *Luehea candida* (*Cebus capucinus,* the white-faced capuchin).

Socially Biased Individual Learning and the Ecological Approach

In keeping with our ethological perspective on social learning and our interest in linking social propensities to niche construction, Fragaszy and Visalberghi (2001) proposed an inclusive model of socially biased learning in natural circumstances. Socially biased learning is framed within the social and physical setting of behavior, as well as influenced by the characteristics of the individual, and all these elements are interrelated. Individual characteristics include behavioral repertoire, general attraction to others, salience of specific partners present at that moment, responsiveness to objects, motivation to engage in new activities, prior experience with the setting, and ongoing experience (e.g., current activities, current internal state). Social elements that bear on an individual's likelihood of learning from others include the composition of social partners, tolerance of these individual for the focal learner, the behavior of the other, the value added to an object or a place from another's actions there as well as its emotional expressions while doing so (e.g., vocalizations associated with food), and enduring changes in the environment that remain from the other's activity (e.g., bits of food, altered substrates—hereafter, physical traces). The physical setting includes the abundance of sites to act and the accessibility of these sites. Finally, the physical setting affects risks for action. For example, monkeys are less likely to explore a new opportunity for action in a setting where perceived risk of predation is high compared to a setting where perceived risk of predation is low.

Foraging for Larvae Hidden inside Bamboo Stalks

Gunst et al. (2010) describe how young wild brown capuchins *(Cebus apella)* learn to find and retrieve beetle larvae hidden inside tough stalks of bamboo *(Guadua latifolia)* in the Raleighvallen Central Suriname Nature Preserve (Suriname). Gunst et al. (2008) characterize the larvae obtained from the interior of bamboo stalks as difficult foods, in contrast, for example, to the new young leaves of bamboo, which the monkeys find and eat in the same area. Obtaining a larva from its tough, concealing substrate requires selecting an appropriate bamboo stalk, locating the larvae hidden inside (both components of searching), and ripping the stalk open and extracting the larvae (handling components) (Figure 6.1 a, b, c). Locating an appropriate stalk and an appropriate site on the stalk is not easy, because the areas of bamboo that contain larvae do not differ in appearance from areas lacking larvae. Choosing the right spot to open is important because ripping the stalk open requires strength and is time consuming. Monkeys reach adult efficiency at this foraging task—obtaining five to six larvae per hour allocated to searching for larvae—at about five years of age, although they devote considerable time to inspecting and opening bamboo stalks from about one year of age.

Gunst et al.'s studies show how social partners' alteration of the physical environment can aid the young monkeys' development of skill in obtaining larvae. Young monkeys are attracted to canes already opened by adults, and at these sites they practice behaviors that contribute to finding and obtaining larvae. Specifically, immature monkeys performed significantly more larvae-related foraging behaviors (rapidly tapping the cane with the fingertips—called tap scanning—inspecting the cane with fingers or nose, biting into and ripping bamboo stalks apart) within two minutes after approaching a ripped bamboo stalk left by a skilled forager than they did in the two minutes before (Figure 6.1d). In contrast, experienced foragers inspected ripped bamboo stalks briefly and did not follow inspection with foraging. Thus, the physical traces left by skilled foragers stimulate in youngsters activities likely to contribute to the acquisition of the foraging skill at hand. In short, skillful individuals "leave

gure 6.1. Bamboo ripping by an adult male (the alpha male) while an infant is watching and b). After finishing ripping, the adult male is about to extract a larva while a juvenile watching (c; the infant is hidden behind the adult male, probably watching too, but es not show on the photo). The adult male having left the foraging spot with its larva, e infant is inspecting the already ripped bamboo stalk (d). Photos courtesy of Nöelle unst.

the landscape littered with prepared 'practice' sites that appeal to younger monkeys" (Gunst et al. 2008, 21). We can think of physical traces as a form of niche construction and the young monkeys' response to physical traces as delayed, indirect social facilitation.

Finding the larvae seems to be the most challenging part of this foraging task. Bamboo grows in dense groves, and the sections containing larvae do not appear visibly different from canes that do not. In direct inspection of all the canes in five five-square-meter quadrats in a large bamboo patch, researchers found no larvae inside rotten stalks with light brown epidermis, internodes (sections of cane between growth nodes, usually about 30 centimeters long) already ripped apart by capuchins, or thin stalks. In contrast, large and medium green stalks contained an average of 0.05 larva per intact internode, and never more than one per internode. These findings indicate the importance of directing ripping activity to where larvae might be present. Accordingly, they suggest that perceiving cues associated with larvae and using them to guide search is important to optimize the time and effort devoted to searching for larvae.

Faced with a vast expanse of bamboo canes in a patch, how do capuchins search for larvae? Before finding a larva, the monkeys commonly tap scan, sniff, and inspect canes visually and manually and bite and rip the stalk apart. Stepwise linear regressions using foraging efficiency as the dependent variable demonstrate that visual inspection and tap scanning predict foraging efficiency, whereas the other behaviors do not. Visual inspection and tap scanning tended to become more frequent with age, whereas manual inspection and biting, which were not predictive of finding a larva, tended to decline with age, although even adults performed these behaviors at low rates.

Whereas younger animals directed extractive behavioral patterns toward small healthy stalks, or already-ripped stalks, the adults focused on large healthy stalks, where larvae are likely to be found. Interestingly, two adult males that had recently immigrated into the study group spent less time than other adults searching for larvae and were no better than older juveniles at finding larvae. These findings suggest that these monkeys were relatively naïve about this particular foraging activity, and highlight the dependence of efficient searching behaviors upon extended practice even for individuals with full physical capabilities.

Social context could help the monkeys learn the perceptual cues that indicate the presence of a larva inside a cane. For example, through their attraction to the sites where others have already opened canes and extracted larvae, youngsters could learn to notice the presence of the tiny hole made by the insect while laying the eggs that develop into larvae, or the odor associated with the larva. Similarly, from watching adults searching, they could learn that tapping serves as a relatively reliable cue about the presence of the larva inside the stalk. Capuchins in other regions than Gunst et al.'s study area tap scan while foraging for insects embedded in dead branches (weeper capuchins, *C. olivaceous*, Fragaszy 1986; brown capuchins, *C. apella*, Phillips et al. 2003). Thus tapping is a genus-typical behavior, performed by capuchins monkeys in many settings, but the monkeys in each location must learn when to use it effectively and for what purpose.

Use of Hammers and Anvils to Crack Palm Nuts

All over their geographical distribution, capuchin monkeys pound objects, such as hard fruits or snails, on hard surfaces in order to get access to the inner parts (for review see Fragaszy et al. 2004). Both in captivity and in the wild, some capuchins learn to use hammer stones and anvils to crack open nuts (for review see Visalberghi and Fragaszy 2006). Young capuchins, like young chimpanzees (Inoue-Nakamura and Matsuzawa 1997), learn to crack nuts over several years (Resende et al. 2008). Why does it take so long to master cracking nuts using a hammer stone and anvil? How does social setting support or hinder learning to crack nuts?

Resende et al. (2008) systematically investigated the ontogeny of manipulative behavior and nut cracking in nine young semi-free-ranging capuchins (*Cebus* spp., probably mostly *C. libidinosus* and *C. nigritus* and hybrids of these species) living in Tiête Ecological Park (near the city of São Paulo, Brazil) over 23 months of observation. The monkeys began to pound objects on surfaces at 2–3 months of age, and at the same time as they began to act directly with objects. Between 6 and 12 months they manipulated stones or nuts separately; banging the nut or stone directly on a substrate was the most common action. During the second year, manipulative activities became both very frequent and more vigorous. In particular, pounding became the most common action linking object and surfaces. Placing an object, such as a nut, on a surface and then releasing it was rare and was the last action necessary for nut cracking to appear; this action was first seen when the monkeys were 19–24 months old. The two young monkeys in the study that cracked nuts did so for the first time at 25 and 29 months. Resende et al. (2008) suggested that placing the nut on an anvil and releasing it posed one of the main difficulties for nut cracking for capuchin monkeys. Perhaps releasing an object in which they are still interested requires overriding a strong proclivity to maintain a secure grip on it. Eventually, the monkeys open nuts by placing them one at a time on an appropriate substrate and striking them forcefully and accurately with a hammer. Young capuchins generally follow the same pattern of acquisition described by Visalberghi (1987) for two captive adult capuchins when they encountered, for the first time in their life, wooden blocks and nuts and learned to use the blocks as hammers to crack the nuts on a concrete floor. One striking difference, however, is that the captive adults first cracked nuts in a few days, rather than a few years.

The research team (including the authors) studying the wild bearded capuchins *(C. libidinosus)* in Gilbuès, Piauí, Brazil, at Fazenda Boa Vista (hereafter FBV; see http://EthoCebus.net) has written several reports about nut cracking in these monkeys (Visalberghi, Spagnoletti, et al. 2009; Liu et al. 2009; Spagnoletti et al. 2011; Fragaszy, Pickering, et al. 2010; Fragaszy, Greenberg, et al. 2010). This site is in the northeast of Brazil, at approximately 9 degrees south and 45 degrees west. We have as yet few developmental data about nut cracking from FBV. We have noticed, however, that the capuchins seem to acquire nut cracking following the same trajectory as the semi-free-ranging capuchins in Tiête Ecological Park as described by Resende et al. 2008. In FBV, youngsters devote most of their efforts, and achieve their first successes, with partially opened nuts that they recover from the vicinity of the anvil, or with the least resistant species of nuts (unpublished data). Some youngsters at FBV can crack open less-resistant nuts by two years of age.

We suggest that social setting can positively bias learning indirectly, when group members are not presently cracking nuts, and directly, when they currently are involved in nut cracking. Indirect influence arises from the previous actions of others that create a supportive physical environment, as was also the case for monkeys learning to forage for larvae in bamboo canes, as reviewed above. At FBV there are three kinds of physical traces of activity that are helpful to youngsters learning to crack nuts. First, capuchins transport stones to anvil sites and leave them there after having cracked nuts. Later, monkeys arriving at the anvil sites use the hammer stones already present there. Second, because anvils are relatively soft sandstone or wood, as they are used repeatedly, pits develop in the area where the monkeys strike the nuts with the hammer. By producing pits, capuchins improve the affordances of the anvil for themselves as well as for future nut crackers (Fragaszy et al. 2010). By leaving hammer stones on anvils and by creating pits in the anvil surfaces through repeated use, capuchins make it easier for youngsters (or other unskilled individuals) to learn to crack nuts with hammers. The hammers are on the anvils, and the pits provide a ready place to put the nut so that it can be struck securely (Fragaszy et al. 2010). Monkeys only need to show up at the anvil with their nuts; the materials to crack them are ready at hand, prepared by others. Finally, monkeys frequently leave bits of nuts and shells at the anvil when they leave the site, after having cracked one or more nuts. The hammer stones and anvils retain oily (and fragrant) traces of the nut kernels where they have been smashed against these surfaces. These features attract young monkeys' attention when they approach the anvil site, whether or not they are able to crack nuts on their own.

Direct positive social influence occurs when group members are cracking nuts. Nut cracking is a noisy, vigorous activity, and the sound and motion attract youngsters. They may watch from some distance, or they may stay near the anvil while another is cracking, sometimes handling smaller stones and nut shells in the vicinity, and they may take pieces of nuts cracked by others while the others are still at the anvil (Figure 6.2). Over many months they spend a long time in this permissive social setting (Figure 6.3). Eventually, when proficient tool users leave their hammer and/or partially opened nuts on the anvil, youngsters use them to

Figure 6.2. The alpha male has extracted a kernel from a piassava nut, and a juvenile is nearby searching for bits of nut remaining on the ground. Photo by Elisabetta Visalberghi.

Figure 6.3. A juvenile closely monitors the nut-cracking behavior of the alpha male. Photo by Elisabetta Visalberghi.

"practice," if they are strong enough to lift the hammer stone (which may weigh more than the young monkey trying to lift it), or when very small, they strike one nut on another. In this way youngsters' exploratory actions with nuts and stones occur in a place with the appropriate elements for success (that is, pieces of nuts, hard stones, and pitted anvil surfaces). Although adults do displace juveniles from anvils, direct agonism is rare (Verderane 2010).

Ottoni et al. (2005) assessed the proficiency of each tool user in semi-free-ranging capuchins in the Tiête Ecological Park near São Paulo. They also recorded the extent to which other monkeys observed each tool user and/or collected remains of nuts afterward (scrounging), while or just after the other monkey cracked the nut. In 76 percent of dyads, the tool-using monkey was more proficient at cracking nuts than was the observing monkey. Scrounging occurred in 35 percent of the episodes in which the monkey using the tool cracked the nut. Since the most proficient nut crackers would tend to yield, on average, the highest payoff for scrounging, it is likely that the proximate cause of the young monkeys' selective observation of particular nut crackers is the opportunity for scrounging that these proficient individuals afford the observer. The nonrandom pattern of observing others enhances scrounging payoffs and, coincidentally, maximizes opportunities to associate places, objects, and actions with obtaining nuts. As Ottoni and co-workers write, "simple associative or re-inforcement processes can underlie the tendency of capuchins—youngsters in particular—to watch nutcrackers at work. This simple mechanism could, by itself, optimize the conditions for the social learning of nut cracking techniques and for the diffusion of tool-aided nut cracking as a behavioral tradition" (p. 218).

Ramos da Silva (2008) obtained parallel results in a group of wild bearded capuchins in FBV. He found that one or more monkeys watched 25 percent of the nut-cracking episodes, with the number of observers ranging between one and four. As in Tiête Ecological Park, the more fre-quently watched capuchins were those that used tools more frequently and with higher rates of success. In 35 percent of the nut-cracking epi-sodes observed by a juvenile, the juvenile manipulated the same hammer or the nut and/or attempted to crack a nut (or part of a nut) with the hammer within a few minutes after watching another cracking with that hammer and anvil. This finding suggests that exploration of anvil sites and hammers is socially biased. Social facilitation of the specific actions associated with nut cracking (e.g., pounding a stone on a surface) is prob-ably part of this package (see also Visalberghi 1987). Experimental studies to determine if pounding an object can be socially facilitated in capuchins would strengthen this interpretation.

In summary, nut cracking by wild capuchins appears to be socially biased in several ways that increase the likelihood of naïve individuals learn-ing to crack nuts with stone tools. First, young capuchins up to about two years of age are well tolerated by adults in feeding contexts and are likely to scrounge bits of food from others (see also Fragaszy et al. 1997). Sec-ond, the repeated use by group members of the same anvil sites and the same hammers, together with enduring traces of nuts cracked at those sites, provides familiar places with appropriate resources to practice ham-mering. Third, juveniles are strongly attracted to watch the activities of others, which is linked with the probable social facilitation of pounding, and with the motivation to explore the hammers and anvils. All these so-cially provided elements increase the likelihood of a juvenile acquiring tool use if it lives in a group that practices this behavior routinely (compared to a group that does not). In other words, through their behavior, adults construct a niche in which youngsters reliably learn to crack nuts. Nut cracking is likely to be a tradition in capuchin groups.

Does this means that the presence of skillful group members is the *conditio sine qua non* for monkeys to learn to crack nuts? Apparently not, since there is plenty of evidence that the behavior can be acquired by naïve individuals (youngsters as well as adults) without social input (e.g.,

Visalberghi 1987; Fragaszy et al. 2004). When they have access to encased food such as nuts, and in a physical setting promoting nut cracking
(with hard substrates and hammers), some capuchins (but not all) discover on their own how to use tools to crack nuts. Interestingly, naïve
adults exhibit the same array of explorative behaviors and (spatially correct and incorrect) actions combining the nut and the tool that have been
described for young capuchins (Resende et al. 2008). On the other hand,
merely living in a group that practices nut cracking does not guarantee
that a monkey will acquire this skill. Of the 24 physically normal capuchins observed at FBV that are old enough to use tools, one individual
was never seen to crack nuts using a stone tool.

Might young monkeys learn to crack nuts, or improve their technique,
from directly copying some aspect of the behavior of others? Field observations cannot answer this question decisively, but we think the answer is
no. Pounding because another monkey is pounding is one entry point for
skill development, but simply pounding a stone on a nut is not sufficient
to crack the nut. Cracking a nut requires skillful placement of the nut followed by skillful handling of a heavy stone, but no specific technique of
placing or handling. It is a motor skill more than a matter of special techniques. Novice alpine skiers must practice on their skis to master skiing
down a snowy slope under control, and no amount of watching a skilled
instructor will substitute for direct practice. It seems that nut cracking is
like skiing: the observable components of the action are straightforward,
but their competent execution requires extended practice. The trajectory
of the stone, the force with which it strikes the nut, the position of the nut
in the anvil—all these aspects affect success and need a lot of individual
practice to optimize. Even after the young monkey reliably produces
all the relevant actions in the correct sequence, it may take years before it
succeeds in cracking a nut. Having proficient group members to watch
could contribute to skill development over this long period, not because
young monkeys learn anything specific from watching others, but because
watching others, like encountering physical traces of their activity, increases their motivation to act and channels their choice of elements with
which to act toward the right ones to learn to crack nuts.

Processing Encapsulated Seeds

Infant capuchins begin to exhibit most species-typical manipulative
actions such as pounding and rubbing objects on substrates in the first
months of life (Adams-Curtis and Fragaszy 1994; Spinozzi 1989). At first,

pounding and rubbing are not differentiated, but gradually they become distinctive, with pounding involving brief force applied intermittently in a plane perpendicular to the substrate and rubbing involving sliding a held object in a plane parallel to, and in contact with, the substrate, usually with cyclical back-and-forth motions. In general, during foraging these actions are used for different purposes: pounding is used to break a rigid surface, whereas rubbing is used to remove a pliant layer (for example, to remove the chemical and mechanical defenses of caterpillars). However, in some cases, as the one illustrated below, the two actions are directed at the same object for the same goal.

Perry (2009) studied how young white-faced capuchins *(Cebus capucinus)* developed the processing techniques used to exploit the seeds of *Luehea candida.* At Perry's site, the Lomas Barbudal Biological Reserve in Costa Rica, as well as elsewhere (Fragaszy et al. 2004), white-faced capuchin monkeys eat luehea seeds. The monkeys devote up to 15.4 percent of their foraging time to these seeds during the peak fruiting season (Perry and Ordoñez-Jiménez 2006). Thus luehea seeds constitute an important part of the diet for the monkeys in Lomas Barbudal in the season when they are available. The luehea trees average 10 meters in crown diameter and 15 meters in height, and typically two to three monkeys forage in one tree simultaneously. The fruits are wooden capsules containing many tiny, nutritious seeds, and when ripe, their five seams slowly open to release the wind-dispersed seeds (Figure 6.4). Capuchins feed on the seed pods before they are fully open.

The monkeys adopt two different techniques to loosen the luehea seeds from their point of attachment deep in the cracks so that they fall out or can be more easily plucked from the tip of the fruit. The pounding technique consists of repeatedly striking the fruit against a substrate; the rubbing or scrubbing technique consists of repeatedly moving the fruit back and forth across a rough surface. Fragaszy et al. (2004, 131) describe the action of rubbing as occurring "when an object is drawn backward and then forward against the substrate"; we believe that scrubbing and rubbing refer to the same behavior. Perry (2009) tried herself to extract the seeds by pounding and by scrubbing. Through pounding she obtained 7.8 seeds per 10 seconds and through scrubbing 5.8 seeds per 10 seconds. Thus these methods produce seeds at roughly equivalent rates.

An unusual and very valuable feature of Perry's study is that it was longitudinal, extending five years. Perry observed that juveniles between one and two years old tried a wide variety of techniques, including pounding and scrubbing and a combination of the two, and they were generally not

Figure 6.4. A white-faced capuchin extracts winged seeds from luehea pods. In the upper right corner: a close view of the pod and its seeds. Santa Rosa National Park, Costa Rica. Photos courtesy of Katherine C. MacKinnon.

able to get the seeds. When three to five years old, these same youngsters gradually abandoned the inefficient variants they had used earlier. By five, they had generally settled on one technique (pounding or scrubbing). In all four groups studied by Perry (2009), pounding and scrubbing were each used by at least one adult. Young individuals are typically exposed to both pounding and scrubbing, although at different rates (because most adults pound).

Perry (2009) examined the technique used by each youngster during its first five years of life, in relation to that individual's estimated exposure to pounding performed by the mother and by other group members during each luehea season. Her findings show that the youngsters with an early bias for one of the two techniques increased their bias for that technique as they aged. Overall, females were likely to use the same technique as their mothers, whereas males were not. Regression analysis revealed that for both males and females the technique most frequently observed significantly predicted the technique adopted by observers, particularly in the second year of life, although the predictive value of the observed technique for the practiced technique was lower for male observers than female observers.

According to Perry (2009), her findings suggest that observation of others' foraging techniques influences which techniques youngsters use, as has been found in captive capuchins (Dindo et al. 2009 and Crast et al. 2010). Perry suggests that the intrinsic pleasure individuals obtain from copying the actions of individuals with whom they have special bonds, as proposed by de Waal's (2001; Bonnie and de Waal 2007) "Bonding—and Identification—based Observational Learning Model," may support this process. Alternatively, perhaps the observer prefers to be in proximity with group members performing the same technique as itself. This scenario is suggested by the finding of Paukner, Suomi, et al. (2009) that capuchin monkeys affiliate more with humans that contingently match their behavior than humans that match their behavior, but not contingently. In other words, being imitated promotes affiliation in captive capuchins. When a monkey is scrubbing, another individual doing the same thing provides more contingent matching of its behavior than does an individual that is pounding, and thus the first monkey might prefer to be near another monkey that is scrubbing. This process would on average produce congruent techniques for opening seeds in youngsters and their near neighbors. As another alternative, it is possible that both scrubbing and pounding are socially facilitated, as eating behavior is (Visalberghi and Addessi 2003; Ferrari, Maiolini, et al. 2005), and as we suggest pounding nuts may be. Even weak social facilitation may be sufficient for most young monkeys to match, eventually, the technique used by most members of their group. In any case, Perry's findings illustrate the complexity of the social context in natural settings, where monkeys see individuals acting on the same objects in variable ways. In such situations, the monkeys do not show strong fidelity to the particular method they observe the most, unlike in some experimental studies with captive capuchins, where the monkeys show strong fidelity to the one technique they observe another individual performing (e.g., Fredman and Whiten 2008; Dindo et al. 2009).

Conclusions

Capuchin monkeys afford an interesting view of how physical, social, and experiential factors contribute to the development of foraging competence in a long-lived primate (Gunst et al. 2008). These monkeys display regionally variable foraging specializations, reflecting local resources. The three examples of distinctive foraging behaviors acquired by capuchin monkeys in natural settings that we have reviewed here demonstrate how social influences can play out differently in different foraging situations. In the first, young monkeys learning to find larvae embedded in bamboo canes must learn how to find a cane worth opening, as most bamboo canes do not contain larvae. Initially, they preferentially look in canes that adults have already opened. In this example the enduring physical traces of adults' foraging activities strongly influence young monkeys' behavior. From already-opened canes, the youngsters can learn how such canes smell, they can practice ripping bamboo, and perhaps they can learn or practice other features of the task. However, looking in the canes that adults have already opened will not, by itself, ever lead to finding a larvae, because the adults remove the single larva in each cane that they open. We do not yet understand how the monkeys eventually learn to detect the presence of larvae before they open a cane, but it is not from straightforward facilitation of action at the same site.

In the second example, social facilitation of action and activity at a prepared site, together with scrounging opportunities, provides a roundly supportive setting for young monkeys learning to crack nuts. The anvils are improved by use, with the formation of pits that serve to reduce the probability that the nut will be displaced from the anvil as it is struck (Fragaszy et al. 2010). Durable hammer stones, which are rare in the landscape (Visalberghi, Spagnoletti, et al. 2009) are routinely left at anvil sites and thus are available to the next user. Thus, as in the previous example, enduring physical traces of others' activity set the stage for effective practice by young monkeys. Additionally, and unlike the previous example, young monkeys have repeated opportunities to scrounge bits of food at the anvil, while or after others have cracked nuts there, and they have ample opportunities to watch others cracking nuts. All these features of the setting promote persistent exploration of the relevant materials and may directly facilitate pounding actions. Collectively these features contribute to young monkeys becoming proficient nut crackers, but we do not yet know their relative contributions.

In the third example, young monkeys learning to open luehea seeds do not have prepared sites or enduring physical traces to explore, nor do they have extensive opportunities to scrounge food from others foraging on these seeds. But, as in the two previous examples, they do have repeated exposure to others performing a noisy and vigorous activity and obtaining food from doing so, and this seems sufficient to motivate persistent efforts by young monkeys to open these seeds themselves. In this case, the food item is manipulated directly (as opposed to being hidden inside bamboo or struck with another object, as in the previous examples). Being with others that use a particular technique seems to bias the technique adopted by young monkeys, although the precise mechanism for the congruence between youngsters and their group mates is not yet clear.

These examples illustrate that foraging presents diverse challenges, and that social influences promoting shared foraging skills, a social form of niche construction (Odling-Smee, Laland, and Feldman 2003), can take diverse forms. In recent years, a lot of attention has been paid to some sources of social influence (for example, visible actions), while others (such as physical traces) have received less consideration. We think that broadening our attention to all the components of the social context will improve our understanding of social contributions to learning. Direct visual observation may not be the primary source of social influence on learning, and copying behavior will often not be a sufficient basis for learning a skill. Many of the relevant parameters of foraging actions are not amenable to observation; they could not be learned by copying the behavior of others.

Many of the studies of social learning in monkeys, particularly in foraging (our own included), carried out in the laboratory are necessarily disconnected to a greater or lesser extent from the species' natural ecology. For example, group size and composition, the nature of the food resource, its temporal and spatial distribution and abundance, etc., do not match those known for the species in natural environments. Often the experimental tasks involve problems the animals do not deal with in nature (for example, transparent containers, or fixed panels or levers), and time constraints are imposed, both on our observation and on the animals' exposure to the problem. We believe this state of affairs has contributed to our blindness to the richness of the social biases affecting the behavior of individuals, particularly more naïve individuals, and especially to biases that become manifest over long (developmental) periods. Foraging skills, for example, develop over years. Understanding social contributions to learning important life skills, such as foraging, will be best accomplished if we keep clearly in mind that a wide range of parameters can affect learning over a long period, and the same parameters may have variable impact over this long period, a point of view expressed some years ago by Russon (2003). Social niche construction impacting learning can affect a range of physical dimensions, and individual learners bring diverse susceptibilities, predispositions, and emerging skills to the setting. Thus it is inappropriate to ask if a given skill is "socially learned" in a natural setting—all skills are acquired, and most slowly, in a social context that, like the learner, changes over time.

Acknowledgments

European project IM-CLeVeR FP7-ICT-IP-231722.

CHAPTER SEVEN

Social Learning and Culture in Child and Chimpanzee

*Lydia M. Hopper, Sarah Marshall-Pescini,
and Andrew Whiten*

Introduction: "Deep Social Mind" and Culture

More than three decades of primate research have accumulated since Nick Humphrey galvanized the study of primate societies with his "social intellect hypothesis" (Humphrey 1976). The essence of Humphrey's idea was that the distinctive complexities of primates' social lives have been responsible for shaping their intelligence more than have the physical problems they face. This idea resonated and interacted with other developments in the field, perhaps most influentially Griffin's efforts to make the study of animal mind a reputable and worthy enterprise (1976) and de Waal's discoveries about "Chimpanzee Politics" (1982), a subject to which repeated quotations from the architect of human intrigue, Machiavelli, were aptly applied. Soon after, the gathering empirical evidence on social intellect was accordingly collated under the title "Machiavellian Intelligence" (Byrne and Whiten 1988; Whiten and Byrne 1997; see also "Social Brain" hypothesis, Dunbar 1998). These developments were followed by growing swaths of related research on social complexity in other mammals and birds (de Waal and Tyack 2003; Emery et al. 2007).

Whiten (1999a, 2006) hypothesized that a distinctive hominid elaboration on this social intelligence—"Deep Social Mind"—underlay the evolutionary success of our own genus in developing the exploitation of a new, cognitive, hunting-and-gathering niche, unique among primates. This mentality was inferred to include four main components, which together made evolving hominins' minds more deeply interconnected than any other species: (1) egalitarianism and cooperation, incorporating division of labor coupled with unprecedented interdependence; (2) mind reading (a.k.a. theory of mind), in which minds interdigitate with each other in increasing depth; (3) culture, in which the contents of minds derive increasingly from social transmission between them; and (4) language, which permits all of these to develop deeper interpenetration between minds.

These developments made humans the most deeply sociocognitive species on the planet, but did not spring from nowhere. Fundamental features of the first three components have been identified in other primates, permitting inferences about their forms in our common ancestors. These provided vital foundations for their subsequent elaboration in hominin evolution. In the present chapter we focus on the cultural strand in this multicomponent adaptive complex (Whiten and van Schaik 2007). The other strands are all discussed elsewhere in this volume.

Our central aim is to establish features shared between our own species and our closest living relatives, chimpanzees, thus indicating the kind of cultural mind at work in our common ancestry, 6–10 million years ago (Patterson et al. 2006). The other side of this coin is identifying key interspecific differences, generated during our diverging evolutionary pathways. For recent reviews of social learning in primates with a broader perspective than the ape focus of our present chapter, see Subiaul (2007); Caldwell and Whiten (2010); Price et al. (2010); Whiten (in press) and Visalberghi and Fragaszy (this volume).

Social Learning

What Is Social Learning?

Social learning can be described most simply as learning from others. The clearest cases are those where one individual learns from its direct perception of another; but situations where there is learning from the results of others' actions (for example, the tools, odors, or debris they leave behind) mean that the concept of social learning has inherently fuzzy edges.

In this chapter we focus on observational learning—learning through watching. One main area thus omitted should be briefly mentioned. This is teaching, or "scaffolding" (Whiten 1999b), where a tutor's actions inform or shape what a pupil learns. Teaching in this functional sense does occur in nonhuman animals, but appears rare in nonhuman primates. It seems notable in predatory species, where such parental effort supports the developmental leap to catching difficult prey. Recent, exciting work on this topic is reviewed by Thornton and Raihani (2008) and Hoppitt et al. (2008). In primates (including human hunter-gatherers—a way of life that typified most of human evolution), the majority of social learning appears to be observational. This typically begins within the mother-infant relationship, which Matsuzawa (1994), talking of chimpanzees, has characterized as a "master-apprentice" relationship, broadening later to encompass learning from others.

The Role and Significance of Social Learning

Learning by observation can be better than learning by oneself for several reasons. Often it can be more efficient to exploit what others already know, such as what items are good to eat, or how to use a tool. Learning from what others already know about predators can be enormously safer than personal discovery! However, there is a limit on the benefits of social learning. Imagine, at one extreme, that members of a community learn *only* by observation. As soon as the environment changes (a new food source becomes available, or a new predator comes on the scene), simply following what others do may leave one in a maladaptive rut (Laland 2004; Kendal et al. 2005). Modeling work by Boyd and Richerson (1985) suggested that the value of social learning follows a U-shaped curve, being unhelpful in environments that show either minimal or very high rates of change, but highly adaptive in intermediate ones (see also Eriksson et al. 2007). Since all primates have at their disposal both social and individual learning capacities, it may be best to think of this in terms of achieving an optimum balance between these, adapted to local circumstances.

Social learning is often beneficial over the short term, as in learning what is a good fruiting tree. However, some social learning has longer-term effects, generating traditions (Imanishi 1957). Traditions, as defined by Fragaszy and Perry (2003, xiii), involve "a distinctive behavior pattern shown by two or more individuals in a social unit, which persists over time and that new practitioners acquire in part through socially aided learning." As noted above, such traditions can be highly adaptive, yet they can also become maladaptive and die out. For some authors, the concept of culture is equated with traditions, but others require additional features to ascribe culture. Whiten and van Schaik (2007), for example, define cultures as more-complex entities composed of multiple traditions, as occurs in humans and, as we discuss below, in other species including the great apes.

The study of social learning in animals has a longer history than that focused on culture, with experimental research getting well under way in the early years of the twentieth century (Whiten and Ham 1992; Tomasello and Call 1997; Whiten et al. 2004 provide reviews). In this chapter we describe how more-recent studies have gone on to change our ideas about the social learning and cultures of apes. Studies of social learning in children have also by now spanned over a century (Whiten and Ham 1992; R. Mitchell 1987; M. Carpenter and Nielsen 2008), and a "Web of Science" search reveals more than 1,000 articles concerning imitation alone, far beyond what any review can now aspire to cover. Below we selectively highlight several recent advances we think significant.

Social Learning Mechanisms: Chimpanzees and Children

Both chimpanzees and children have the social prerequisites for observational learning (Tomasello and Call 1997), but what form does this learning take? To what sources of information do learners attend? In broad terms, individuals can replicate either the bodily actions of others (imitation) or the outcomes and goals of their actions (emulation) (see Table 7.1). This is a broad-brush distinction that we dissect further below. Complementary, detailed learning taxonomies are offered by Hopper

Table 7.1   Mechanisms of social learning defined

Social learning mechanism	Definition
Imitation	Learning an intrinsic part of a novel act (cf. Whiten and Ham 1992)
Response facilitation	"The presence of a conspecific performing an act . . . increases the probability of an animal which sees it doing the same. Response facilitation crucially differs from imitation in that only actions already in the repertoire can be facilitated" (Byrne 1994, 237)
Program-level imitation	Individuals combine novel and extant behaviors to copy a hierarchically organized sequence of actions
Emulation	Object movement reenactment—the observer replicates the movements of the physical artifact that the model interacted with
	Goal emulation—observers are motivated to achieve the same goal as that demonstrated, but may ignore the means they witness and instead use their own method to attain it
	Affordance learning—an individual learns about the physical properties of the environment and relations among objects

and Whiten (in press), Call and Carpenter (2002, 2003), Zentall (2003, 2006), and Whiten et al. (2004, 2009). We start first with emulation, considered by some to be a simpler process, and then describe imitation.

Emulation

"Emulation" was first used to label the responses of children who achieved the same end-state as a demonstrated action but used a novel way to reach that same goal (D. Wood 1989) and has since been particularly identified with chimpanzees (Nagell et al. 1993; Tomasello 1999; Call et al. 2005). Emulative learning can be further subdivided into "object movement reenactment," "goal-emulation," and "affordance learning" (Whiten et al. 2004).

OBJECT MOVEMENT REENACTMENT (OMR)

OMR (Custance et al. 1999) is a form of emulative learning in which the observer replicates the movements of the physical artifact that the model interacted with. The upshot is that the observer copies what the object, not the model, does. Often, this kind of emulation may offer a more parsimonious explanation of observational learning than imitation (Tomasello 1996).

GOAL EMULATION AND END-STATE EMULATION

Goal emulation is what was referred to by D. Wood (1989) when he first described emulative learning; the observers are motivated to achieve the same goal as that demonstrated, but may ignore the means they witness and instead use their own method to attain it (Tomasello 1990).

The distinction between "goals" and "end states" also warrants attention (Gattis 2002). From an early age, children recognize others' goals (Bekkering et al. 2000). As described in their theory of goal-directed imitation (GOADI), Wohlschläger et al. (2003) proposed that it is perceived goals that drive the observational learning of children, rather than rote imitation. GOADI, although developed to describe the imitative behavior of children, can perhaps be applied to the goal emulation of other species, including chimpanzees.

AFFORDANCE LEARNING

Affordance learning describes what an individual learns about the physical properties of the environment and relations among objects (Byrne 1998a). Byrne argued that "having seen a change of state effected, the observer knows more about the physical nature of an object: that nuts crack

[and] that rocks are heavy and hard" (pp. 604–605) and that an observer may similarly be able to learn the physical relationships between objects and their structural makeup.

Imitation

Thorndike (1898) defined imitation as simply "doing an act from seeing it done." Whiten and Ham (1992, 250) defined it more specifically as a process in which "B learns some aspect(s) of the intrinsic *form* of an act from A." One form of imitation we skirt in this chapter but should note nonetheless is vocal learning. There is minimal evidence for this in primates, including apes (Crockford et al. 2004), in contrast to its pervasive occurrence in songbirds (Catchpole and Slater 1995) and some mammals, notably cetaceans (Janik and Slater 1997) and bats (Knörnschild et al. 2009).

Byrne (2002) distinguished contextual imitation, where the subject applies an already-known behavior to a new situation (cf. Byrne's 1994 definition of "response facilitation") from production imitation, where the individual adds new actions to its repertoire. However, such novelty may be graded, insofar as the imitation of any "new" action in some way builds on what the imitator already has in its repertoire (Whiten et al. 1996). From observations of wild gorillas and rehabilitant orangutans, Byrne and Russon (1998) suggested that individuals may combine novel and extant behaviors to copy a hierarchically organized sequence of actions at a "program-level" of imitation—for example, the specific manner by which gorillas fold nettle leaves in order to make them safe to eat. It should be noted, however, that naïve, captive gorillas also appear to process nettles before eating them in ways that other apes do not (Tennie et al. 2008). This suggests that this behavior is, to some degree, genetically predetermined; whether social learning adds to this remains to be tested (Sawyer and Robbins 2009).

Some consider that, to imitate, the observer must have an understanding of the model's intentions (Tomasello 1999; Horowitz 2003; Huang et al. 2006), or be able to take the model's perspective (Boyd and Richerson 1996). However, as imitative learning has been reported for a variety of species, including rats and pigeons, others suggest it is unlikely that these levels of social cognition are necessarily required (Zentall 2001). It has further been suggested that the evidence for imitation that suggests intention reading (for example Behne et al. 2005; Call and Tomasello 2008) could also be explained by a simpler algorithm in which the behavioral responses of the demonstrator are monitored (see Baldwin and Baird 2001 for a review).

"OVER-COPYING" VERSUS SELECTIVE COPYING

Both chimpanzees and children have been shown to copy actions "rationally," taking their context into account. In a classic study (Gergely, Bekkering, and Király 2002a), a child observed an adult use her head to switch on a light. Children imitated this odd action if they saw the person had both hands free and could have used them instead. By contrast, if the model had both hands wrapped in a blanket, children then (rationally) used their hands and not their head to operate the light. Selective imitation of this kind has also been shown in "enculturated" (human-reared) chimpanzees (Buttelmann et al. 2007), non-enculturated orangutans (Buttelmann et al. 2008), and in a seemingly more fleeting form in dogs (Range et al. 2007) on several similar tasks. In other studies children were shown to selectively copy intentional actions in preference to accidental ones (M. Carpenter et al. 1998; Behne et al. 2005).

Further studies have shown selective imitation of functional actions (using a stick to extract a reward from a side hole in a box) rather than causally redundant ones (initially inserting the tool ineffectually in a top hole). Chimpanzees working with an opaque version of the box, where this difference was not visible, reproduced both the observed behaviors in the order seen; but when working with an equivalent but transparent apparatus where causal connections were perceptible, they tended to carry out only the second action, directly relevant to the task (Horner and Whiten 2005). By contrast, young children tend to repeat all the actions in the transparent conditions as well as the opaque one (Horner and Whiten 2005; McGuigan et al. 2007). A thorough investigation of this issue in young children (Lyons et al. 2007) showed that even after experimenters' efforts to train children to be more selective, they continued to copy redundant actions, leading the authors to attribute to them an automatic disposition for "over-imitation" (Lyons et al. 2007)—although since this tendency might include emulative elements it might be more safely called "over-copying" (Whiten et al. 2009). Whiten et al. (2005, 280) suggested that "a plausible explanation [for over-copying] is simply that we are such a thorough-going cultural species that it pays children, as a kind of default strategy, to copy willy-nilly much of the behavioral repertoire they see enacted before them." Other authors note that the cognitive opacity of many human tasks can make selective imitation difficult, and blanket-copying a more effective strategy (Gergely and Csibra 2005, 2006).

By contrast, Gergely and Csibra (2006) identified a mechanism they described as "natural pedagogy," in which behaviors such as direct eye contact, gaze alternation, talking and use of the child's name alert children to actions directed toward them, from which they may learn. The importance of these signals is supported by several studies showing that social learning is strengthened by their presence (Brugger et al. 2007; Nielsen 2006; Topal et al. 2008). Authors of such studies have suggested that the overimitation phenomenon may be a by-product of such ostensive communicative signals being exhibited by the demonstrator, inducing children to ignore their own individually observed "reality" and give more weight to demonstrators' actions.

Over-copying appears to be distinctive to humans rather than apes, but some recent studies with chimpanzees have hinted at dispositions akin to it. Chimpanzees were found to "perseverate" with a socially learned action even when a more successful and costly alternative was repeatedly witnessed (Marshall-Pescini and Whiten 2008a; Hrubesch et al. 2009). These are not examples of overimitation *sensu* Lyons et al. (2007), since subjects were not copying additional "useless actions." However, what had been initially copied was resistant to replacement by new information, similarly rendering the subject's behavior relatively inefficient.

In another intriguing example, chimpanzees who, via observation of a conspecific, had socially acquired a tool-combination skill to rake some out-of-reach food continued to use this laborious method even when the proximity of the food did not require the combination of two tools (Price et al. 2009). Chimpanzees who had individually learned the combination task, however, used it only when necessary. Other cases in which social learning has overridden individual experience, broadly conceived of as social "conformity," come from species as different as guppies (Laland and Williams 1998), dogs (Pongrácz et al. 2003), rats (Galef and Whiskin 2008), capuchin monkeys (Dindo et al. 2009), and chimpanzees (Whiten et al. 2005; Hopper, Schapiro, et al. 2011).

The potency of social over individual learning can thus lead individuals to (i) copy redundant or even maladaptive actions (children: Horner and Whiten 2005; Lyons et al. 2011; dogs: Prato-Previde et al. 2008), (ii) stick to a socially influenced/learned behavior that is "visibly" inefficient (chimpanzees: Price et al. 2009; dogs: Pongrácz et al. 2003; guppies: Laland and Williams 1998) and (iii) mask or slow down the subsequent acquisition of a more efficient strategy (chimpanzees: Hrubesch et al. 2009; Marshall-Pescini and Whiten 2008a). Notwithstanding such pitfalls, imitation can be a powerful means of learning from the accumulated wisdom and skilled actions of others in a complex cultural world, likely explaining its very potency in leading learners "astray" in the contexts summarized above (Hopper, Schapiro, et al. 2011).

Distinguishing Forms of Learning in Child and Chimpanzee

Any of the different social learning mechanisms might potentially explain the observational learning abilities of chimpanzees and children, with much possible overlap between them. To distinguish these, a number of test paradigms have been developed.

Testing for Bodily Imitation: "Do as I Do" and Spontaneous Copying

Given the general agreement that imitation involves the replication of an observed bodily movement, a number of studies have tested apes using a "do as I do" (DAID) paradigm. The procedure involves training an animal to repeat a set of observed actions on command and later presenting novel (untrained) actions to be imitated. The first report of a DAID test comes from the early work of Hayes and Hayes (1952), in which their home-raised chimpanzee, Vicki, was reported to reproduce novel actions such as stretching her mouth with her fingers.

A more systematic investigation of chimpanzees' imitative abilities in a DAID task was reported by Custance et al. (1995), who found that two juvenile nursery-reared chimpanzees matched (although rarely faithfully) approximately a third of a set of 48 novel actions. The same set of target behaviors was later used with a human-reared, language-trained orangutan, Chantek (Miles et al. 1996; Call and Tomasello 1995; Call 2001). Chantek showed a high level of behavioral matching on most behaviors, particularly those involving contact between body parts.

Myowa-Yamakoshi and Matsuzawa (1999) tested five human-reared chimpanzees, initially on their spontaneous tendency to imitate and later on the more classic DAID test. Results showed very few incidents of spontaneous imitation (only of actions involving contact between two objects), although performance noticeably improved after training. The actions chimpanzees found easiest to reproduce were those involving object-to-object interaction. Byrne and Tanner (2006) tested a female zoo-living gorilla with a set of seven body movements, with no prior training. Results showed that the gorilla exhibited recognizable matching, although rarely with high fidelity. The authors concluded that, given none of the actions presented could truly be considered novel, the most parsimonious explanation for the results was response facilitation or mimicry, whereby the observation of a known behavior elicits the subject's own action.

A final example of spontaneous, but in this case also real-time, mimicry (the observer performing actions in concert with the demonstrators' actions) comes from a study demonstrating observational learning of nut

Figure 7.1. Three frames from video footage showing a sequence of nut-cracking mimicry by Baluku (on the left) watching Mawa (on the right) nut cracking. This example of spontaneous bodily imitation shows Baluku matching with his own body the movements made by Mawa when nut cracking (Marshall-Pescini and Whiten 2008a). The video footage, in supplementary information to Marshall-Pescini and Whiten (2008a) can be viewed at http://dx.doi.org/10.1037/0735–7036.122.2.186.supp.

cracking in a group of juvenile chimpanzees (Marshall-Pescini and Whiten 2008b). This is a particularly interesting case because of its spontaneity—it was not based on "DAID" or any other training experience. Two of 11 chimpanzees made rhythmic hitting motions with their hand on the ground, roughly in time with an older chimpanzee pounding a nut with a stone (Figure 7.1). This behavior occurred on seven separate occasions for one of the chimpanzees, and once for a second chimpanzee. A third chimpanzee carried out a slightly different behavior: her whole body rocked, following the motion of the observed pounding action. In all cases the observer chimpanzees had already attempted to crack a nut following demonstrations, although none had been successful in doing so.

These results are of particular interest in relation to the operation of mirror neurons in primates. Mirror neurons, which fire both when an individual performs certain actions or sees them done by another individual, were first identified in monkeys, but lack of evidence for imitation in monkeys led the discoverers to suggest that mirror neurons' function instead lay in more general abilities to read goal-directed behavior in others, through simulation in the self (for reviews see Ferrari, Bonini, and Fogassi 2009 and this volume). In humans, by contrast, mirror neurons have been implicated in imitation (Iacoboni, this volume). So far, little is know about mirror neurons in apes, because of a lack of the neural recording done with monkeys (typically for ethical reasons) and the fMRI approaches used with humans (apes will not lie still in a scanner!). The mirroring behavior recorded by Marshall-Pescini and Whiten suggests apes may represent an intermediate position in which mirror neurons are also being incorporated into imitative and perhaps other "like me" cognitive processes. It will thus be interesting to see if such "mirroring" as is shown in the videos sampled in Figure 7.1 are to be seen in capuchin monkeys engaged in nut cracking. Visalberghi and Fragaszy (this volume) emphasize the extent to which capuchins appear not to copy the nut-cracking actions of others, but rather learn from the environmental affordances of anvil sites. The chimpanzee mirroring discussed here, together with the results of the ghost studies (see our discussion below), leads us to conclude that for apes, the actions of others are more important than the portrayal by Visalberghi and Fragaszy of what occurs in capuchins.

Involvement of mirror neurons has also been suggested in relation to the reproduction of familiar body actions by a gorilla (Byrne and Tanner 2006). Custance et al. (1995) noted that when presented with a novel DAID action, chimpanzees sometimes tended to first produce the closest trained (familiar) behavior and then modify it to achieve a closer match. When attempting to acquire a novel action, the closest available known behavior may be activated through the mirror neuron system, but additional neural systems, perhaps prefrontal ones, may be necessary to "parse the behavior and reconstruct it in a novel sequence" (Ferrari, Bonini, and Fogassi 2009).

To imitate others, subjects are faced with the "correspondence problem" (Heyes 2001): they must visually recognize the behavior being carried out and transform it into a corresponding motor output. It may thus be problematic that a DAID demonstrator has invariably been a human, not a conspecific (Boesch 2007, 2008). Although ape and human morphology is similar, the difference may still be strong enough to affect results. However, in Marshall-Pescini and Whiten's (2008b) study, the chimpanzees saw both human and conspecific models, and no difference was apparent in the time spent observing the human versus a conspecific demonstrator (Marshall-Pescini, unpublished data). The real-time mimicking described above occurred both with a familiar human and a chimpanzee as demonstrator.

Another relevant behavior was observed during this study. On three occasions, the two chimpanzees who carried out real-time mimicking placed their hand or foot on the human demonstrators' own hand while she was pounding a nut. Such co-action is thought to help the acquisition of novel behaviors by creating a "motor" memory of the action being performed. Instances have been described among both wild (McGrew 1977) and captive chimpanzees (Horner et al. 2006) as well as capuchin monkeys (Westergaard and Fragaszy 1987). The fact that in the experiment it occurred only with a human demonstrator might be explicable because these chimpanzees were attached to very familiar human caregivers. They may have identified more with such caregivers than the chimpanzee demonstrator involved, a sometimes intolerant and boisterous young male.

This example highlights the need to consider multiple factors when assessing the influence of a model. Perhaps not only conspecificity (Boesch 2007, 2008), but also familiarity (de Waal 1998), dominance (Nicol and Pope, 1999; Pongrácz et al 2008), and friendship relationships (Coussi-Korbel and Fragaszy 1995) can be equally influential. Only future studies including both human and conspecific demonstrators, matched for other potentially important variables, will be able to ascertain the weight of each.

Testing for Emulation: Ghost Display Conditions

To distinguish emulative learning from bodily imitation, ghost conditions have been employed. Byrne (2002, 90) noted that, in emulation, "the important distinction from other kinds of imitation is that motor behavior per se is not copied. Thus, in principle, emulation could be as effective if the behavior were not seen." Ghost displays fit this rationale; in a ghost display the movements of an apparatus are shown to a naïve observer through the use of a hidden mechanism, without a live demonstrator operating it. Ghost displays were first developed in the 1970s and have been employed with humans, primates, dogs, rats, and birds (Hopper 2010 provides a review).

When chimpanzees were presented with a tool-use task (the "panpipes"), we found that, although able to learn how to operate it after having seen a demonstration by a conspecific, they were unable to learn the same action after a ghost display (Hopper et al. 2007). In the social learning conditions, a model used a tool to raise a blockage on the top of the panpipes to release a reward inside (see Figure 7.2). Seeing the blockage raised and lowered discreetly with fishing-line, with no model present, was not sufficient for the chimpanzees to learn how to release the reward themselves, despite seeing over 200 of such "demonstrations." This was even the case when the tool was incorporated into the display.

In a direct comparison of the abilities of children and chimpanzees to learn from ghost displays, Hopper et al. (2008) presented three- to four-year-old children and captive chimpanzees with a simpler, bidirectional task, in which a reward could be gained by sliding a small door on a box to either the left or right. The subjects were presented with one of three forms of demonstration: (1) a conspecific (child or chimpanzee) moving the door, (2) a ghost display in which the door was moved discreetly by the experimenter using fishing line, or (3) an "enhanced ghost" display condition in which the subjects saw a ghost display with a passive conspecific

Figure 7.2. A side view of the panpipes apparatus showing (a) a chimpanzee using lift to retrieve a reward from the panpipes; (b) the ghost display shown to chimpanzees in which the blockage was raised and lowered discreetly by the experimenter using fishing line; (c) the same ghost display but with the tool included. Even with the tool added to the display, none of the chimpanzees presented with this task were able to retrieve the reward themselves (Hopper et al. 2007), in contrast with learning from a conspecific (Whiten et al. 2005).

"model" sitting in front of the box, collecting the rewards. This third experimental condition was run to rule out a mere presence effect as a possible cause of greater copying in the full demonstration, compared to a ghost display condition (Klein and Zentall 2003). A fourth group of subjects was presented with the task in a no-information control condition.

Children and chimpanzees copied the direction that the door was pushed with their first response after seeing an active model, ghost, or enhanced ghost display, providing the first evidence of emulative learning by chimpanzees from a ghost display (Hopper et al. 2008). They learned not only that the door could move, but matched its direction of movement in the ghost conditions, consistent with either object movement reenactment, or affordance learning (i.e., the door afforded shifting to left, or right, to reveal the food reward). However after this first action, the behavior of the children and chimpanzees differed. In their responses overall, the children matched the active model and the enhanced ghost display (passive model) above chance levels, thus continuing to display emulation in the latter ghost condition (Figure 7.3).

By contrast, after their first trial, chimpanzees no longer matched what they saw in the ghost conditions; they instead tended to explore pushing the door in both directions. The chimpanzees thus showed only fleeting evidence of emulation in their first trial. Nevertheless, they persisted in matching what the active model chimpanzee did in the full demonstration condition, just as the children did. In combination with the earlier finding that chimpanzees did not learn at all from ghost displays in a more-complex tool-use task (Hopper et al. 2007), these results emphasize that for chimpanzees to learn by observation, it is generally important they can watch an active agent perform that act, a conclusion that resonates much with the theme of the present volume, of "connecting" with others.

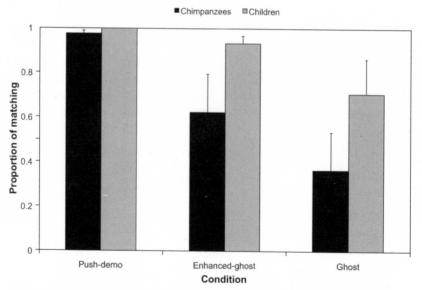

Figure 7.3. The proportion of matching responses made by chimpanzees and children after seeing each of the three forms of "demonstration" of opening a slide box. Children and chimpanzees continued to match the demonstrated action of a conspecific with high fidelity (100 percent and 99 percent respectively), but this level of matching only continued for the children, not the chimpanzees, in the two ghost-display conditions (Hopper et al. 2008).

Overall, children thus seem better able than chimpanzees to gain information from ghost conditions (Hopper 2010; Hopper et al. 2010) and in this sense are better emulators. This difference may be because children are more familiar with interacting with automated mechanical devices. Subiaul et al. (2007, 240) suggested that the reason children they tested were able to learn from a ghost display was that, despite being an inanimate object, the task "behaved as if it were animate and displayed the characteristics of social agents including, agency and goal-directedness [and] . . . children, from infancy, are sensitive to cues that index animacy and goal-directedness." However, given accumulating evidence for chimpanzees' sophistication in discriminating similar cues in the general domain of "mindreading" or "theory of mind" (Call and Tomasello 2008), it is difficult to see why the same should not apply to their perception of ghost displays as well, if Subiaul et al.'s speculation is correct.

Testing for Emulation: End-State Conditions

A second method used to tease apart imitative from emulative learning is the end-state condition. In this, individuals are shown the initial state of the test object, which is then manipulated out-of-sight, and they are then shown its resultant form. If the observer is able to complete the task after such a "demonstration," goal emulation is inferred.

The end-state condition was first used with birds to examine how milk-bottle-top piercing behavior, observed among blue tits *(Parus caeruleus)*, spread across Britain (for reviews see Bonner 1980; Reader and Laland 2003). Sherry and Galef (1984, 1990) tested black-capped chickadees *(Parus atricapillus)* and found that exposing the birds to bottles previously opened by the experimenter (the end state) was sufficient for the naïve birds to learn how to pierce the bottle tops themselves. More recently, end-state conditions were used to test how black rats *(Rattus rattus)* learned to strip open pinecones to eat the kernels, as they are observed doing in the wild in Israel (Zohar and Terkel 1991; Terkel 1996). Zohar and Terkel reported that young rats who were raised with expert mothers were the most likely to learn the stripping behavior; however, of rats raised by naïve mothers, those that were exposed to pinecones at various stages of opening (end-state conditions) were more likely to learn how to strip the pine kernels than those who only saw intact pine kernels.

Meltzoff (1985) ran the original end-state experiment with (14-month-old) human infants. His study revealed no significant evidence that infants could deconstruct a dumbbell after having seen only the start state (object intact) and the end state (ends removed). The majority of end-state studies have replicated and extended upon Meltzoff's study, using the same dumbbells, sometimes in conjunction with other test objects (Bellagamba and Tomasello 1999; Huang et al. 2002, 2006). To date, infants from 12 to 41 months old have been tested following Meltzoff's design, but only 19-month-olds tested with the end-state condition performed with success equal to infants that saw a full demonstration by an adult (Huang et al. 2002).

One study has implemented the end-state condition with chimpanzees (Call et al. 2005) as well as children. The test object used by Call et al. had similar physical properties to Meltzoff's dumbbell; it was a tube that had to be deconstructed in one of two ways (broken in half or remove the ends) to retrieve a reward inside. Chimpanzees were no better at opening the tubes after having seen a model perform the action than after an end-state "display." Chimpanzees and children were tested with both opaque and clear tubes, and the chimpanzees were more likely to open the clear tubes, probably because they could see the reward inside the tube. There was no difference between the full demonstration and end-state conditions in the likelihood of chimpanzees copying the method used to open the tube.

Unlike the chimpanzees, the children that saw a demonstration by an adult had greater success in opening the tubes than those that saw an end-state condition, but Call et al. (2005) noted that for any condition the children were "reluctant" to break the tube. Call et al. concluded that the responses of the chimpanzees provided evidence for emulative learning, whereas children were more likely to imitate. However, as both chimpanzees and children showed success in a no-information control, it is also possible that their successes were shaped by asocial, rather than social, learning.

Given Chimpanzees' and Humans' Social Learning Capacities, Do They Develop Cultural Traditions?

It is unquestioned that humans are cultural beings, but whether animals show comparable traditions is a hotly contested debate, as recently well illustrated in *The Question of Animal Culture* (Laland and Galef 2009). Culture in apes, the subject of this chapter, has been seriously studied for only a few decades. Just fifty years ago we knew nothing of chimpanzee culture—indeed, scarcely anything of their behavior in the wild—but now we know much (Whiten 2011).

Are Chimpanzees Cultural?

In "Cultures in Chimpanzees," Whiten et al. (1999) offered evidence for behavioral traditions among wild communities of chimpanzees. These behavioral variants included different ways of grooming each other (e.g., social scratch, grooming handclasp), gathering food (e.g., ant dipping, pestle pounding), and behavioral practices (e.g., rain dance). It was postulated that social learning was the underlying mechanism for the spread and maintenance of the observed patterns, insofar as alternative environmental and genetic explanations can likely be excluded. For example, tool use to extract nutrients from the central growing point of palm trees (pestle pounding) is found at Bossou (Guinea) but not in the Taï Forest (Côte d'Ivoire) just a few hundred kilometers away, where similar palm trees and the same subspecies of chimpanzee can be found. However, one can never be sure that some subtle factor has not been missed in this approach (Laland et al. 2009). One way to test whether chimpanzees' social learning is capable of sustaining such traditions is to conduct experimental studies in captivity, where more control can be exerted by the researcher (Whiten and Mesoudi 2008).

Behavioral Traditions in Captivity

Two main forms of diffusion experiment have been used with chimpanzees (and also for comparative purposes with human children): "diffusion chains" and "open diffusion" studies.

A diffusion chain allows for tight control over the spread of an introduced behavior and follows the format of the "telephone game" (a.k.a. "Chinese whispers"). An individual (A) is introduced to a novel method of solving a task, which the individual then demonstrates in the presence of a conspecific (B). After observing A, individual B is allowed to work on the task independently and, if proficient, perform in front of a third conspecific (C), and so on. This method was originally used by Bartlett in 1932 to study human memory in the transmission of folktales, but it was 1972 before a comparable "replacement method" was pioneered by Menzel and colleagues, in a study of chimpanzees' habituation to novel objects (E. Menzel, Davenport, and Rogers 1972).

Only much more recently were chimpanzees tested with a true diffusion chain (Horner et al. 2006). Horner and colleagues began one chain of chimpanzees with a model trained to operate a "Doorian" puzzle box by sliding a door panel open, and a second chain with another chimpanzee who instead lifted a small door set into the sliding panel. Chimpanzees copied the method used by the individual before them in the chain, creating two distinct, unfolding traditions. The same high fidelity of matching was also found in children tested in the same manner.

In a diffusion chain, the subjects have no control over the model they observe, and all observation episodes are dyadic. To remedy this, some experimenters have used an open diffusion method to examine the development and spread of behavioral traditions within whole groups of chimpanzees. These studies with captive chimpanzees support the inferences drawn about traditions in the wild, showing that chimpanzees are capable of social learning that supports the spread and maintenance of new behaviors both within and between groups (Whiten et al. 2005, 2007; Horner et al. 2006; Bonnie et al. 2007; Hopper et al. 2007).

Cumulative Culture

Despite the traditions identified among wild chimpanzees, there is a stark contrast between the complexity of human and chimpanzee culture. This difference in richness, we believe, is because humans may be better able to build and improve upon extant technologies, languages, and other behavior ("ratcheting") and in this way develop cumulative cultures (Tomasello 1999). Not only does such an ability enable individuals and communities to develop rich and varied cultures, it also allows them to rise above outdated and potentially maladaptive habits (Laland 2004).

Although inhabiting a less complex cultural world than humans, chimpanzees have been documented using "tool sets"—using more than one tool in succession to achieve a goal (Brewer and McGrew 1990). Bermejo and Illera (1999, 626) concluded that the use of tool sets showed "chimpanzees used the tools in a flexible, insightful way . . . [recognizing] the functions of the tools and the hierarchical nature of the multiple action involved in the use of the tool-set." More-recent observations of wild chimpanzees in the Goualougo Triangle, Republic of Congo, have shown that chimpanzees attribute multiple functions to a single tool and that they manufacture tools with specific physical characteristics relevant for each particular task (Sanz et al. 2004; Sanz, Call, and Morgan 2009; Sanz, Schoning, and Morgan 2009). For example, chimpanzees create one tool to perforate the ground and a second, modified to have a brush tip, to insert into the tunnel to catch termites. These tools can be distinguished both in the plant material they are constructed from and their average dimensions; perforating tools are longer and thicker than fishing tools.

Sanz, Call, and Morgan (2009, 295) concluded that it is "likely that simple tool technologies (involving a single unmodified tool) were the precursors of techniques with more tool modifications and more complex tool sets," suggesting evidence for ratcheting and, therefore, some small degree of cumulative culture in the wild. However, experimental evidence for such ratcheting with captive chimpanzees has proved less conclusive. Results indicate that chimpanzees are limited in their ability to build upon, or change from, previously held knowledge to alternatives presented in their social world (Hrubesch et al. 2009; Marshall-Pescini and Whiten 2008a).

To determine whether chimpanzees are capable of switching from a previously learned food-gathering technique to a more complex but efficient method, eleven sanctuary-housed chimpanzees, aged two to seven years, were tested with a honey-dipping apparatus that could be opened in either of two ways (Marshall-Pescini and Whiten 2008a). The chimpanzees were first shown a simple "dipping" technique. The human demonstrator held back a door on the apparatus with one hand, revealing a hole. The model then inserted a rod tool into the hole, which contained honey. If successful in learning the dipping method, the chimpanzees were then shown a second, more complex, "probing" technique. For this, the demonstrator used the rod tool to release a hidden catch, which then allowed a second defense to be pulled back, revealing the chamber containing the honey and peanuts. So, although probing was motorically more challenging, it gave a larger payoff.

Chimpanzees aged three to seven years socially learned the dipping technique, but only one of them upgraded from the dipping to the probing technique, and this individual had already discovered the probing technique during baseline trials through asocial learning. Thus, although Marshall-Pescini and Whiten (2008a) showed the chimpanzees were capable of social learning, they found no evidence that the chimpanzees could switch to a novel behavior after learning an already successful technique (see also Hrubesch et al. 2009 and Tennie et al. 2009). The reports of flexibility in tool use from the wild, noted above, encourage more varied experimental research on this topic.

Conclusions

In this chapter we have illustrated how the lives of both chimpanzees and human children incorporate some rich forms of social learning and culture. Phenomena they share, indicating the cultural nature of our last common ancestor, include the existence of cultures constituted of multiple traditions of diverse kinds, including such behaviors as tool use and social customs; likely underwriting these, capacities for bodily imitation and real-time action mirroring; selective imitation according to intent and rationality in others, and perceptible causal connections in the physical world; context-dependent potency of social over individual learning; and emulation in elementary physical tasks in the stripped-down context illustrated by a ghost experiment. Among these capacities, both species show remarkable degrees of interconnectedness with the actions and even the psychology of others.

Principal ways in which in childhood we already see cognitive correlates of the enormous cultural gulf between our own species and chimpanzees include a greater capacity for high-fidelity imitation and even "overimitation" (two phenomena that are probably related, and represent the high end of a continuum that runs from high conformity and fidelity to the actions of others, to more flexible and selective strategies; Whiten et al. 2005, 2009), a greater capacity for emulation in respect of more-complex artifacts, and a receptivity to informal pedagogy. In these respects children connect culturally ever more closely with others' actions and minds. However, the evolution of these extraordinary abilities now appears less mysterious than in the past, given the shared ancestral precursors we can now infer through such comparative studies as we outlined here.

Acknowledgments

All three authors thank the Biotechnology and Biological Sciences Research Council (BBSRC, UK) for supporting their research summarized in this chapter. AW also thanks the Royal Society and the Leverhulme Trust for support of research reported here. LH also thanks the National Science Foundation, which she was funded by when writing this chapter through a CAREER grant (SES 0847351) awarded to her postdoctoral mentor Sarah Brosnan. We thank Frans de Waal and Ludwig Huber for comments on drafts of this chapter.

Empathy, Perspective Taking, and Cooperation

CHAPTER EIGHT

A Bottom-Up View of Empathy

Frans B. M. de Waal

Introduction

Dictionaries typically define "empathy" as the ability to share and understand someone else's feelings and situation. In the scientific literature, however, a more mentalistic definition, closer to theory of mind, has become popular. Accordingly, empathy is a way of gaining access to another's mind by pretending to enter that individual's "shoes." For example, Goldman (2006) sees empathy as a combination of simulation and projection: inside its own head, the subject simulates how it would feel being in the other's situation and proceeds to assign mental states of its own to the other. Similarly, Baron-Cohen (2005, 170) describes empathy as involving "a leap of imagination into someone else's headspace." Most of these definitions sound so cognitively demanding that it is hardly surprising that animal empathy is rarely considered.

But what if the beginnings of empathy are much simpler? What if empathy does not require the subject to sort through information gained from the other as well as digging inside itself to arrive at an evaluation of what might be going on with the other? What if subjects share in the other's state of mind via bodily communication? The immediacy of the empathic response hints at this possibility. If we see a child fall and scrape its knee, we flinch and exclaim "ouch!" as if what happened to the child happened at the same instant to ourselves. This was already known to Theodor Lipps (1903), who developed the concept of empathy and aptly called it *Einfühlung* (German for "feeling into"). We are in suspense watching a high-wire artist, Lipps wrote, because we vicariously enter his body and thus share his experience. It is as if we are on the rope with him. We obviously cannot feel anything that happens outside of ourselves, but by unconsciously merging self and other, the other's experiences echo within us as if they are our own. Such identification, argued Lipps, cannot be reduced to any other capacities, such as learning, association, or reasoning. Empathy offers access to "the foreign self."

Empathy as feeling one with another's state, rather than some sort of cognitive deduction, was already a major point of discussion in early twentieth-century philosophy, from Ludwig Wittgenstein to Max Scheler (Zahavi 2008). This bottom-up view has the advantage of explaining the unconscious reactions demonstrated by Dimberg et al. (2000), which are unexplained by the more cognitive view. With small electrodes registering facial muscle movements, investigators presented human subjects with pictures of angry and happy faces on a computer screen. Even if the pictures flashed too briefly for conscious perception, subjects still mimicked the faces and experienced corresponding emotions. Subjects exposed to happy faces reported feeling better than those exposed to angry ones, even though neither group was aware of what they had seen. Clearly, their empathy with the perceived emotions was brought about unconsciously, without cognitive simulations or projections. Interpersonal emotional connections seem to run as much via bodies as minds (Niedenthal 2007).

If this is true for humans, it is probably equally true for other animals. But unfortunately, attention to executive cognitive function and a hierarchical view of the mind, which puts the neocortex in control, have turned comparative studies of cognition into interspecies contests. The emphasis on cognitive complexity pits humans against apes, apes against monkeys, primates against corvids, and primates against any other large-brained species, while neglecting the role of subcortical processes and the shared neural background of the capacities in question (Parvizi 2009; Chapter 1, this volume). This shared neural background explains why cognitive traits that are considered uniquely human, or uniquely hominid, are often later also found in monkeys, canines, even rodents. Here I will pursue a bottom-up perspective on empathy, which ties it to motor imitation and bodily synchronization without denying more-advanced forms of empathy requiring greater cognitive complexity.

Bodily synchronization is as adaptive for prey as it is for cooperative predators. Social animals need to coordinate movement, collectively respond to danger, communicate about food and water, and assist others in need. Responsiveness to the behavioral states of conspecifics ranges from a flock of birds taking off all at once because one among them is startled by a predator, to a mother ape returning to a whimpering youngster to help it from one tree to the next by draping her body as a bridge between the two. The first is a reflexlike transmission of fear, which may not involve any understanding of what triggered the initial reaction but which is undoubtedly adaptive. The selection pressure on paying attention to others must have been enormous. The mother-ape example is more discriminating, involving anxiety at hearing one's offspring whimper, assessment of the reason for its distress, and an attempt to ameliorate the situation.

At the core of the empathic capacity lies a mechanism that provides the subject with access to the subjective state of another through the subject's own neural representations. When the subject attends to the other's state, the subject's neural representations of similar states are automatically activated. This lets the subject get "under the skin" of the other, bodily sharing its emotions and needs. This neural activation, which Preston and de Waal (2002) dubbed the "perception-action mechanism" (PAM) of empathy, fits with H. Damasio's (1994) somatic marker hypothesis of emotions, Prinz's (1997) common coding theory of perception and action, as well as evidence for a link at the cellular level between seeing and doing, such as the mirror neurons first discovered in macaques (Chapters 2 and 3, this volume).

Imitation Follows Synchronization

If PAM underlies both imitation and empathy, as argued here, we predict correlations between both capacities. Highly empathic persons are indeed more inclined to unconscious mimicry (Chartrand and Bargh 1999), and humans with autism spectrum disorders are deficient not only in empathy but also in imitation (Charman et al. 1997). Functional magnetic resonance imaging (fMRI) studies neurally connect contagious yawning, a typical form of motor mimicry, with empathic modeling (Platek et al. 2005). Other primates, too, yawn when they see conspecifics yawn (Anderson et al. 2004), and do so even in response to the animated yawns of an apelike drawing (M. Campbell et al. 2009; Figure 8.1). Other forms of spontaneous behavioral copying are well known from both monkeys and apes, including feeding facilitation (Addessi and Visalberghi 2001; Dindo and de Waal 2006), contagious self-scratching (Nakayama 2004), and neonatal imitation (Bard 2007; Ferrari 2006).

De Waal (1998) proposed "identification" with the other as a precondition for imitation and empathy. Identification entails bodily mapping the self onto the other (or the other onto the self), and as such relates to the capacity for shared neural representation (Decety and Chaminade 2003). Body mapping has been tested in do-as-I-do experiments with

Figure 8.1. Since yawning is an involuntary reflex, yawn contagion is closer to empathy than imitation. Chimpanzees are so sensitive to the yawns of others that even a three-dimensional animation of a yawning head, of which this drawing shows eight stages, induces yawns in chimpanzees watching it on a computer screen (M. Campbell et al. 2009).

primates (Custance et al. 1995) and is also reflected in synchronized movements between model and apprentice chimpanzees (Marshall-Pescini and Whiten 2008b). Bodily similarity—such as with members of their own sex and species—and social closeness likely enhance identification, hence imitation. Yawn contagion in gelada baboons, for example, occurs mostly between individuals that are socially close (Palagi et al. 2009). Identification or a lack thereof also explains the variation in outcomes of imitation experiments. Initial skepticism about imitation in nonhuman primates was based on their failure to copy human models (Tomasello, Kruger, and Ratner 1993), but two kinds of studies have yielded more-promising results, namely those concerning a) human-raised apes watching a human model (Tomasello, Savage-Rumbaugh, and Kruger 1993; Bjorklund et al. 2000) or b) apes raised by their own kind watching a conspecific model (see below). In both cases, rearing history guaranteed identification with and attention to the model species.

This perspective does not exclude ape imitation of human models, provided the apes have grown up with or are socially close to the humans who test them. Most apes, however, will be at a disadvantage in such tests, such as in a recent eye-tracking study in which chimpanzees failed to follow human gaze direction, whereas the same apes had no trouble following the gaze of conspecifics (Hattori et al. 2010). Such bodily factors should be considered before concluding anything about cognitive differences if apes and children have both been tested by humans (de Waal, Boesch, et al. 2008). Exposed to models of their own species, chimpanzees faithfully copy tool use and foraging techniques, arbitrary means to achieve rewards, and necessary (but not unnecessary) actions to reach a goal (reviewed by Horner and de Waal 2009; Chapter 7, this volume). Given the role of identification, which is greater with same-sex partners, it is further unsurprising that when young chimpanzees learn to use a wand to fish for ants, sons copy their mothers not nearly as precisely as do daughters (Lonsdorf et al. 2004).

The key role of identification and body mapping urges a return to the classical definition of imitation (i.e., "learning to do an act from seeing it done"—Thorndike 1898), since the newer meanings, which stress the understanding of another's goals and methods to achieve this goal (Whiten and Ham 1992), may be overly cognitive. Even for adult humans, imitation is probably not as complex as generally assumed (Horowitz 2003). If unconscious body mapping is at the core of imitation, as suggested here, the focus shifts away from the understanding of goals, intentions, and rewards. Capuchin monkeys, for example, do not need any rewards to be influenced by the foraging choices of conspecifics, nor do they need to see others obtain rewards (Bonnie and de Waal 2007). Rather, the rewards of imitation may be intrinsic, making synchronization and conformism part and parcel of socially positive relations (de Waal 2001). Intrinsic rewards are also suggested by how being imitated is positively valued by both humans and other primates (van Baaren et al. 2003; Paukner, Suomi et al. 2009).

Basic Forms of Empathy

Since emotional connectedness in humans is so common, starts so early in life (e.g., M. Hoffman 1975; Zahn-Waxler and Radke-Yarrow 1990), and shows neural and physiological correlates (e.g., Adolphs et al. 1994; Decety and Chaminade 2003) as well as a genetic substrate (Plomin et al. 1993), it would be strange indeed if no continuity with other species existed. Emotional responses to displays of emotion in others are in fact so commonplace in animals (Plutchik 1987; de Waal 1996) that Darwin (1871, 77) already noted that "many animals certainly sympathize with each other's distress or danger." For example, rats and pigeons display distress in response to perceived distress in a conspecific, and temporarily inhibit conditioned behavior if it causes the other pain (Church 1959; Watanabe and Ono 1986). A recent experiment demonstrated that if mice perceive other mice in pain, this intensifies their own responsiveness to pain (Langford et al. 2006).

The next evolutionary step occurred when emotional contagion became paired with appraisal of the other's situation and attempts to understand the cause of the other's emotions, a step dubbed *cognitive empathy* by de Waal (1996). A first sign of cognitive empathy is concern for others. Thus, Yerkes (1925) reported how a young bonobo showed intense concern for his sickly chimpanzee companion, and Ladygina-Kohts (1935) noticed similar tendencies in her home-reared chimpanzee toward herself. She discovered that the only way to get her chimpanzee off the roof of her house (better than rewards or threat of punishment) was by acting distressed, hence by inducing concern for herself.

Consolation is perhaps the best-documented nonhuman primate example of what in humans is known as sympathetic concern. Consolation is defined as reassurance provided by an uninvolved bystander to one of the combatants in a previous aggressive incident. For example, a third party goes over to the loser of a fight and gently puts an arm around his or her shoulders (Figure 8.2). After the initial studies of chimpanzee con-

Figure 8.2. Consolation behavior is common in humans and apes but largely absent in monkeys. A juvenile chimpanzee puts an arm around a screaming adult male, who has been defeated in a fight. Photo by Frans de Waal.

solation (de Waal and van Roosmalen 1979; de Waal and Aureli 1996), other studies have confirmed this behavior in different ape species (Palagi et al. 2004; Cordoni et al. 2004; Mallavarapu et al. 2006). However, when de Waal and Aureli (1996) set out to apply the same observation protocol to detect consolation in monkeys, they failed to find any, as did others (Watts et al. 2000). The consolation gap between monkeys and the hominids extends even to the one situation where one would most expect consolation to occur: macaque mothers fail to comfort their own offspring after a fight (Schino et al. 2004). Similarly, O'Connell's (1995) content analysis of hundreds of reports confirms that reassurance of distressed others is typical of apes yet rare or absent in monkeys.

The issue of how widespread consolation is remains unresolved, especially now that similar behavior has been reported for canines (Cordoni and Palagi 2008; Cools et al. 2008) as well as corvids (Seed et al. 2007). This behavior may require a separation between internally and externally generated emotions, as also required for perspective taking (see below). Recent studies confirm that chimpanzee consolation serves to reduce arousal, is biased toward socially close individuals, and shows a sex difference, with females showing more of it than males (Fraser et al. 2008; Romero et al. 2010; Chapter 15, this volume), which is consistent with other empathy expressions in both humans and other animals (Preston and de Waal 2002).

Empathic Perspective Taking

E. Menzel (1974) was the first to investigate whether chimpanzees understand what others know, setting the stage for studies of nonhuman theory of mind and perspective taking. After several ups and downs in the evidence, current consensus seems to be that apes, but probably not monkeys, show some level of perspective taking both in their spontaneous social behavior (de Waal 1986, 1996) and under experimental conditions (B. Hare et al. 2001, 2006; Shillito et al. 2005; Hirata 2006).

An important manifestation of empathic perspective taking is so-called "targeted helping," which is help fine-tuned to another's specific situation (de Waal 1996). For an individual to move beyond being sensitive to others toward an explicit other-orientation requires a shift in perspective. The emotional state induced in oneself by the other now needs to be attributed to the other instead of the self. A heightened self-identity allows a subject to relate to the object's emotional state without losing sight of the actual source of this state (M. Hoffman 1982; Lewis 2002). The required self-representation is hard to establish independently, but one common avenue is to gauge reactions to a mirror. The *co-emergence hypothesis* predicts that mirror self-recognition (MSR) and advanced expressions of empathy appear together in both development and phylogeny.

Ontogenetically, there is compelling support for the co-emergence hypothesis in human children (Bischof-Köhler 1988, 1991; Zahn-Waxler et al. 1992). Gallup (1983) was the first to propose phylogenetic co-emergence, a prediction empirically supported by the contrast between monkeys and apes, with compelling evidence for both MSR, consolation, and targeted helping only in the apes. Apart from the great apes, the animals for which we have the most-striking accounts of consolation and targeted helping are dolphins and elephants, which are also the only mammals other than the apes to pass the mark test in which an individual needs to locate a mark on itself that it cannot see without a mirror (Reiss and Marino 2001; Plotnik et al. 2006; Figure 8.3).

Figure 8.3. An Asian elephant with a visible X-shaped mark on the right side of her head and an invisible sham mark on the left side touches the visible mark with the tip of her trunk, thus indicating self-recognition, which is thought to correlate with perspective taking. This still image was collected by a lipstick video camera embedded in the mirror (Plotnik et al. 2006).

It should be added, however, that self-representation is unlikely to have appeared *de novo* in a few large-brained animals. The framework of developmental psychologists, according to which self-representation emerges in small incremental steps (Lewis and Brooks-Gunn 1979; Rochat 2003), may also apply to phylogeny. Instead of adhering to an all-or-nothing division, monkeys may reach an intermediate stage of self-awareness similar to that of pre-MSR human infants (de Waal et al. 2005).

Empathic perspective taking overlaps with but is not identical to theory of mind in that its focus is not so much on the other's state of knowledge as on the other's situation, emotional state, and needs. It is a broader capacity that arises earlier during human ontogeny than theory of mind (Wellman et al. 2000). In animals, empathic perspective taking is strongly suggested by occasional high-cost helping, which cases are mostly anecdotal. The reason is that high-cost helping (e.g., dolphins forming a protective ring around human swimmers threatened by sharks or a chimpanzee biting through a poacher snare around another's limb; for other examples see de Waal 2009a) does not lend itself to experimentation. The same is true for human high-cost altruism, however. If one accepts qualitative accounts of human heroism—which are the only accounts we have—what would be the reason to dismiss similar accounts in relation to other animals? Animal observations are further in line with research on low-cost altruism, for which we do now have experimental support.

Altruistic Behavior

An old female, named Peony, spends her days with other chimpanzees in a large outdoor enclosure near Atlanta, Georgia. On bad days, when her arthritis is acting up, she has great trouble walking and climbing. But other females help her out. For example, Peony is huffing and puffing to get up into the climbing frame in which several chimpanzees have gathered for a grooming session. An unrelated younger female moves behind her, places both hands on her ample behind and pushes her up with quite a bit of effort, until Peony joins the rest.

Even though there are abundant examples of spontaneous helping among primates, the modern literature still depicts humans as the only truly altruistic species, since all that animals care about are return benefits (e.g., Dawkins 1976; Kagan 2000; Fehr and Fischbacher 2003). The problem with this view is that the evolutionary reasons for altruistic behavior are not necessarily the animals' reasons. Do animals really help each other in the knowledge that this will ultimately benefit themselves? To assume so is cognitively demanding in the extreme, requiring animals to have expectations about the future behavior of others and to keep track of what they did for others versus what others did for them. Thus far, there is little or no evidence for such expectations. Helpful acts for immediate self-gain are indeed common, but it seems safe to assume that future return benefits remain largely beyond the animal's cognitive horizon.

Once evolved, behavior often operates with motivational autonomy— that is, its motivation is relatively independent of evolutionary goals (de Waal 2008). An example is sexual behavior, which arose to serve reproduction. Since animals are, as far as we know, unaware of the link between sex and reproduction, they must be engaging in sex (as do humans much of the time) without progeny in mind. Just as sex cannot be motivated by unforeseen consequences, altruistic behavior cannot be motivated by unforeseen payoffs such as inclusive fitness or return benefits in the distant future.

The motivation to help must therefore stem from immediate factors, such as sensitivity to the emotions and/or needs of others. Such sensitivity would by no means contradict self-serving reasons for the evolution of behavior, as long as it steers altruistic behavior into the direction predicted by theories of kin selection and reciprocal altruism. In humans, the most commonly assumed motivation behind altruism is empathy. We identify with another in need, pain, or distress, which induces emotional arousal that may translate into sympathy and helping (Batson 1991). The same hypothesis may apply to other animals (de Waal 2008). This can be tested by evaluating how animals perceive another's situation, and under which circumstances they try to ameliorate this situation.

Apart from assisting an aging female in her climbing efforts, chimpanzees occasionally perform extremely costly helping actions. For example, when a female reacts to the screams of her closest associate by defending her against a dominant male, she takes enormous risks on behalf of the other. She may very well get injured. Note the following description of two longtime chimpanzee friends in a zoo colony: "Not only do they often act together against attackers, they also seek comfort and reassurance from each other. When one of them has been involved in a painful conflict, she goes to the other to be embraced. They then literally scream in each other's arms" (de Waal 1982, 67). This kind of cooperation, expressed in alliances and coalitions, is among the best-documented in primatology (Harcourt and de Waal 1992).

Another well-known form of assistance is food sharing. Outside the mother-offspring relation or immediate kin-group, sharing is rare in the primate order (Feistner and McGrew 1989), yet common in callithrichid monkeys, capuchin monkeys, and chimpanzees (Figure 8.4). The two main hypotheses to explain this kind of food sharing are (1) the sharing-under-pressure hypothesis, and (2) the reciprocity hypothesis. According to the sharing-under-pressure hypothesis, individuals share in order to be left alone by potentially aggressive beggars (Stevens and Stephens 2002; Gilby 2006). This hypothesis is contradicted, however, by the fact that the most generously sharing individuals are often fully dominant (de Waal 1989; Nishida et al. 1992), aggression is more often shown by food possessors than nonpossessors (de Waal 1989), food transfers occur even if negative behavior is prevented by physical separation (Nissen and Crawford 1932; de Waal 1997b), and many primates vocally announce the presence of sharable food, thus attracting beggars. In fact, chimpanzee begging behavior is rarely of a threatening nature, as it derives from infant and juvenile expressions of need aimed at the mother (e.g., pouting, whimpering, holding out a hand). None of the above observations fits the sharing-under-pressure hypothesis.

The reciprocity hypothesis, on the other hand, predicts that food is part of a service economy, hence exchanged for other favors. It has indeed been shown that adult chimpanzees are more likely to share with individuals who have groomed them earlier in the day. In other words, if A groomed B in the morning, B was more likely than usual to share food with A in the afternoon. Food-for-grooming exchanges among chimpanzees have been shown to be partner specific (de Waal 1997a). Of all examples of reciprocal altruism in nonhuman animals, these exchanges come closest to fulfilling the requirements of calculated reciprocity, that is, exchange with the same partner after a significant time delay reflecting memory of previous events and a psychological mechanism that Trivers (1971) described as "gratitude" (Bonnie and de Waal 2004).

Spontaneous helping behavior may be partly based on learned contingencies between help given and received, yet since these contingencies are highly probabilistic and occur over intervals lasting days, weeks, or longer, it is hard to see how they might explain high-risk helping, such as when Washoe, the world's first language-trained chimp, heard another female scream and hit the water. Fouts and Mills (1997, 180) describe how Washoe raced across two electric wires, which normally contained the apes, to reach the victim and waded into the slippery mud to reach the wildly thrashing female and grab one of her flailing arms to pull her to safety. Washoe barely knew this female, having met her only a few hours before.

Even if contingent reciprocity were to play a role, it is good to realize that it is impossible to learn behavioral contingencies without spontaneously engaging in the behavior in the first place. We must therefore assume an impulse that propels individuals to defend, share with, or rescue others. In the case of Washoe, this impulse needed to be strong enough to overcome her species' hydrophobia (chimpanzees cannot swim). Empathy has the potential to provide such an impulse, as it produces a stake in the recipient's well-being through shared representations. In the words of M. Hoffman (1981b, 133), empathy has the unique property of "transforming another person's misfortune into one's own feeling of distress." Inasmuch as both humans and other animals are most empathic toward past cooperators and socially close individuals, empathy biases altruistic behavior precisely as predicted by theories of kin selection and reciprocal altruism (de Waal 2008).

For both practical and ethical reasons, however, there is a scarcity of experiments on emotionally charged situations that could trigger costly altruism. This is true not only for animal altruism, but equally so for human altruism. Instead, experiments concern low-cost altruism, sometimes called "other-regarding preferences." A typical paradigm is to offer one member of a pair the option to either secure food for itself by manipulating part A of an apparatus or secure food for both itself and the other by manipulating part B of the same apparatus. In the first such experiment, Colman et al. (1969) found one of four tested macaques to be consistently other-regarding. When replications failed to find the same tendency in chimpanzees, however, this led to the suggestion that other-regarding preferences may be uniquely human (Silk et al. 2005; Jensen et al. 2006). It is impossible to prove the null hypothesis, however, and recent studies with different methodologies have yielded results more in line with expectations based on naturalistic primate behavior.

In one study, investigators tried to rule out reciprocity by having apes interact with humans they barely knew and on whom they did not depend for food or other favors (Warneken et al. 2007). The investigators also ruled out the role of immediate return benefits by manipulating the availability of rewards. In this experiment, chimpanzees spontaneously assisted persons regardless of whether or not this yielded rewards and were also willing to open a door for conspecifics so that these could reach a room with food. One would think that rewards for the actor, even if not strictly necessary, at least stimulated helping actions, but in fact rewards proved irrelevant. The decision to help did not seem based on a cost/benefit calculation, therefore, consistent with predictions from empathy-induced altruism.

Spontaneous helping has also been experimentally demonstrated in both capuchin monkeys (de Waal, Leimbruger, and Greenberg 2008; Lakshminarayanan and Santos 2008) and marmosets (Burkart et al. 2007). In our study, two capuchin monkeys were placed side by side separated by mesh. One of them needed to barter with us with small plastic tokens, which we would first give to a monkey, after which we would hold out an open hand to let them return the token for a tidbit (Figure 8.5). The critical test came when we offered a choice between two differently colored tokens with different meaning: one token was "selfish," the other "prosocial." If the bartering monkey picked the selfish token, it received a small piece of apple for returning it, but its partner remained unrewarded. The prosocial token, on the other hand, rewarded both monkeys with apple at the same time. Since the monkey who did the bartering was rewarded either way, the only difference was in what the partner received.

Monkeys preferentially bartered with the prosocial token. This preference could not be explained by fear of future punishment, because dominant partners proved to be more prosocial than subordinate ones.

Figure 8.4. Spontaneous food sharing among chimpanzees may be entirely voluntary, such as here, showing an adult female putting food in the mouth of an unrelated adolescent male. Photo by Frans de Waal.

Figure 8.5. One capuchin monkey reaches through an armhole to choose between differently marked pieces of pipe while her partner looks on. The pipe pieces can be exchanged for food. One token feeds both monkeys, the other feeds only the chooser. Capuchins typically prefer the "prosocial" token (de Waal et al. 2008). Drawing from a video still by Frans de Waal.

Familiarity biased the choices in the predicted direction: the stronger the social tie between two monkeys, as measured by how much time they associated in the group, the more they favored the prosocial token. Moreover, choices were reflected in accompanying behavior, with greater orientation toward the partner during prosocial choices (de Waal, Leimgruber, and Greenberg 2008).

In short, there is mounting evidence from both naturalistic observations and experiments that primates care about each other's welfare and follow altruistic impulses in some contexts, probably based on empathy, which in both humans and other animals increases with familiarity. The empathy mechanism automatically produces a stake in the other's welfare—that is, the behavior comes with an intrinsic reward, known in the human literature as the "warm glow" effect. Actions that improve another's condition come with pleasant feelings (Andreoni 1989), so that humans report feeling good when they do good and show activation of reward-related brain areas (Harbaugh et al. 2007). It will be important to determine if the same self-reward system extends to other primates.

Russian Doll Model

Empathy covers all the ways in which one individual's emotional state affects another's, with simple mechanisms at its core and more-complex mechanisms and perspective-taking abilities as its outer layers. Because of this layered nature of the capacities involved, I have referred to it as a Russian doll in which higher cognitive levels of empathy build upon a firm, hardwired basis, such as the PAM (Figure 8.6). The claim is not that PAM by itself explains sympathetic concern or perspective taking, but that it underpins these cognitively more-advanced forms of empathy and serves to motivate behavioral outcomes. In the absence of emotional engagement, perspective taking would be a cold enterprise that could just as easily lead to torture as helping (de Waal 2005).

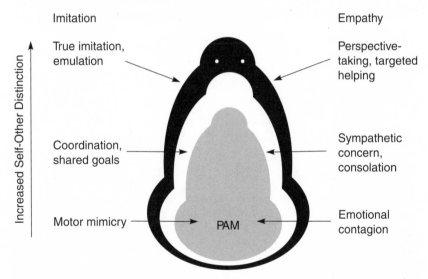

Figure 8.6. The "Russian doll model" of empathy and imitation. Empathy (right) induces an emotional state in the subject similar to that of the object, with at its core the perception-action mechanism (PAM). The doll's outer layers, such as sympathetic concern and perspective taking, build upon this hardwired socio-affective basis. Sharing the same PAM, the doll's imitation side (left) starts with motor mimicry, followed by coordination, shared goals, imitation, and emulation. Even though the doll's outer layers depend on learning and prefrontal functioning, they remain fundamentally linked to its inner core.

In accordance with the PAM as outlined by Preston and de Waal (2002), the motivational structure of both imitation and empathy includes a) shared representation, b) identification with others based on physical similarity, shared experience, and social closeness, and c) automaticity and spontaneity. This, then, is the bottom-up view according to which basic matching mechanisms between the bodies of individuals led to the capacity to mimic actions and share emotions, which mechanisms were refined ever more during evolution to lead to the capacity to spread entire cultures and to mentally trade places with others. The latter, more fully formed capacities obviously require more than matching. They require an appraisal of the situation of the other and an understanding of the other's goals and intentions, but all these outer layers developed out of and were never fully separate from the Russian doll's core.

Postscript: Empathy as Umbrella Term

After having written the above, the author entered into a discussion with fellow authors Tania Singer and Marco Iacoboni about what we should call empathy and what not. In this discussion, lines were drawn around empathy as distinct from emotional contagion, mentalizing, sympathy, and compassion. In response, let me clarify my own use of the term, which is considerably broader (cf. Preston and de Waal 2002), and also how it relates to the mirror neurons discussed in Chapters 2 and 3 of this volume. Since I am quoting e-mail exchanges, the tone of this postscript is rather informal.

For me, empathy is the umbrella term that encompasses all levels of the Russian doll model. Adult humans may show the whole spectrum, whereas many animals show only a few, as do human infants. The following definition of empathy comes from de Waal (2008, 281):

> The capacity to a) be affected by and share the emotional state of another individual, b) assess the reasons for the other's state, and c) identify with the other, adopting his or her perspective. This definition extends beyond what exists in most animals, but the term "empathy" applies even if only (a) is met.

There exists a tendency in psychology to slice and dice phenomena until we are left with little pieces without connections between them. This is not how a biologist thinks. Nursing and bonding are obviously different things, but we also know that nursing promotes bonding, and vice versa, and that the two probably evolved together in mammals, which explains their shared physiology (e.g., oxytocin). This is also true for empathy and imitation. There has been a tendency in the literature to isolate "true" imitation from the rest, but such distinctions are hopeless, since even the imitation of birdsong and primate tool use may rely on the same neural substrate (Chapters 1–3, this volume). The same holds for empathy: many layers can be recognized, yet they remain connected. I am for empathy as the larger term to cover all these integrated aspects, from the emotional to the cognitive.

I have trouble with empathy as "shared feelings," because feelings are self-perceived emotions. I would prefer just the word "emotions." I don't think you need to know your own emotions to be able to share them with others or to be affected by those of others. The emotion literature has good reason to keep feelings separate from emotions. I should also add that the distinction between emotional contagion and empathy drawn by Singer and Hein in Chapter 10 does not strike me as useful, since it seems to imply that one can achieve empathy without being affected by the other's emotions, which I find hard to imagine.

Asked whether motor mimicry is a necessary and sufficient condition for empathy, or just one route that may help initiate empathy, I doubt that motor mimicry is always needed. The perception-action mechanism model speaks of shared neural representations, not all of which need to involve motor processes. We stayed deliberately neutral on the role of mirror neurons. They may be involved in facial mimicry or bodily transmissions, but there may be other channels to affect others.

As for the distinction between empathy and cognitive perspective taking (such as theory of mind and mentalizing), I like to be careful. I see perspective taking as another level of empathy, not as separate. In normal development (and also in evolution), perspective taking is added to the emotional process. The child begins to wonder what causes emotions in others, rather than just being affected by them, and hence begins to focus on the situation of the other. The same applies to some large-brained animals. They add perspective taking to the emotional process, without in any way replacing it. So, rather than viewing perspective taking as separate, it is fully integrated with the emotional one, which is why I speak of "empathetic perspective taking."

Only in psychopaths do we see a disconnect, where perspective taking is applied without the emotional core. This is very interesting: they adopt the viewpoint of the other without probably ever having felt empathy (de Waal 2009a describes this as an empty Russian doll). But this is not how it normally works, and why psychopathy is a psychiatric condition. Normally, the child progresses from emotional contagion and personal distress to empathy that includes the self-other distinction and perspective taking.

I try to avoid the term "theory of mind" as much as possible since it has a very special meaning (knowing what others know) that has little to do with empathy. Theory of mind is such an emotion-less, body-less term (and it should be noted that having an actual "theory" of another's mind has never been demonstrated) that I really cannot see it apply to young children or animals. It seems also a rather narrow angle for intersubjectivity among human adults. It is unclear how relevant the false belief task is to a discussion of empathy: I would prefer to know more about the understanding of another's needs, intentions, and desires.

Finally, with respect to a possible contrast between human and nonhuman empathy, let me just say that there is a long history of cognitive distinctions between humans and other animals that have fallen by the wayside. Currently, experiments indicating human-ape cognitive differences are overwhelmingly based on children being tested by their own species and apes being tested by us, a different species. If bodily connections matter the way I think they do, this amounts to a fatal design flaw. What chimpanzees understand of the human mind is irrelevant, but often used as the gold standard of their cognition. My personal guess is that the role of perspective taking in empathy is substantially similar between humans and apes, and likely also between humans and other large-brained mammals.

CHAPTER NINE

What Does the Primate Mind Know about Other Minds? A Review of Primates' Understanding of Visual Attention

April M. Ruiz and Laurie R. Santos

Introduction

Despite wide-ranging differences in group size, social structure, and mating systems, all primates are inherently social creatures. Primates grow up in a world full of conspecifics, a perceptual environment filled with many complicated social interactions. In addition to being competitors for resources, group mates can also serve as models, sentinels, informants, mating partners, and playmates. In order to survive in this socially pressured world, individual primates must quickly learn to make sense of the actions of others and exploit them for their own ends.

One of the most important cues provided by conspecifics comes from where individual group members happen to be looking. A primate's gaze direction is informative of what he knows and doesn't know, what he happens to be thinking about, and how he is likely to act in the near future. Indeed, individual primates have the potential to interact with others in sophisticated ways simply by recognizing and interpreting cues such as head orientation and eye gaze. For this reason, the capacity to recognize the direction of others' attention has long been considered a critical capacity for socially living species.

In this chapter, we will discuss how primates are able respond to others' attentional states. Specifically, we review behavioral work on how primates read and react to others' attention direction and explore what such evidence tells us about primates' capacities to understand others. We begin by discussing behavioral evidence suggesting that primates recognize and respond to gaze cues in surprisingly sophisticated ways, allowing them to locate objects of interest in their environment and to interact with others in appropriate communicative modes. We then turn to the more complicated question of what primates' understanding of visual attention means for their ability to represent others' mental states, and how this ability may vary across primates. We then propose possible neural mechanisms that would support these cognitive capacities in an effort to understand, from the bottom up, how complex social behavior may arise. We argue that the ability to recognize the direction of visual attention of other individuals is the base upon which more-complex sociocognitive abilities are built, and that the neural mechanisms supporting gaze-following are, as such, also essential in supporting capacities such as the understanding of intentions and beliefs.

Following Gaze Direction

One of most valuable skills a social animal can possess involves recognizing where another individual is looking. The ability to follow others' gaze direction, however, requires an individual to be able to pick out relevant cues in the environment (e.g., social vs. nonsocial entities, heads and faces vs. legs and feet), attend to them more closely, and respond appropriately. Recognizing and following the gaze of others offers many opportunities for a social animal: others' gazes can provide hints about where to locate food sources, detect predators, and witness important social interactions (Emery 2000). Gaze direction also provides information helpful in predicting another individual's behavior (Seyama and Nagayama 2005). As the ability to recognize targets of gaze direction can provide an observer with qualifying information (such as whether a target is dangerous or safe), psychologists have long hypothesized that the development of visual co-orientation in humans is necessary for the emergence of language and social cognition (Bruner 1983; Csibra and Gergely 2006) and also for the development of theory-of-mind capacities (e.g., Whiten and Byrne 1988; Santos and Hauser 1999), that is, the ability to reason about other individuals' behavior in terms of mental states.

Researchers tracing the evolution of visual co-orientation among primates have typically asked whether primate subjects were able to follow the line of gaze of a human experimenter. In most of these studies, an experimenter would stand in front of a subject and direct his own attention to a distant location using cues such as head orientation and eye-gaze direction. In this type of situation, great apes and Old World monkeys are able to follow human gaze (e.g., Itakura 1996 and Bräuer et al. 2005 for great apes; Ferrari et al. 2000 for Old World monkeys); in contrast, prosimian primates have typically failed at this task (Itakura 1996).

Although most experimental tests of primate gaze-following involve a human experimenter, visual co-orientation in the real world more likely involves the gaze of conspecifics. As such, some researchers have argued that conspecific gaze may provide a more valid test of primate gaze-following capacities (Neiworth et al. 2002; Ruiz et al. 2009). Fewer studies have investigated whether primates spontaneously follow a conspecific's gaze, but such studies often demonstrate these abilities in a wider range of species. In a study with chimpanzees and four Old World monkey species, Tomasello and colleagues (1998) enticed one individual to look at an attractive piece of food and then measured the reaction of a second conspecific (the subject) who had been looking at the first. Tomasello and colleagues found that all five species spontaneously followed the gaze of conspecifics. Using similar logic, Neiworth and colleagues (2002) demonstrated that cotton-top tamarins, a New World monkey species, are able to follow the gaze of a group mate but not of a familiar human experimenter (see Burkart and Heschl 2007 for a similar result with common marmosets). Hattori and colleagues (2010) used an eye-tracking technique and found that chimpanzees follow conspecific cues better than human cues. Finally, using an innovative method involving wearable cameras, Shepherd and Platt (2008) found that ring-tailed lemurs follow the gaze of conspecifics when navigating their natural environment.

Although the use of conspecific models has shown that gaze-following extends broadly across the primate order, studies using live conspecific models present their own methodological problems. For example, it is often difficult to determine whether the subject saw the object of the other individual's attention independently of following the other individual's gaze (see Tomasello et al. 1998 for one way to control for this issue). In addition, researchers are required to train conspecific models to look to certain locations in order to carry out more-controlled tests (Neiworth et al. 2002; Horton and Caldwell 2006), and such training can limit the types of questions an experimenter can investigate using live conspecific models. To circumvent some of these issues, some researchers have turned to the use of photographic stimuli. In these studies, subjects are presented with a static image of a conspecific, either on video screen or paper. This paradigm allows the experimenter to strictly control what cues are made available to the subject, in what specific direction the model is "attending," and how much time the subject is exposed to the cue. Tests exploring visual co-orientation using photographs instead of live models have shown that lemurs (Ruiz et al. 2009), Old World monkeys (Lorincz et al. 1999; Scerif et al. 2004) and apes (Horton and Caldwell 2006) can successfully follow the direction of a conspecific's visual attention.

Using Others' Gaze Direction

Given all the evidence that primates can follow others' gaze direction, one might expect that many primates are also able to use others' gaze direction for practical purposes, such as locating a hidden piece of food. Surprisingly, however, primates have generally performed poorly on tasks that require the use of others' gaze direction to locate hidden objects. Much of this work has used a task known as the object-choice method. In object-choice tasks, primate subjects must follow the visual attention cues (direction of gaze, pointing, etc.) provided by an experimenter in order to find a piece of food that has been hidden in one of several potential hiding places. Historically, primates' performance on this task has been inconsistent at best (e.g., Call et al. 1998, 2000; Herrmann et al. 2006; Anderson, Montant, and Schmitt 1996; Vick and Anderson 2000). For example, even though orangutans and gorillas follow human gaze, they appear unable to spontaneously use the cues provided in object-choice tasks, although some improvement with extensive training has been reported (Peignot and Anderson 1999; Byrnit 2004). Chimpanzees have performed similarly poorly on this task (see Call et al. 2000), with successful performance limited to specific methodological circumstances. For example, chimpanzees perform better when they are allowed to watch the experimenter approach and search the correct location before they make their choice (Buttelmann et al. 2008) or when vocalizations are used in conjunction with gaze cues (Call et al. 2000). In another example, Barth and colleagues (2005) showed that chimpanzees performed better on this task when they were required to leave the testing area between trials, rather than remain for multitrial blocks. In addition, Call and colleagues (1998) found that chimpanzees succeeded on an object-choice task when food was hidden inside tubes or behind barriers, rather than underneath cups, suggesting that chimpanzees succeed more when experimenters had true visual access to the food item. Finally, Hare and colleagues demonstrated that chimpanzees perform better when the task is performed in a competitive context in which an "evil" experimenter always takes food from the subject (B. Hare and Tomasello 2004).

Although the success of some methodological variations provides important insight into the types of ways in which researchers should frame experimental questions, performance on the object-choice task as a whole suggests that most primates do not routinely use human visual attention to discover hidden resources. This pattern of performance poses a puzzle, as it is unclear why primates are able to follow another individual's gaze but not use it in a communicative way. Some researchers have argued that this disconnect might stem largely from the use of human gaze cues as stimuli in these studies. It is possible that primates do not reliably interpret human gaze as conveying information, even though they automatically follow human gaze. Indeed, some have argued that the ability to recognize the informational content of human gaze might involve a rerouting of cognitive resources already in existence for conspecific communicative skills in order to read human cues (Tschudin et al. 2001), or might require extensive contact with humans in order to enhance possible predispositions to attend to human cues (Scheumann and Call 2004). In fact, the nonhuman primate subjects who tend to perform best on these tasks are those individuals that are highly enculturated (e.g., Itakura 1996), lending support to the latter proposal. Further, various studies have shown that aquarium dolphins (Tschudin et al. 2001) and fur seals (Scheumann and Call 2004), who spend their days with human trainers, and also domesticated dogs (B. Hare et al. 2002) are able to use human gaze cues in object-choice tasks. Other researchers have argued that primates' performance in an object-choice task is actually limited by their success in gaze-following. Ruiz and colleagues (2009) hypothesized that primates might have difficulty realizing that the gaze cue being presented by the experimenter is relevant to the food search task; as such, primates might perform better in the object-choice test in cases in which they successfully attend to and follow gaze. To get at this possibility, Ruiz and colleagues presented lemurs with an object-choice test and compared subjects' gaze-following and performance on a trial-by-trial analysis. Although the lemurs' choice performance appeared to be only slightly above chance, the analyses revealed that when subjects followed the direction of attention of the model, they were more likely to choose the correct location of the hidden item. In this way, one could interpret limitations in object-choice performance in terms of limitations on gaze-following.

Other researchers, however, have tried to explain primates' object-choice failures by emphasizing motivational rather than cognitive deficits. Tomasello and his colleagues have argued that many of primates' deficits on object-choice tasks result from the fact that they are missing "a special kind of shared motivation in truly collaborative activities" (Tomasello et al. 2005, 11). As Tomasello and colleagues (2005) explain, "although apes interact with one another in myriad complex ways, they are not motivated in the same way as humans to share emotions, experiences, and activities with others of their own kind." Although Tomasello and colleague's motivational hypothesis fits nicely with the available data, some researchers have criticized this account on the lines that it is somewhat difficult to test empirically. Lyons and colleagues (2005), for example, have pointed out that it is not clear what study could falsify this hypothesis, making this hypothesis and motivational accounts in general hard to address operationally. Unfortunately, the problem of why primates perform poorly on object-choice tasks is still a very open question, one that continues to generate much debate among researchers today.

Tailoring Behavior to the Attentional State of Others

In addition to using other individuals' direction of gaze to learn more about the environment, primates must use others' visual attention when attempting to communicate successfully. Primates often find themselves in a position of wanting to be seen or heard by another individual—mating solicitations, threatening expressions, and other communicative gestures can only be effective when the recipient is able to perceive them (Bruner 1975; Gómez 1991). For this reason, primates who engage in these communicative acts should have an understanding of the attentional states of others and should use gestures in a manner directly related to the attentional state of a recipient (Tomasello 1995): gesturing more when the recipient can see the gesture, and less (if at all) when the recipient cannot.

Much empirical work has begun to address whether primates do in fact use an understanding of visual attention during communicative acts. In a landmark initial study, Povinelli and Eddy (1996) examined whether chimpanzees spontaneously modify their use of a begging gesture depending on the attentional state of a human experimenter. After an extensive training period in which chimpanzees learned to use a begging gesture to receive a reward from an experimenter, chimpanzees were presented with a choice to gesture toward one of two experimenters. Across a number of different conditions, the experimenters varied in that one individual could see the chimpanzee and one could not (e.g., full body oriented toward versus away from the subject, head turned toward or away from the subject, a bucket held over or just to the side of the experimenter's head, a blindfold covering the experimenter's eyes or the experimenter's mouth). In an initial condition, Povinelli and Eddy found that chimpanzees used more manual gestures when a human experimenter was oriented toward them than when the experimenter had his back turned. In all other conditions, however, chimpanzees chose randomly between the two experimenters, ignoring visual attention cues. Povinelli and Eddy interpreted this pattern of performance by hypothesizing that chimpanzees had learned a "face rule"—namely, gesture toward the person with the most "face" visible. Consistent with this view, over time chimpanzees chose correctly in conditions where the presence of the experimenter's face could be used as a discriminative cue. Further, some subjects even learned an "eye rule" and so, over time, chose correctly when the presence of eyes could be used as a discriminative cue. Reaux and colleagues (1999) replicated this study a few years later with the same subjects to see if sensitivity to attentional states improved with development. Once again, chimpanzees chose correctly when discriminating between an experimenter whose body was oriented forward and one whose back was turned. Chimpanzees performed well when their "face rule" could be applied, as in the original study.

Povinelli and Eddy (1996) speculated that chimpanzees may understand what a human experimenter can see but are unable to use this information in creating a gesture strategy. One important aspect of their study to consider, however, is that the experimenter to whom the chimpanzee should have gestured never established attentional contact by looking directly at the chimpanzee's face (see discussion in Gómez 2005). If mutual attention is a key aspect of successful gestural communication, as has been suggested (Leavens et al. 1996), then perhaps chimpanzees would do better in cases in which mutual attention was established (see Povinelli and Eddy 1996 for some evidence that chimpanzees do better when an experimenter makes direct eye contact).

The results obtained by Povinelli and Eddy (1996), however, did not end the story. In the decades that followed, other labs provided conflicting insight into primates' understanding of attentional states. In a recent study, Bulloch and collaborators (2008) tested chimpanzees using a paradigm similar to that of Povinelli and Eddy (1996). However, the results obtained from their replication were strikingly different; Bulloch and colleagues found that chimpanzees begged preferentially toward the experimenter who could see them right from the beginning of the experiment without having to learn this during the course of the experiment. It seems, then, that the extensive training and behavioral shaping Povinelli and Eddy employed before their key experimental manipulations could have overshadowed their chimpanzees' true sociocognitive capacities. These new findings suggest that primates can, in fact, recognize others' visual attention cues in a begging task.

Chimpanzees also appear to use other individual's visual attention cues when deciding when to gesture in a naturalistic context. Tomasello and colleagues (1994) observed that chimpanzees use visual gestures flexibly, adding tactile signals to get an observer to reorient. Liebal and colleagues (2004) provide further evidence to support this claim, reporting that chimpanzees in a natural context use more visually based gestures when the recipient is already attending to them (but see Hattori et al. 2007 for failures in capuchin monkeys). These findings together suggest that chimpanzees adapt their gestural signals to the attentional states of others, but are unfortunately silent on the specific cues that chimpanzees use to determine whether another individual can see them. To get at this question, Kaminski and colleagues (2004) systematically varied cues related to the observer's body and face orientation. They found that when the experimenter's body was oriented toward the subject, ape subjects used more gestures when the experimenter's face was also oriented toward the subject than when it was oriented away. However, when the experimenter's body was oriented away, apes' use of gestures did not change whether the experimenter's face was oriented toward or away from the subject. Subjects in Kaminski's study differentiated between an experimenter with her face oriented toward or away from them right from the beginning of the experiment and did not need to learn a "face rule" (these results also stand in contrast to those of Povinelli and Eddy [1996], demonstrating that it is important to investigate null results, as they can be due to many factors, of which a cognitive deficit is only one). However, apes' sensitivity to face orientation was limited to those cases in which the experimenter's body was oriented toward the subject, as they did not exhibit this sensitivity when the experimenter's body was oriented away. But why would chimpanzees use face orientation correctly in one instance but not in the other? The researchers proposed that apes could reason about the two cues in different ways: perhaps body orientation indicates an observer's likelihood to perform an action, but face orientation indicates an observer's perceptual access. As such, apes should gesture when a human experimenter can see them and is likely to respond, but not when an experimenter can see them but is unlikely to respond.

Using Visual Attention Cues in Competition for Resources

Another domain in which visual attention cues are helpful is in competition for scarce resources. From covert mating acts to exploiting food sources, an individual primate faces many situations in which it would be better off if other individuals were ignorant of its actions. For this reason, understanding whether other individuals are knowledgeable or ignorant can offer an important competitive advantage for primates deciding when to engage in a potentially deceptive act. But do primates use others' visual access cues to decide when to behave deceptively?

B. Hare and colleagues (2000) addressed this question in what is now widely considered to be a landmark study. In their study, Hare and colleagues explored what chimpanzees understood about the visual access of their competitors. They placed dominant and subordinate chimpanzees into an arena in which two individuals competed for access to contested

pieces of food. The question of interest was whether subordinate chimpanzees would spontaneously approach foods that the dominant individual
could not see. Impressively, subordinate chimpanzees did just this, selectively approaching foods that were hidden from the dominant chimpanzee's visual access. Chimpanzees also can use another individual's previous
visual access to make predictions about what they currently know. B. Hare
et al. (2001) presented subordinate chimpanzees with a situation in which
a dominant individual saw food hidden in one location. The dominant
chimpanzee's visual access was then blocked as the food was moved to a
new location. Hare and colleagues observed that subordinate chimpanzees
preferentially approached foods that the dominant individuals had not
seen move, suggesting that they recognize when the dominant individual is
ignorant of food's new location. In addition, chimpanzees also understand
that different individuals can have differential visual access. B. Hare et al.
(2001) presented subordinate chimpanzees with the same contested food
task but swapped the first dominant individual for one who failed to witness the baiting event. Subordinate chimpanzees were more likely to approach the food when an ignorant dominant individual was present than
when a knowledgeable dominant individual was present, suggesting that
this species is capable of taking into account both the present and past visual access of multiple individuals.

Primates are also able to recognize the visual attention cues of a human
experimenter in a competitive context. Flombaum and Santos (2005) examined whether rhesus monkeys would selectively steal food from human experimenters that could not see them. In their studies, two experimenters approached a free-ranging monkey and then placed a desirable
piece of food on the ground. One of the two experimenters could see this
piece of food, while the other could not. Across a number of different conditions (see Figure 9.1), monkeys selectively stole food from the experimenter who couldn't see them. Like chimpanzees, rhesus monkeys in this
task successfully recognized and interpreted the attentional states of their
competitors, reacting in a way that allowed for successful deception (see B.
Hare et al. 2006 for a similar result in chimpanzees).

igure 9.1. The five comparisons presented to rhesus monkeys in Flombaum and Santos (2005).
n all cases, monkeys selectively stole food from the experimenter who could not see them.

Other recent studies demonstrate that primates' understanding of their competitors' attention is not restricted to the visual domain. Santos and colleagues (2006) explored whether rhesus monkeys also take into account a competitor's auditory access (see Figure 9.2). They allowed monkeys to steal one of two containers from a human competitor who was facing away. One of the two containers was silent, but the other made a sound when it was moved—it was covered in small, noisy bells that jingled when moved. Monkeys selectively attempted to steal the silent container over the noisy one. Importantly, this effect only held in cases where the experimenter wasn't looking. In trials in which the experimenter watched the monkey, thereby making auditory information irrelevant, subjects approached the silent and noisy boxes equally. Melis, Call, and Tomasello (2006) found a similar result in apes; chimpanzees selectively

Figure 9.2. A photograph of the Santos et al. (2006) stealing study. Monkeys watched as an experimenter placed food into silent and noisy boxes. Santos and colleagues found that monkeys selectively stole food from the silent box.

choose a quiet over a noisy door when removing contested pieces of food. Taken together, these results suggest that several primates have a sophisticated understanding of their competitors' attentional access, even across different modalities.

Representing Attentional States as Mental States?

As should be evident from our review thus far, comparative researchers now have abundant behavioral evidence that Old World monkeys and apes are able to interpret the attentional cues of others and behave appropriately in reaction to these cues. They follow the gaze of others, and they use this information to make informed decisions about when to gesture to others and when deceptive acts are likely to work. In light of these findings, what can we conclude about the way in which primates represent this attentional information at a cognitive or information-processing level? Put differently, what can primates' performance on these tasks tell us about the representations they use to solve these social problems?

The question of how primates represent the attentional states of others is one that has long generated much debate among psychologists interested in the nature of information processing in nonhuman animals (e.g., Tomasello et al. 2005; Povinelli and Vonk 2003). Much of this theoretical fervor began with Premack and Woodruff's (1978b) landmark study of their chimpanzee Sarah's understanding of others' actions. Rather than focusing merely on chimpanzees' performance in this study, Premack and Woodruff were interested in the cognitive representations that Sarah used to reason about others' actions. Specifically, they wondered if Sarah was able to think about the actions of another individual not only in terms of that individual's behaviors, but also in terms of his unobservable mental states—his perceptions, beliefs, desires, and intentions. This cognitive process—often referred to as a theory of mind—is one that the human primate engages in quite automatically. Indeed, it's hard for adult humans to watch any social interaction and not consider how such actions bear on the desires of those involved, what they know about the world, and so on. Nonetheless, the cognitive computations required to process another individual's mental states are incredibly complex. The process of mental state attribution requires an individual to go beyond the available behavioral data and postulate states that are not directly observable (see discussion in Heyes 1998). In the case of visual attention, for example, an organism would need to go beyond the directly observable behavioral features of a problem (e.g., a competitor's eyes are facing away from me) and represent a set of properties that are not directly observable in the world (e.g., the competitor doesn't see me and, there-fore, is ignorant of what I am doing). While these steps could lead an in-dividual to perform an act when a dominant individual is unlikely to punish him for it, an individual can get to that choice more simply. In-stead of reasoning about whether or not the dominant individual can see him and is ignorant of his actions, he could instead merely associate the dominant's forward-oriented face with a negative result, and the opposite orientation with a positive result. Since these two scenarios lead to the same behavior in the acting individual, it is important to conduct con-trolled experiments in order to determine whether primates reason, for example, in terms of seeing and ignorance or in terms of a more simple association.

Faced with the difficulty of representing unseen mental states, research-ers have long wondered whether nonhuman primates share humanlike theory-of-mind abilities (see Heyes 1998 for review). Indeed, primate re-searchers have debated the answer to this question almost since the mo-ment Premack and Woodruff originally posed this question in 1978 (for more historical perspective on primate theory-of-mind questions, see Menzel and Menzel, Chapter 16 in this volume). Much of the debate sur-rounds the question of what would count as evidence for a mentalistic (rather than merely behavioral) understanding of others' actions. In Premack and Woodruff's (1978b) original commentaries, several philos-ophers noted that real mental-state reasoning requires an individual to realize that mental states are psychological in nature and thus can differ from real states of the world (see discussions in Dennett 1978; Pylshyn 1978). Using this logic, many researchers became interested in whether primates understood a set of special cases in which an individual's mental state differed from the true state of the world. One such case occurs when an individual possesses a false belief. Take, for example, a case in which I know that another person falsely believes that it's raining out-side. Note that the content of this individual's belief state (i.e., "it is rain-ing outside") differs both from what is really occurring in the world (i.e., it's not actually raining outside) and from what I myself believe about the world (i.e., I think it's not raining outside). My understanding of this in-dividual's false belief, then, involves an explicit realization that there is a difference between this individual's mental representation of the world and the real-world state, as well as an understanding of the difference between the content of this individual's mental state and the content of my own. In this way, false-belief scenarios provide a fertile test ground for studying the extent to which an individual's understanding of others' actions is indeed mental (as opposed to behavioral) in content.

Although developmental psychologists easily devised scenarios to test false-belief understanding in human children (see Wellman et al. 2001 for reviews), primate researchers had a more difficult task in developing methods that could be used to test false-belief understanding in nonverbal primates. Call and Tomasello (1999) reported an initial attempt testing chimpanzees and orangutans. In this study, experimenters with true and false beliefs attempted to communicate the location of hidden food to ape subjects. The logic was that the apes should ignore the experimenter's communicative cue in cases in which the experimenter had a false belief. The apes performed quite poorly on this task, failing to behave differently when the experimenter had a false belief. Unfortunately, apes also did rather badly on this task in general; on the whole, apes perform poorly when reading an experimenter's communicative cues (see review in Tomasello and Call 1997), so it's possible that apes performed poorly due to task demands rather than an inability to understand false beliefs. Kaminski and colleagues (2008) dealt with these task demand issues in a more recent study; they presented apes with a false-belief task in the context of competition, exactly the kind of situation in which primates do best on other sociocognitive tasks (see reviews in B. Hare 2001; Santos et al. 2006). In this study, chimpanzees played a role-switching game in which they alternated searching in buckets for food. In the critical condition, the experimenter gave one chimpanzee a false belief about the location of a piece of food. The question of interest was whether the other chimpanzee would exploit this chimpanzee's false belief in order to get more food. Although chimpanzees perform well on versions of the task that do not involve false beliefs, the same apes did poorly when required to use another individual's false belief. Although chimpanzees are able to use another individual's ignorance to deceive him, they fail to use another individual's false belief to do the same. In this way, chimpanzees differed from human children, who easily used false beliefs to deceive others in an identical task (Kaminski et al. 2008).

Marticorena and colleagues (in press) observed similar failures in rhesus macaques on a different measure of false-belief understanding. The researchers adapted a looking-time measure (see Onishi and Baillargeon 2005 for the same method in human infants) to explore whether rhesus monkeys represented the false beliefs of a human experimenter. Monkeys watched as a human experimenter witnessed the hiding of a piece of food in one of two locations. In one case, the experimenter had a true belief about the location of the food—it remained in the original location in which the experimenter had seen it originally placed. In another case, the experimenter had a false belief—the experimenter's view was hidden while the food switched locations. If monkeys represent the experimenter's false belief in this case, they should expect her to search for this food in the original location, because she falsely believes it is in the location in which she last saw it. Monkeys should then show a reliable difference in looking time between the expected and unexpected conditions, as such a difference would indicate monkeys view the events as qualitatively different. Like human infants (Onishi and Baillargeon 2005), monkeys made correct expectations in the true belief condition; macaques looked significantly longer when an experimenter with a true belief searched for the food in the wrong location than when he searched in the correct location. In contrast to human infants, however, macaques made no prediction in the false-belief condition. When the experimenter had a false belief about the location of the food, monkeys looked equally at events in which he searched the correct and incorrect box. Even in this simpler looking-time task, macaques appear to represent another individual's knowledge and ignorance, but not his false beliefs. Taken together, this pattern of performance in chimpanzees and macaques suggests that primates may represent some, but not all, aspects of others' mental lives. Across several studies (e.g., B. Hare et al. 2000; B. Hare et al. 2001; Kaminski et al. 2008; Marticorena et al., in press; Santos et al. 2006), the performance of chimpanzees and macaques is consistent with the idea that these species reliably recognize the behavioral cues that are relevant for determining when other individuals are knowledgeable and when they are ignorant. Such understanding appears to fall short of the performance of older human children, however, in that to date chimpanzees and macaques appear to lack the additional ability to reason about cues related to others' false beliefs. Indeed, primates appear to perform poorly on such false-belief tasks even when task demands are low (e.g., Marticorena et al., under review).

Apes' and monkeys' poor performance to date on false-belief studies complicates the question of how primates actually represent social cues. Historically, researchers have used an individual's understanding of others' false beliefs as evidence that that individual represents social cues in mentalistic terms (see discussions in Call and Santos, in press; Wellman et al. 2001). The fact that primates perform poorly on false-belief tasks has caused some researchers to argue that primates do not represent others social behaviors in mentalistic ways, as humans do; instead, primates may merely only represent and generalize over the observable behavioral components of others' actions (e.g., Penn and Povinelli 2007b). For example, some researchers have argued that primates may understand another individual's visual attention without understanding his "mental perspective" (Byrne and Whiten 1992). Under this view, primates can appreciate the relationship between gaze direction and the location of some visual stimulus without understanding anything about the mental experience of another individual (Horton and Caldwell 2006). This is achieved simply by understanding that attention is directed toward targets, that attention is determined by gaze direction, and that individuals usually act upon (or react to) objects to which they are attending (Gómez 1996). On the contrary, other researchers have argued that primates can, in fact, represent some kinds of mental states (e.g., knowledge versus ignorance) but merely fall short of representing others' beliefs (see Tomasello et al. 2003; Call and Santos, in press). Tomasello and colleagues (2003), for example, have suggested that theory of mind is not black or white, yes or no. Instead, it could be viewed as an umbrella term that covers a wide range of sociocognitive processes. Similarly, Santos and her colleagues (e.g., Santos et al. 2006; see also discussion in Call and Santos, in press) have claimed that primates can represent some concepts—such as knowledge and ignorance—mentalistically, but fall short of real mentalistic understanding of others' beliefs.

The Future of Primate Mental State Attribution: An Interdisciplinary Approach

In the spirit of the interdisciplinary nature of this volume, we will end our review with another open area in the field of primate theory of mind: the question of how different neural architectures may support these processes. Although much work has explored how the brain mediates visual perspective taking in our own species (see reviews in Allison et al. 2000; Gallagher and Frith 2003; Saxe et al. 2004), less work has examined whether homologous areas are at work in primate visual perspective taking. Indeed, neuroscientists have only recently begun looking at the ways in which primate brains process social stimuli and, specifically, information related to the attention and actions of others (see reviews in Fogassi and Gallese 2002; Ghazanfar and Santos 2004; Platt and Ghazanfar 2009; Shepherd 2010).

Some of the earliest work on this topic explored how macaque monkey brains processed the direction of other individuals' gaze (see reviews in Grosbras et al. 2005; Nummenmaa and Calder 2009; Shepherd 2010). In an important series of studies, Perrett and his colleagues showed that areas near the macaque superior temporal sulcus (STS) selectively respond to the direction of another individual's gaze, independent of that individual's head orientation (Perrett et al. 1982; Perrett et al. 1985; Perrett et al. 1992; see Allison et al. 2000 for findings in a homologous area of human cortex). This work suggests that specialized areas of the macaque STS process information related to where another individual is looking, but it is silent on whether these same regions are employed during higher-level perspective-taking tasks. Put differently, do STS regions merely encode gaze direction, or are they critical for more-complicated computations related to what other individuals can see? Unfortunately, little work to date has directly addressed this question. Interestingly, however, there are at least a few hints that neural mechanisms for gaze detection may be more complex than previously thought. First, K. Hoffmann and colleagues (2007) observed gaze-sensitive neurons in the amygdala, an area known to process emotionally valenced stimuli. In addition, Hoffman and colleagues found that gaze sensitivity in this region was modulated by other socially relevant cues, such as what expression the stimulus monkey was making. The flexibility observed in gaze-selective areas hints that gaze processing in this region may involve processes more complex than mere direction detection. Similarly flexible responses to gaze cues have also been observed in STS areas. Jellema and colleagues (2000) explored whether STS regions that respond to gaze direction take into account the broader context in which gaze is employed. To do so, they presented macaques with an event in which an actor varied his gaze direction while making a goal-directed reach toward a target. Impressively, they observed that gaze-sensitive neurons in the STS only respond selectively when the actor's gaze is oriented at the target object. This result suggests that STS neurons flexibly encode gaze direction in a way consistent with a sophisticated understanding of attention-following.

Another system implicated in primate perspective taking is the so-called mirror neuron system (see reviews in Fogassi and Gallese 2002; Iacoboni, this volume; Rizzolatti and Craighero 2004). Mirror neurons, which were first identified in macaque motor area F5 by Rizzolatti and colleagues (1988), have the unique property of responding both when the monkey subject makes a goal-directed action and when the same subject watches the same goal-directed action being performed by another individual. This close coupling of representations for action execution and representations for action perception led Rizzolatti and colleagues to speculate that mirror neuron regions may be critically involved in making sense of others' actions and attentional states. Under this view, macaques process the action of another person using the same circuitry that they would to execute that action themselves. This link therefore allows macaques to watch the goal-directed actions of others while simultaneously accessing the representations they themselves would employ when performing goal-directed action. Recent work suggests that this action-perception coupling may also work in the context of gaze-following. Shepherd and colleagues (2009) have recently observed areas in macaque parietal cortex (LIP) that respond both to gaze shifts and the perception of gaze shifts in others. Shepherd and colleagues hypothesize that this LIP "gaze mirroring" network could potentially support eye gaze understanding in the same way the F5 motor systems support action understanding.

But are mirroring systems actually necessary for representing the goal-directed actions and attentional states of others? Again, little work has directly addressed this question, particularly in macaque monkeys. Nevertheless, much recent evidence is consistent with the view that mirror neuron systems are critical for perspective-taking processes. First, there is a growing body of work using fMRI that suggests a link between the mirror system and perspective-taking abilities in humans (see reviews in Gallese and Goldman 1998; Iacoboni and Dapretto 2006; Oberman and Ramachandran 2007; J. H. G. Williams et al. 2001). For example, populations with known perspective-taking deficits, such as children with autism spectrum disorders, appear to have less activation in mirror neuron networks during perspective-taking tasks (Dapretto et al. 2006). Second, there is mounting evidence that macaque mirror neurons are tuned less to the kinematic movements of actions and more to the goals associated with those actions. For example, Umiltà and colleagues (2001) showed macaques events in which a hand grabbed a peanut either out in the open or behind an occluder. The researchers observed that the same mirror neurons were active regardless of whether the monkey could actually see the exact grasp motion. In another example, Fogassi, Ferrari, et al. (2005) observed that neurons in this region appear selective to particular goals, rather than particular movements. They presented macaques with identical hand movements in different contexts (e.g., grasping an object to move it versus grasping an object to eat it) and found that neurons in this region were selective more to the goal of the action than the particular movement involved. Taken together, mirror neurons appear to represent action not in terms of mere kinematics, but instead in terms of goals.

Despite some progress on these issues, we feel the stage is now set for an even more thorough exploration of primate perspective taking at a neural level, one that takes advantage of recent insights from behavioral work on primate social cognition. One outstanding question concerns whether gaze-selective regions can track where eyes have been (and thus what an individual has seen) over time. As reviewed above, a number of researchers have observed that primates are capable of tracking cues related to what an individual has seen over time in order to get an accurate sense of what that individual currently knows. Such behavioral results hint that gaze-selective neurons may respond differently in cases in which an individual's gaze is directed at new information (of which the individual is currently ignorant) versus old information (about which the individual is currently knowledgeable). To date, however, little work has explored how gaze-sensitive neurons are affected by an individual's information over time.

Another open issue concerns the extent to which gaze-sensitive neurons are modulated by information from other modalities. As reviewed above, several studies have demonstrated that primates take into account both what individuals can see (i.e., where their eyes are pointed) and what they can hear to determine what they know (e.g., Melis, Call, and Tomasello 2006; Santos et al. 2006). Such behavioral findings ask whether gaze-directed neurons can also be modulated by representations of what other individuals can hear. If so, these pathways may provide a neural basis for modality-independent representations of others' knowledge and ignorance.

Finally, little work to date has addressed the connection between the mirroring regions that respond to action perception and regions that represent eye-gaze direction. The existing behavioral work on primate perspective taking suggests that action understanding is intimately connected with information related to what other individuals can and cannot see. But is neural activity in mirror regions modulated depending on whether others are looking? Note that this is exactly the kind of modulation one would expect if these mirror regions were involved in more-complex representations that arise from true perspective taking. Consider, for example, the events presented by Umiltà and colleagues, in which an agent reached for a peanut behind a hidden barrier. In this study, mirror regions responded based on the agent's perceived goal—they fired only when a peanut was present as a target of the agent's reach. But what would happen in a slightly different case in which the agent had previously seen a peanut go behind a barrier? If the peanut was later removed but the agent thought the peanut was present, how would this region respond? The answers to such questions would provide insight into the question of how these regions treat cases of knowledge and ignorance, cases that behavioral data suggest macaques understand at some level (e.g., Flombaum and Santos 2005; Santos et al. 2006).

Although many questions remain for researchers interested in primate perspective taking—both at the behavioral and at the neural level—there is much reason to be excited by progress in this area. First, researchers have gained much insight into the ways that primates make sense of others at the behavioral level. Though undoubtedly much work is left to be done, our current insights into primate perspective-taking behaviors provide a rich body of work on which to base hypotheses about the cognitive and neural representations underlying these behaviors. In addition, there appears to be a growing enthusiasm for interdisciplinary approaches that attempt to synthesize insights across different fields. As the other chapters in this volume can attest, the last five years have seen an unprecedented increase in collaborations across disciplines, ones that have led to new insights in primate neuroscience and cognition. Our hope is that the many advances researchers have made on primate perspective taking at the behavioral level can now be translated into new advances at the neural and theoretical levels. In doing so, researchers will have embraced the promise of a truly interdisciplinary approach to the primate mind, one nicely championed by others in this volume as well.

CHAPTER TEN

Human Empathy through the Lens of Psychology and Social Neuroscience

Tania Singer and Grit Hein

HERE IS a long tradition of research and writing on human empathy and emotions, crossing the disciplinary boundaries between philosophy, social psychology, anthropology, and neuroscience. Part of this tradition has always been the question concerning whether empathy and emotions are uniquely human, or whether they are also part of the reality of nonhuman species (Darwin, Ekman, and Prodger 2002). Starting in the 1970s, empirical approaches were developed that allowed researchers to investigate emotions in animals, in particular their ability to empathize with a conspecific. The goal of this chapter is to give a concise overview of the most prominent approaches and results of empathy research in humans and to establish a link to animal findings. To clarify the terminology, in the first part of our chapter we provide definitions of empathy that were developed in social and developmental psychology and in social neuroscience. In the second part, we review empirical approaches and results from these fields that focus on empathy and theory of mind in humans. In the last part, we provide selected examples of empathy-related research in animals in an attempt to link those findings to human empathy research, and we discuss possible future research integrating research streams from psychology, neuroscience, and evolutionary anthropology.

Definitions of Human Empathy

Human empathy has been defined in different ways by different fields. Here we focus on prominent definitions of empathy derived from social and developmental psychology and social neuroscience research (for comprehensive reviews, see Batson 2009; de Vignemont and Singer 2006; Eisenberg 2000; M. Hoffman 2000; Singer and Lamm 2009). M. Hoffman (1981a), for example, proposed a relatively broad definition: "an affective response appropriate to someone else's situation rather than one's own." According to the "Russian doll model" (de Waal 2008), empathy is the result of a developmental sequence, beginning with babies crying when they hear another baby's cry and arriving, after considerable development, at a clear sense of others as distinct from the self. Distress perceived in the other is passed on to the observer and can elicit helping behavior, mainly driven by the urge to relieve the aversive arousal caused by one's own distress.

Modifying Hoffman's (1981) approach, Eisenberg and Miller (1987) pointed out that one's own distress can also lead to withdrawal rather than helping behavior. According to their definition, seeing a person in need elicits empathy, defined as "an affective state that stems from the apprehension of another's emotional state or condition, and that is congruent with it." With further processing, empathy either turns into sympathy or personal distress, or a combination of both (Eisenberg 2000). Sympathy is defined as "an emotional response stemming from another's emotional state or condition that is not identical to the other's emotion, but consists of feelings of sorrow or concern for another's welfare" (p. 92) and motivates prosocial behavior, unlike personal distress, which leads to withdrawal behavior (Eisenberg and Miller 1987).

The term "empathic concern," introduced by Batson, is similar to Eisenberg's definition of sympathy. Empathic concern is defined as an other-oriented response, which is congruent with the distress perceived in another person (e.g., Batson et al. 1995) and motivates altruistic helping (Batson 1991).

Based on these distinctions, which are made in social and developmental psychology, de Vignemont and Singer (2006) suggested that empathy can be further differentiated from emotional contagion and cognitive perspective taking. They describe empathy as follows: we "empathize" with others when we have (a) an affective state (b) that is isomorphic to another person's affective state and (c) that was elicited by observing or imagining another person's affective state (d) when we know that the other person's affective state is the source of our own affective state.

The definition of empathy as an affective state is important because it differentiates empathy from its cognitive counterpart that has been termed "theory of mind (ToM)," "cognitive perspective taking," and "mentalizing" (Baron-Cohen et al. 2000; Frith and Frith 2003; Premack and Woodruff

1978b; Wimmer and Perner 1983). The term "mentalizing" (Frith and Frith 2003) connotes a person's ability to cognitively represent the mental states of others, including beliefs and thoughts as well as affective states. Importantly, while engaging in cognitive perspective taking, a person may understand the affective states of another purely on the basis of knowledge but without becoming emotionally involved, that is, without sharing the emotional states of others. In contrast, the term "empathizing" connotes the capacity to share other people's feelings. Accordingly, when one empathizes with another person who is in pain, one feels the other person's negative affective state in one's own body. In contrast, when one understands someone else's thoughts, one does not feel that person's thoughts in one's own body. There are no qualia attached to the representation of the other person's thoughts. This difference and its significance become clearer when we consider psychopaths: they do not appear to have an impaired ability to understand other people's wishes, beliefs, intentions, and desires, as they are known to be manipulative. However, psychopathy is characterized and diagnosed as a lack of empathy. Thus, they probably lack the embodied feeling of empathy, which allows nonpsychopaths to anticipate and appreciate others' suffering, thereby often preventing them from harming others (for a similar argument, see J. Blair et al. 2005).

Part (b) of de Vignemont and Singer's (2006) description of empathy is important in distinguishing "empathy" from "sympathy," "empathic concern," and "compassion" (see Batson 2009; Eisenberg and Miller 1987; Singer and Steinbeis 2009, for a similar distinction). The latter three constructs involve the vicarious feeling for the other person. But when we "empathize," we share the other person's feelings: we feel as the other person (see also Figure 10.1a). When we "sympathize" or show "compassion," we do not necessarily share the same feeling. For example, to use the first person for a moment, when I empathize with a person who is sad, I feel sad myself. When I sympathize with or feel compassion for a sad person, I feel pity or concern for the person, but I am not sad myself. Also, when I notice that someone is jealous or envious of me, I can sympathize with or show compassion for that person, but I am not jealous or envious myself. Further, empathy is not necessarily linked to prosocial motivation—that is, a wish to maximize the other person's happiness or alleviate the other person's distress. In contrast, sympathy, empathic concern, or compassion are believed to be associated with prosocial motivation and prosocial actions or action tendencies. For example, empathy can be misused by a torturer who empathizes to find his victim's weakest point, but he is far from showing compassion for the suffering person.

Finally, de Vignemont and Singer's (2006) conception of empathy distinguishes between "empathy" and "emotional contagion." The latter refers to a reaction in which one shares an emotion with another person without realizing that the other person's emotion was the trigger. For example, babies start crying when they hear other babies crying, long before they develop a sense of a self separate from others. These reactions might be a precursor of the development of the capacity for empathy (see Sagi and Hoffman 1976), but they are not considered empathic responses per se because the babies are not aware that they are vicariously feeling with another person's distress. The distinction between empathy and emotional contagion is specifically important with respect to

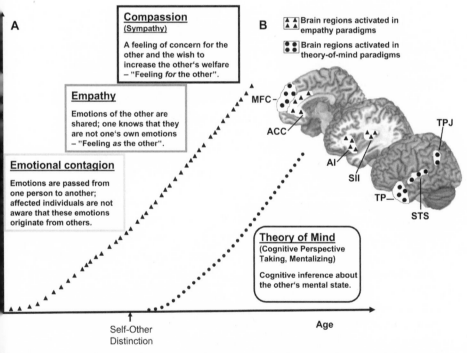

Figure 10.1. A. Definition of emotional contagion, empathy, compassion, and theory of mind and illustration of their suggested developmental sequence. B. Schematic overview of brain regions typically involved in understanding others based on theory-of-mind dots and empathy triangles. The latter represents regions identified using the cited empathy-for-pain and touch paradigms. Note, however, that the depicted empathy-related regions are not exhaustive. MPC = medial prefrontal cortex; ACC = anterior cingulate cortex; AI = anterior insula; SII = secondary somatosensory cortex; TP = temporal poles; STS = superior temporal sulcus; TPJ = temporoparietal junction.

the question concerning whether nonhuman species have empathy or just its precursor, emotional contagion (see the last section for discussion of this point).

Figure 10.1a depicts a schematic illustration of the relevant concepts and their suggested developmental sequence (see also Singer 2006). Accordingly, emotional contagion is a precursor of empathy, which is a vital first step in a chain of emotional responses that lead to feelings of compassion, empathic concern, or sympathy. Whereas the latter three vicarious affective responses involve feeling for someone, empathy involves feeling as someone else. Nevertheless, all four concepts refer to affective responses. In contrast, cognitive perspective taking, ToM or mentalizing refer to abstract propositional thoughts such as the inference of beliefs and intentions of others. Note that even though psychology and social neuroscience have provided accumulating empirical evidence for the need to distinguish between these concepts and routes of social cognition, in a normally developed adult human brain, these routes usually interact with each other in an integrated fashion to produce an understanding of the other's mind and behavior.

In the next section, we will review empirical evidence from psychology and neuroscience that provides support for the validity of the outlined conceptual distinctions and the notion of different neural circuitries underlying our ability to empathize and to mentalize (theory of mind).

Empirical Evidence for Empathy in Social Psychology

In psychological research, empathy is usually assessed using self-report scales (e.g., Batson et al. 1997; Davis 1983; Mehrabian and Epstein 1972). Self-reports of empathy have been found to be reliable across cultures (Siu and Shek 2005); in different populations, such as those of prisoners (Lauterbach and Hosser 2007), physicians (Yarnold et al. 1996), and patients (Alterman et al. 2003); and in different age groups (Funk et al. 2008; Grühn et al. 2008).

Going one step further, in an impressive series of experiments, Batson and colleagues investigated the behavioral relevance of self-reported empathy with respect to actual helping behavior. According to Batson (1991), observing a person in need can elicit empathic concern or personal distress (see also Eisenberg and Miller 1987). Both feelings can trigger helping, but the underlying motivation is very different for the two outcomes. A distressed person helps the other in order to reduce his own distress (i.e., based on a self-regarding motivation). In contrast, empathic concern produces altruistic helping with the other-regarding motivation to increase the other's welfare. Personal distress felt when seeing a person in need can be reduced by escaping the distressing situation without helping. In contrast, empathic concern triggers helping, even if there is an easy escape option. Based on this reasoning, Batson and colleagues predicted that distressed people would help only if there was no other way out of the distressing situation, but would be less likely to do so if they could escape easily, whereas potential escape options would have no impact on helping behavior in empathic people. To test this, Batson and colleagues had participants observe a person receiving electric shocks in an ostensible learning experiment. After a few trials, subjects were given the opportunity to relieve the other person's suffering by taking the remaining shocks themselves (helping behavior). In the "easy escape" condition, participants were informed that they would only have to observe the other person suffering during the first two trials. In the "difficult escape" condition, they were told that they would have to watch the other person suffering until the end of the session. The degree of empathic concern versus distress was manipulated with an emotion-specific misattribution technique (Batson 1991). Confirming their predictions, subjects who reported a high level of distress when observing the person in need helped significantly less in the "easy escape" condition as compared to the "difficult escape" condition, whereas there was no such difference for subjects reporting empathic concern. This and many other experiments (for a review, see Batson 2009) confirm the hypothesis that empathic concern acts as a precursor of altruistic motivation, whereas personal distress leads to a self-oriented response. As will be outlined in the next section, parallel research in developmental psychology has extended these findings to children.

Empirical Evidence for Empathy from Developmental Psychology

In the domain of developmental psychology, Eisenberg and colleagues (Eisenberg et al. 1989) studied how children's responses to needy others differed from adults' responses. One of their studies used a paradigm similar to that developed by Batson and colleagues: the paradigm focused on the easy escape condition, since it differentiates best between altruistic and selfish motives. In addition to measuring subjects' individual rates of helping behavior, Eisenberg and colleagues also recorded facial expressions and heart rate while subjects watched a video clip of a hospital scene depicting a woman and her two injured children, who were suffering from the consequences of a car accident. After viewing the film, subjects reported their emotional reactions and were given the opportunity to offer to help the woman in the clip by assisting her with some household tasks (in the case of adults) or to help the injured children by bringing them their homework materials (in the case of children). In line with the hypothesis that sympathy promotes prosocial behavior, this experiment showed that for adults, self-reports and facial expressions of sympathy predicted helping. Interestingly, children's verbal reports of distress and sympathy were not related to their prosocial behavior, suggesting that children's capacity to report self-experience may be underdeveloped. However, children's facial display of distress tended to be negatively related to their helping behavior: children who showed more facial signals of distress tended to help less. Taken together, the general pattern of results supports the claim that empathic concern promotes helping, whereas personal distress is linked to self-oriented motivation. Several other developmental psychological studies have confirmed with various measures (ranging from facial and behavioral to physiological reactions to the distress of others) that children's increased empathic concern is positively correlated with higher rates of helping (for a review see Eisenberg 2000).

Empirical Evidence for Empathy in Social Neuroscience

With the emergence of the field of social neuroscience, researchers have also embarked on studies of the neural underpinnings of empathy using neuroscientific methods such as functional magnetic resonance imaging (fMRI). Most of these empathy studies have focused on empathy for pain suffered by another person (Bird et al. 2010; Cheng et al. 2007; Gu and Han 2007; Lamm, Batson, and Decety 2007; Lamm et al. 2010; Lamm, Nusbaum, et al. 2007; Morrison and Downing 2007; Saarela et al. 2007; Singer et al. 2004, 2006, 2008; for a recent meta-analysis, see Lamm et al. 2009). Two early fMRI studies by Singer and colleagues (2004, 2006) investigated empathy "in vivo" with an interactive empathy-for-pain paradigm. In this paradigm, the volunteer in the fMRI scanner either receives pain herself or perceives pain in another person, delivered via pain electrodes on the back of the volunteer's or the other person's hand. The other person sits next to the fMRI scanner, and a mirror system allows the participant inside the scanner to see her own as well as the other person's hand lying on a tilted board. Differently colored flashes of light on a screen behind the board point to either the volunteer's or the other person's hand, indicating which of them will receive painful and which will receive nonpainful stimulation. This procedure permits us to measure pain-related brain activation when pain is applied to the scanned volunteer (felt pain) or to the person outside the scanner (empathy for pain). Singer et al. 2004 used this paradigm to assess empathy in couples. Here, the female partner was the volunteer in the scanner who received pain herself or perceived her husband suffering from pain. The results suggest that parts of the so-called "pain matrix"—bilateral anterior insula (AI) and the rostral anterior cingulate cortex (ACC)—were activated when she experienced pain herself as well as when she saw the arrow cue indicating that her husband was experiencing pain (Figure 10.1b, triangles, for a schematic illustration). These areas are involved in the processing of the affective component of pain, that is, how unpleasant the subjectively felt pain is. Interestingly, activation of areas involved in the sensory-discriminative component of our pain experience (activation of primary and secondary somatosensory cortex contralateral to the stimulated hand) was not involved in empathic responses to the suffering of the other, but unique to the pain experience in oneself. Thus, both one's own experience of pain and the knowledge that the other person is experiencing pain activated parts of the same affective but not the sensory-discriminative pain circuit, suggesting a neural simulation of the emotional suffering of the other person in absence of pain stimulation to one's own body. A more recent study (Singer et al. 2006) with the interactive empathy-for-pain paradigm showed that empathic brain responses in AI and ACC are not restricted to a beloved partner, but also occur when an unknown but likeable person is in pain (see also Figure 10.1b).

Imaging studies on empathy for pain in which participants viewed pictures or videos of unknown faces displaying expressions of pain (Lamm, Batson, and Decety 2007; Saarela et al. 2007) or of body parts in painful situations (Cheng et al. 2007; Gu and Han 2007; Lamm, Nusbaum, et al. 2007; Morrison and Downing 2007; Figure 10.2a) have revealed a similar pattern of results, emphasizing that the neural simulation of another person's pain occurs independently of the affective link between the empathizer and the person in pain. In a recent meta-analysis, Lamm et al. (2009) compared four cue-based "online" empathy paradigms such as the one by Singer et al. (2004) described above (e.g., Bird et al. 2010; Hein et al. 2009; Singer et al. 2004, 2006, 2008) with four picture-based empathy-for-pain paradigms (Jackson et al. 2006; Jackson et al. 2005; Lamm et al. 2010; Lamm, Nusbaum, et al. 2007). The results identified a core network of empathy for pain including bilateral AI cortex and dorsal ACC during empathy for pain. Depending on the paradigm used, this core network of empathy can, however, be engaged by different neurofunctional pathways. Pictures of body parts in painful situations recruit neural structures underpinning sensory-motor functions and action understanding (inferior parietal/ventral premotor cortices), whereas abstract information about the other's affective state engages areas associated with mentalizing (pre-cuneus, ventral medial prefrontal cortex, superior temporal cortex, and temporoparietal junction; see also Figure 10.1b). In addition, primary so-matosensory areas are activated rather unspecifically by the picture-based paradigms but not by the cue-based online paradigms, indicating that previously reported discrepant findings of somatosensory involvement in empathic pain might be related to differences in experimental paradigms. Images of body parts being touched may unspecifically activate somato-sensory areas, but this activation does not seem to be functionally in-volved in the simulation of how painful or aversive something feels for another, that is, does not seem to be involved in shared networks under-lying pain empathy.

Evidence for shared neural networks underlying the processing of affec-tive states of self and of others was also found in domains besides the pain domain. There is, for example, evidence that the observation of touch and the firsthand experience of touch activate similar regions in secondary so-matosensory cortex (Keysers et al. 2004; see also Figure 10.1b, triangles, for a schematic illustration). In another recent study, participants watched video clips showing people sampling pleasant and unpleasant tastes and then experienced the different tastes themselves (Jabbi et al. 2007). In line with the results of empathy for pain, Jabbi and colleagues found neural activation in AI cortex when people passively watched disgust in another person and when they were disgusted themselves (for an earlier study in the domain of disgust see also Wicker et al. 2003).

Moreover, the amplitude of empathic brain responses was found to be modulated by several factors, such as the intensity of the displayed emo-tion (Saarela et al. 2007), the participant's appraisal of the situation (Lamm, Batson, and Decety 2007), characteristics of the suffering person such as perceived fairness (Singer et al. 2006), previous experience with pain-inflicting situations (Cheng et al. 2007), and group membership (Hein et al. 2009). The results of the latter study further showed that in-creased AI activation when observing the suffering of an in-group mem-ber as compared to an out-group member predicts in-group favoritism in later helping behavior, that is, participants' willingness to receive painful shocks in order to help the other person (Hein et al. 2009).

Empirical Evidence for Theory of Mind in Social Neuroscience

Using fMRI allowed researchers to distinguish empathy (i.e., the sharing of emotions perceived in another person) from perspective taking (i.e., the cognitive apprehension of others' minds). As we reviewed above, neu-

A Empathy for pain

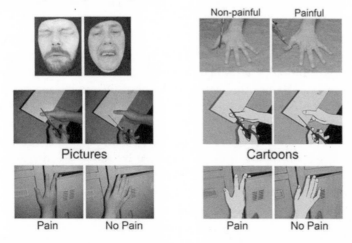

B Theory of Mind

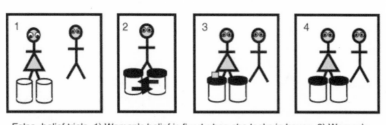

False–belief trials. 1) Woman's belief is fixed when she looks in boxes. 2) Woman's belief becomes false when man swaps boxes. 3) Woman makes her belief manifest by indicating box. 4) Participant is asked to identify box containing object.

Figure 10.2. A. Examples of faces displaying expressions of pain (with kind permission from M. Saarela) and pictures and cartoons of body parts in painful situations (top, with kind permission from C. Lamm; bottom, with kind permission from S. Han) used to investigate empathy for pain in humans. B. Schematic diagram of participants' view of the event sequence for the false-belief task used to measure theory-of-mind abilities (with kind permission from I. A. Apperly).

roimaging studies of cognitive perspective taking are also often referred to as "ToM" or "mentalizing" experiments. In most cases, they are conducted with healthy adults who are asked to understand the intentions, beliefs, and desires of a protagonist in a story or a cartoon (see Figure 10.2b for an example of a "false-belief task" used by Apperly and colleagues

[2007]; see also Apperly et al. 2004; for a review see Gallagher and Frith 2003). ToM studies have consistently revealed a neural network comprising the posterior superior temporal sulcus (STS), extending into the temporoparietal junction (TPJ), the medial prefrontal cortex (mPFC), and sometimes the temporal poles (TP) (see Figure 10.1b, dots, for a schematic illustration). Interestingly, the mPFC is not only involved when people mentalize about other people's thoughts, intentions, and beliefs, but also when people reflect on their own mental states (J. Mitchell et al. 2005).

Recent studies indicate a dissociation between empathizing and mentalizing in psychopathology (for a more extensive review see also Singer and Leiberg 2009). Psychopathy is characterized by a lack of empathy and guilt and poor behavioral control, a precursor of aggressive behavior (R. Hare 1991). Interestingly, people with psychopathy are known to be rather good at understanding other people's mental states, that is, at mentalizing, but lack the ability to empathize with others. Thus, behavioral and imaging studies on psychopaths have found selective emotional dysfunction (Birbaumer et al. 2005; R. Blair et al. 2001; Flor et al. 2002; Levenston et al. 2000; Veit et al. 2002). In contrast to psychopathy, autism is mostly associated with a lack of mentalizing (for a review see U. Frith 2001), but not necessarily a lack of emotion sharing (Bird et al. 2010; Silani et al. 2008). Supporting this claim, neuroimaging studies have revealed reduced activation in mentalizing-related brain areas such as the mPFC, STS, TPJ, and TP when autistic subjects make inferences about others' mental states (Castelli et al. 2002; Happé and Frith 1996).

Recent results suggest that a lack of empathy in autism is only found in specific subgroups of autistic patients with interoceptive deficits (Bird et al. 2010; Silani et al. 2008). One important determining factor for the prevalence of an empathic deficit in autism is the level of alexithymia, a condition marked by impaired identification and description of feelings and impaired differentiation of feelings from bodily signals, which is found to a high degree in 50 percent of the high-functioning patients with autism or Asperger syndrome (E. Hill et al. 2004). According to the shared network hypothesis of empathy introduced above, representations of one's own emotional states are a necessary precondition for the ability to share others' emotions. This would predict that deficits in understanding one's own emotions, as present in alexithymia, lead to empathy deficits and are correlated with reduced insula activation. Confirming this assumption, Silani et al. (2008) showed that alexithymia, but not autism, was associated with reduced activation in the AI. Moreover, individual differences in the degree of alexithymia correlated negatively with individual differences in trait empathy, and levels of both alexithymia and empathy were predictive of AI activation during introspection. These results were replicated and extended in a recent follow-up study testing autistic and non-autistic participants with high and low alexithymia scores using the fMRI empathy-for-pain paradigm by Singer et al. (2004) outlined above. Results revealed that the intensity of empathic brain responses in AI varied as a function of the degree of alexithymia, but were not predictive of autism per se. These findings are in line with the prediction that deficits in understanding one's own emotions result in empathy deficits and that both capacities are supported by insula functions (see also Singer et al. 2009). Furthermore, the diagnosis of autism per se does not seem to be associated with deficits in emotional interoception and empathy. This notion awaits further support by studies aiming to dissociate empathizing from mentalizing abilities by testing the same individuals using appropriate paradigms.

The Mirror Neuron System and Its Relation to Empathy and Mentalizing

In the previous section, we introduced accumulating evidence for two different routes of social cognition: the empathizing and the mentalizing route. There is, however, a third route to the understanding of others' intentions and actions: the so-called mirror neuron system, which underlies the understanding of others' motor actions, intentions, and goals.

As this stream of research is reviewed in detail in other chapters of this book (see Chapters 2 and 3), we will refrain from providing an extensive summary of this line of research and try instead to link mirror neuron research to the concepts introduced in the present chapter so far. In short, the mirror neurons were discovered in the premotor cortex of macaque monkeys (Rizzolatti and Craighero 2004) and are activated when self-generating an action as well as when observing an action by another. In humans, a common neural response to the execution and observation of actions was found in inferior parietal lobus (IPL), ventral premotor cortex (PMC), and posterior inferior frontal gyrus (IFG) using functional magnetic resonance imaging (fMRI) (Grezes and Decety 2001), transcranial magnetic stimulation (TMS) (Fadiga et al. 1995), and magnetoencephalography (MEG) (Hari et al. 1998; for a review see Iacoboni and Dapretto 2006). Note, however, that there is still a controversial debate over whether such findings can be interpreted as evidence for the existence of mirror neurons in the human brain (Grafton 2009, for a review). Findings that show that the neural response when observing others' actions is somatotopically organized (Gazzola et al. 2006; Goldenberg and Karnath 2006), stronger for familiar (Calvo-Merino et al. 2005, 2006) and well-trained (Cross et al. 2006) actions, and modulated by contextual information (Iacoboni et al. 2005) speak in favor of the assumption that we use our own action knowledge to infer the action-related intentions and goals of others. However, given that mirror neurons are motor neurons, their function might be restricted to inferring others' states in the context of motor actions and neither necessary nor sufficient for inferring others' emotions. Supporting this view, the observation of gestures with emotional content was found to be more likely to activate brain regions related to emotion processing, such as the amygdala and orbitofrontal cortex, whereas instrumental gestures mainly activated classic mirror neuron regions in premotor and parietal cortex (Gallagher and Frith 2004). In line with these results, more-recent work suggested that inferior frontal and parietal regions are related to the processing of others' facial muscle movements, whereas inferring emotions from facial expressions is associated with activation of emotion-related brain regions such as the insula (Jabbi and Keysers 2008). Moreover, a meta-analysis (Lamm et al. 2009) focusing on different types of empathy-for-pain paradigms showed that, depending on the task used, different routes seem to elicit activation in a core empathy-related network. Thus, the observation of actions causing pain (e.g., pictures of needles penetrating the skin) co-activate inferior parietal and ventral premotor cortices, that is, action-related circuitries, whereas empathy triggered by abstract cues is associated with activation in the STS, the TPJ, and the mPFC, which are known to be involved in inferring others' mental states ("ToM"; Figure 10.1). Importantly, AI and dorsal ACC are activated irrespective of the task context. This suggests that mirror neuron systems can but do not necessarily play a functional role in activating core empathy-related networks.

In summary, the present review suggests that there are different neural routes that allow humans to understand and predict others' actions, feelings, and beliefs. The degree of interaction between these routes depends on individual characteristics as well as on the task context and the information provided in the given situation (see also Lamm and Singer 2010).

Measures of Empathy in Animals and Integration with Human Results

Measures of empathy in nonhuman species are described in detail in other chapters of this book. Therefore, we will only discuss findings relevant to the possible link between measures and findings of empathy in animals and those in humans.

Discussions of empathy in nonhuman species are closely related to the mark test (Gallup 1970), which is used to investigate an animal's self-recognition in a mirror. Animals that show mirror self-recognition (MSR) use their reflections in a mirror to inspect parts of their bodies, including marked body parts that normally cannot be inspected. MSR is interpreted as a sign of self-recognition (Gallup 1977, 1983; but see Heyes 1994 for a critical discussion). Self-recognition requires a preexisting self-awareness or self-concept (Gallup 1977), which itself is the prerequisite for a distinction between self and others, that is, conspecifics in the social environment. As stated above, a self-other distinction enables one to understand another's emotional state without confusing it with one's own emotions and thus to move from pure emotional contagion to empathy. So far, there is evidence for the presence of MSR in primates (e.g., de Waal et al. 2005), elephants (Plotnik et al. 2006; but see Povinelli 1989), and bottlenose dolphins (Reiss and Marino 2001).

A recent study testing empathic responses in animals focused on rodents (Langford et al. 2006). Mice received noxious stimulation either alone or in dyads with a cage mate, a genetically related stranger, or a stranger. The results revealed that rodents showed significantly stronger pain behavior when they saw the other mouse receiving pain, but only if it was a cage mate and not if the other mouse was a stranger. The authors interpreted this result as consistent with the perception-action model of empathy (Preston and de Waal 2002), proposing that we automatically share the somatic and emotional responses of others.

In the following, we will explore the similarities and dissimilarities of empathy research and findings in animals and humans.

Both human and animal researchers have suggested that the capacity to distinguish between self and others as reflected by self-recognition in a mirror is an essential precondition for the development of empathy (Eisenberg and Strayer 1987; but see de Waal 2008 for a critical view). Developmental studies in humans, for example, have shown that MSR is indeed closely linked to the ontogeny of empathy in children (Bischof-Köhler 1988; Zahn-Waxler et al. 1992). Although the validity of the mark test in animals has been the subject of a controversial debate (Gallup et al. 1995; Heyes 1994), such findings nevertheless support the view that MSR in animals and humans is associated with empathy, although the direct test of a correlation between the ability to pass the MSR test and the ability to empathize is still lacking in animals. Furthermore, it would be interesting to expand the range of species tested with the MSR test to provide further evidence for an association between the capacity for self-other distinction and the development of social emotions such as empathic concern.

It is important to note that, according to the above definition, self-other distinction is a critical precondition for empathy, but not for emotional contagion. Despite the fact that there is no evidence for MSR in rodents yet, Langford et al.'s (2006) results nevertheless showed that the mice's pain reaction was modulated by the pain of their cage mates. Even though the authors interpreted these results in terms of evidence for empathy in rodents, we suggest that such a reaction to others' emotions without a clear self-other distinction is likely to be driven by emotional contagion rather than empathy. Due to the known effects of enhanced attention toward familiar conspecifics as compared to strangers, the emotional contagion might still be stronger when the mice are exposed to the suffering of familiar conspecifics as compared to strangers. Other experimental manipulations are needed in animal research to disentangle emotional contagion from other forms of other-regarding feelings such as empathy, empathic concern, and sympathy or compassion. Clearly, the distinction between these concepts is more difficult to draw in animals than in humans, as subjective ratings that allow us to assess the subjective awareness of vicarious feelings and the exact nature and quality of the experienced feelings are not available in nonhuman species. The underlying nature of the feeling or motivation has to be inferred from observations of behavior. Thus, similar to the reviewed behavioral studies in social and developmental psychology in humans, in animal research consolation or helping behavior toward a suffering conspecific has typically been taken as indirect evidence for the existence of empathy in nonhuman species (de Waal and Aureli 1996; Fraser et al. 2008; Schino et al. 2004).

We suggest that the introduction and adaptation of empathy paradigms developed by social and developmental psychologists (see above) into animal research would represent an elegant way to go a step further and disentangle distress-driven from empathic-concern-driven helping behavior. Using paradigms similar to those developed by Batson and colleagues, for example, one could study whether primates stop pressing food levers when pressing them results in another primate in the opposite cage receiving pain or other unpleasant experiences. Refraining from pressing a food lever could be motivated by wanting to alleviate one's own distress caused by watching and hearing the other in distress or by empathic concern for the other's welfare. To disentangle the two motivations, one could add an "easy escape" condition to the experiment in which apes learn to switch another lever causing the other ape's cage to be enclosed so that no visual or auditory distress cues could be seen or heard anymore. If, under this "easy escape" condition, the empathizing primates start pressing the food lever again despite knowing that the other primate is still suffering pain, empathic concern as a motivation to help would have to be excluded and we would have to assume that helping behavior is merely driven by their motivation to reduce their own distress. Some early single-case reports by Masserman and associates (1964) suggest that primates indeed choose to refrain from pressing food levers when this action is coupled with witnessing another primate suffering pain.

In contrast to empathy research, a variety of well-controlled experimental studies on theory of mind have been conducted in nonhuman species in the last few decades. In a recent review summarizing studies performed over the last 30 years in nonhuman species, Call and Tomasello (2008), for example, concluded that "there is solid evidence from several different experimental paradigms that chimpanzees understand the goals and intentions of others, as well as the perception and knowledge of others. Nevertheless, despite several seemingly valid attempts, there is currently no evidence that chimpanzees understand false beliefs. Our conclusion for the moment is, thus, that chimpanzees understand others in terms of a perception-goal psychology as opposed to a full-fledged, human-like belief-desire psychology." (For a more in-depth discussion of ToM in nonhuman species, see also Chapters 9 and 11 in this volume.)

From a neuroscientific perspective, this observation is interesting in many ways: As reviewed above, neuroscience has accumulated evidence that there are at least three distinct routes to the understanding of other minds that usually interact with each other when humans engage in reading the feelings, action intentions, and abstract beliefs of other humans. Thus, we can distinguish between an empathic route that mostly relies on somatosensory, limbic, and paralimbic cortices (e.g., AI, ACC, SII), a cognitive perspective-taking route (ToM) that mostly relies on PFC, TPJ, and STS, and a mirror neuron system that mostly relies on IPL, PMC, and IFG. It has been suggested that nonhuman species mostly rely on their mirror neuron system, and not on the cognitive perspective-taking route, when inferring the intentions and goals of others (Call and Tomasello 2008; see above). Such a suggestion may explain why primates have difficulty inferring abstract beliefs, an ability that develops late in human ontogeny and has been specifically associated with functions of the TPJ (Saxe and Kanwisher 2003; Saxe and Wexler 2005). In turn, the TPJ is a higher-order association cortex that does not develop and mature until very late in ontogeny (Gogtay et al. 2004) and seems to be crucial for self-other distinction (Decety and Lamm 2007). To our knowledge, however, so far no rigorous in-depth experimental research has been performed in nonhuman species that would allow us to disentangle these three different routes of mind reading as well as possible motives underlying other-regarding behavior in nonhuman species. Adopting an interdisciplinary research focus that integrates methods from neuroscience, evolutionary anthropology, and psychology may help us to shed more light on the evolution and ontogeny of our ability to understand others' minds.

CHAPTER ELEVEN

How Much of Our Cooperative Behavior Is Human?

Brian Hare and Jingzhi Tan

WHETHER IT IS our technological, cultural, institutional, or linguistic capabilities, almost every behavior that we might call "human" is ultimately a product of cooperation. Characterizing human cooperation and explaining its evolution will be central to explaining the evolution of our species. One of the tools we have available for testing hypotheses regarding human cooperation is the comparison of cooperative abilities among hominoids. Humans share a common ancestor with the Panins (bonobos and chimpanzees) some 5–7 million years ago. In comparing our own species with bonobos and chimpanzees we can identify traits we share and those that arose as derived traits during human evolution. Therefore, comparisons between the cooperative behavior of chimpanzees, bonobos, and humans allow us to test hypotheses regarding the unique nature of human cooperation. We review a number of studies on chimpanzee and bonobo cooperation in recent experiments that simultaneously help explain the cooperative behavior these species demonstrate in the wild while also testing hypotheses about the unique nature of human cooperation. (See Figure 11.1.)

Preconceptions and Observations of Ape Cooperation

We now know so much about wild bonobos and chimpanzees that it is easy to forget that the first long-term research on wild ape behavior began less than 50 years ago. This means for 90 years after the publication of *The Origin of Species,* little, if anything, was known about the naturally occurring behavior of apes. Because the first attempt to study wild

Figure 11.1. a) Bonobos and b) chimpanzees are our species' two closest living relatives, with both sharing almost 99 percent of our genome through common descent. This means that the genomes of chimpanzees and bonobos are more similar to that of humans than to that of gorillas. The two species diverged from each other around 1 million years ago.

chimpanzees in French Guinea was very short-lived (Nissen 1931), the majority of information on chimpanzee behavior was provided by the pioneering work of Robert Yerkes and Wolfgang Köhler with small captive populations of young chimpanzees (Köhler 1925; Yerkes and Yerkes 1929). As a result, while it was known that chimpanzees were highly social, before the early 1970s little else was understood about the character of chimpanzee social life. Based on such scant information, some in this period suggested chimpanzees might only show cooperative behavior in experiments conducted in captivity:

> Nissen has made the point . . . that in the native habitat, where the necessities for life are very easy to obtain (food, water, simpler shelter, and protection from enemies), the chimpanzee's full capacities are not taxed to the limit. The necessity for well integrated cooperative activity in order to satisfy biological needs perhaps has never arisen, and hence in the field one should not expect to see the type of team work which might be demonstrated in the laboratory, where such behavior could be put at a premium. (Crawford 1937, 4)

Therefore, before field-workers began to unlock the secrets of chimpanzee social life, few might have guessed the extent to which chimpanzee survival and reproduction depends on their ability to work together with other group members in a variety of contexts. We now know that chimpanzees and bonobos cooperate in a wide range of situations in the wild and that they show a fair amount of plasticity in their cooperative behaviors across a range of ecologies (see Muller and Mitani 2005 for a comprehensive review). For example, both bonobos and chimpanzees have long-term relationships in which they support each other through grooming, coalitionary support, and food sharing (de Waal 1982, 1997b; Nishida 1983; Goodall 1986; Kano 1992; Parish 1994; Vervaecke et al. 2000; Watts 2002; Hohmann and Fruth 2002). In addition, male chimpanzees have been observed to regularly hunt monkeys and patrol their territory borders in groups (Nishida 1979; Wrangham 1999; Boesch and Boesch-Achermann 2000; Watts and Mitani 2001; Mitani and Watts 2001; J. M. Williams et al. 2004).

The variety, frequency, and adaptability of Panin cooperation, along with the reduced role of kinship in explaining it (e.g., Gerloff et al. 1999; Langergraber et al. 2007), raise many questions regarding the underlying motivation and cognitive skills that allow for (or constrain) these species' cooperative behavior. The flavor of the questions are the same: is it that Panin cooperation is a by-product of individuals egocentrically pursuing their own interest with little understanding of how their success depends on the behavior of others, or is it that nonhuman apes have a richer understanding of the social nature of their joint successes (Boesch 1989; Watts and Mitani 2002; Noë 2006)? Do they know when they or someone else needs help? Can they work together with anyone within their group? Do they know that some group members are better helpers than others? Will they only help if it is a mutualistic endeavor, or is their helping sometimes more costly in nature? Will they help a stranger? Can they detect, shun, or even actively punish cheaters?

It is here that experiments will make an important contribution to our understanding of Panin cooperation. Many of the questions above are very difficult to answer without careful experimentation in addition to detailed observations. For example, because observation conditions can be so difficult during hunting, there is disagreement regarding the psychology that might underlie this behavior. In this case it may be most productive to examine the collaborative potential of chimpanzees in the laboratory (Muller and Mitani 2005; Mitani 2006). So while Nissen and Crawford's initial speculation regarding the extent to which chimpanzees cooperate has proven incorrect, the spirit of Crawford's suggestion that laboratory work will be crucial for the study of chimpanzee cooperation is indeed valid. It is now the job of experimentalists to test how flexibly the natural cooperative skills of Panins generalize to novel (and sometimes artificial) situations. This way we should be able to expose not just the cooperative potential of other apes, but also their limitations.

Experimental Investigations of Cooperation

Although chimpanzees and bonobos have been observed to cooperate in a diverse number of contexts, until recently there has been little experimental evidence to support the idea that their cooperation relies on sophisticated cognitive mechanisms. When given a novel instrumental task that required joint action (i.e., pulling a heavy tray baited with food), chimpanzees had to be explicitly trained to solve the problem, and the newly learned cooperative skills did not generalize to a slightly different cooperative task (Crawford 1937, 1941; Savage-Rumbaugh et al. 1978; Chalmeau 1994; Povinelli and O'Neill 2000). However, in all these previous experiments, a handful of chimpanzees were tested, no bonobos were studied, and the performance of the chimpanzees seemed constrained by intolerance between potential partners (Chalmeau 1994). This suggests that larger samples of subjects from both species where social relationship variables can be measured and controlled might provide an opportunity to observe spontaneous cooperation in captivity while gaining insight to its evolution.

Tolerance and Cooperation

Work with other primates and even birds has demonstrated that the ability to cooperate is constrained by tolerance (Petit et al. 1992; Chalmeau 1994; Mendres and de Waal 2000; Visalberghi et al. 2000; Werdenich and Huber 2002; de Waal and Davis 2003; Seed et al. 2008). Thus, one possible explanation for the discrepant findings between observational and experimental studies of chimpanzee cooperation is that noncognitive factors constrain the ability of chimpanzees to cooperate in experimental settings (e.g., Chalmeau 1994). A number of studies with Panins now support this hypothesis.

Melis et al. (2006a) presented a novel instrumental task to 16 pairs of chimpanzees but only after testing them for their ability to share food (Figure 11.2a). Once the level of food tolerance within each pair was assessed, a long platform was placed in front of their testing room but out of their reach. Food was placed in two separate piles on either side of the food platform. A single rope was threaded through two different loops on either end of the food platform so that the food tray could be retrieved if both ends of the ropes were pulled simultaneously. If the rope ends were spread beyond the reach of a single individual, then success in retrieving the food required cooperation. However, pulling one of the rope ends would only pull the rope out of the loops like a shoestring in a shoe—resulting in failure. (Because this apparatus was initially designed by Hirata 2003, we will hereafter refer to it as the Hirata paradigm; it is a much more viable technique for comparative studies of cooperation than that deployed previously by Crawford because it does not require the use of heavy weights—which when working with nonhuman apes can be dangerously heavy and bulky.) When pairs in each dyad were released simultaneously, many spontaneously solved the problem. In addition, their performance in a single six-trial session correlated with their feeding-tolerance score from the pretest (which measured the tendency of one individual to monopolize all the available food). Moreover, individuals from pairs with intolerant dyads were re-paired with individuals they tolerated and subsequently solved the same problem almost immediately (Figure 11.2b). The reverse was also done with the initial tolerant pairs, with the opposite effect. They no longer could solve the problem they had already been capable of solving with a tolerant partner. Overall, then, cooperation could occur only if both partners were able to resist scaring the other subject away from the test area (i.e., in an attempt to monopolize the food before it was even obtained—thus resulting in failure) or in monopolizing all the food after successful cooperation.

Melis et al. (2006a) provides direct evidence that chimpanzee cooperation is highly constrained by interindividual tolerance levels and also suggests that species differences in tolerance may explain variance in cooperative abilities across species (including that seen in humans compared with other animals). In order to test this hypothesis, B. Hare et al. (2007) replicated the previous method with a group of bonobos and directly compared their performance to chimpanzees. In the tolerance tests, bonobos were less likely to monopolize food than chimpanzees, showing higher levels of sociosexual and play behavior (also see de Waal 1992; Palagi 2006; but see Jaeggi et al. 2010, who uses a very different measure of sharing). Importantly, the higher level of sociosexual behavior seen in bonobos was completely absent in chimpanzees (Figure 11.2c). Bonobos had a sociosexual interaction in only one of six trials on average. But chimpanzees simply avoided physical contact in this context.

Bonobos were then compared to chimpanzees in the Hirata pulling paradigm in six trials. Crucially, the age and sex of the subjects from both

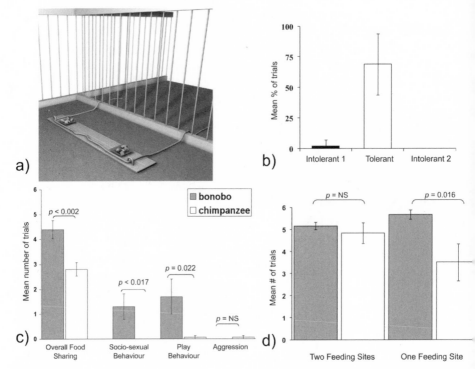

Figure 11.2. a) The cooperation apparatus designed by Hirata (2003). Two food boxes (2.7 meters apart) on a 3.4-meter wooden plank are baited with fruit pieces. One rope is threaded through loops on the plank, and the ends are placed inside the subjects' test room (2.7 meters apart). Reprinted with permission from AAAS. b) Mean percentage of trials out of six that a subject ($n = 12$) was able to cooperatively retrieve the food platform when paired with an intolerant partner before and after being paired with a tolerant partner (Melis et al. 2006a). c) Mean number of trials in which 10 pairs of bonobos and 16 pairs of chimpanzees co-fed and engaged in sociosexual, play, and aggressive behavior in the sharing test (B. Hare et al. 2007). d) Mean number of trials that naïve bonobos and highly tolerant and experienced chimpanzee dyads were able to cooperate to obtain the out-of-reach food tray when food was present in two food sites versus one easily monopolizable feeding site (B. Hare et al. 2007). Error bars indicate standard error of the mean.

species were carefully matched to control for the effect of these variables. The two species were equally skillful at cooperating to retrieve the food when food was placed as in Melis et al. (2006a), with two separate food piles at either end of the food platform. However, in a second condition, when food was placed only in a single pile in the center of the food platform, the bonobos outperformed the chimpanzees (Figure 11.2d). This was also the case when subjects were re-paired with a new, cooperative partner in another round of six trials. When food was easily monopolizable, the more-tolerant species was more skillful at cooperating with multiple partners—even though the chimpanzees tested were very experienced with the task and the bonobos were completely naïve before testing (i.e., the chimpanzees tested here had previously participated in Melis et al. 2006b, 2009, described below). Tolerance is such a powerful constraint that even pairs who are expert at the cooperation task will not succeed when they perceive food as monopolizable.

The species differences found in B. Hare et al. (2007) suggest that bonobos and chimpanzees differ in their ability to cooperate as a result of tolerance constraints on chimpanzees. Thus, understanding the mechanism allowing for these species differences in tolerance will help us to understand how ape cooperation evolves. Comparisons of the physiological response during dyadic food-sharing trials reveal differences in males of the two species in particular (saliva samples were collected before and after a food-sharing trial to examine differential reactivity in steroid hormones). Wobber, Hare, et al. (2010) found that when a pair of chimpanzee males are unable to share food, they show an increase in testosterone but not cortisol in anticipation of being released together into a room full of food. This response is consistent with being primed for a competitive social interaction and avoidance of physical contact (Salvador and Costa 2009). In contrast, male bonobos who are unable to share show a dramatic anticipatory increase in cortisol but not testosterone (see also Hohmann et al. 2008). This response is consistent with a coping style that promotes social interactions in the face of competition that might reduce stress (i.e., play and sociosexual behavior). Wobber, Wrangham, and Hare (2010) also tested the hypothesis that higher levels of feeding tolerance in bonobos result from a shift in the development of their social psychology. Bonobos and chimpanzees ranging in age from young juveniles to adults were again tested for their ability to share food. Subjects of both species were tested in age-matched dyads that controlled for sex. Bonobos again were more tolerant than chimpanzees, showing more play and sex behavior; but in addition an interesting developmental pattern was revealed. While both chimpanzee and bonobo juveniles are equally tolerant, chimpanzees become increasingly intolerant as they age, while bonobos maintain juvenile levels of tolerance into adulthood. Wobber, Wrangham, and Hare (2010) also showed that in two contexts simulating feeding competition, chimpanzees developed inhibitory control skills at a much younger age than bonobos. Even adult bonobos were less inhibited than adult chimpanzees when begging for food.

Taken together, the results show how tolerance can constrain cooperation even in individuals who understand a great deal about the cooperative problem they must solve. Comparisons between Panins show that species differences in cooperation can be a result of differences in tolerance and that developmental shifts in social psychology and differential physiological reactivity play a role in mediating these species' differences in tolerance. Therefore, one of the first changes that had to occur to promote human levels of cooperation in our own species' evolution was an increase in tolerance (i.e., likely bringing it beyond what is seen in either Panin in terms of within-group tolerance). Selection may have acted on the development of our social psychology and physiology to produce a more tolerant hominid that could then more flexibly apply its cooperative skills (B. Hare 2007).

Collaboration

Considerable controversy remains around the question of what degree chimpanzee hunting (or any other form of cooperation) depends on coordinated and intentional collaboration or a simpler cooperative mechanism, such as joint action, in which multiple individuals pursue the same goal independently but simultaneously without awareness of their partners' role in their success (Tomasello and Call 1997; Boesch and Boesch-Achermann 2000; Muller and Mitani 2005; Noë 2006). Moreover, while some have emphasized the strategic nature of chimpanzee meat-sharing, others have argued that sharing might be used as a technique to reduce harassment and piracy by other chimpanzees (Stevens 2004; Gilby 2006; see also Fruth and Hohmann 2002).

In order to test between the rich and lean interpretations of the naturally occurring cooperative behavior of chimpanzees, Melis et al. (2006b) tested whether chimpanzees understood the role of their partner in their success in the Hirata pulling task. In retrieving the food in Melis et al. (2006a), chimpanzee dyads could succeed if they were simultaneously attracted to and willing to pull one of the two rope ends. As in observations of wild cooperation, they may have simply solved the problem through joint action without an understanding of their partner's role. Therefore, Melis et al. (2006b) released only one individual from a highly tolerant pair of chimpanzees into a testing room. The potential partner was locked in an adjacent room with a one-way key that only the subject could remove to free her partner. All subjects had previously learned they could open the door to enter the adjacent room to retrieve food, but they had never been shown they could open the door for another chimpanzee. Instead of pulling the rope out of the food platform and failing to retrieve the food, the subjects spontaneously opened the door for their partner and waited until the partner arrived before they pulled in the platform together successfully. In a control in which the two rope ends were placed close together so that the subject did not need help in retrieving the food, the subjects did not release their partner from the room. Therefore, chimpanzees do spontaneously show an appreciation of the role of another individual in their success solving a novel instrumental task. Chimpanzees were able to use another chimpanzee as a social tool when they needed help solving a problem, but avoided using the other's help when it was not needed (i.e., when they could retrieve the food themselves).

Recognizing the utility of a social partner is just one component of collaboration, because collaborators must also find a way to cooperate even when interests may initially conflict. Melis et al. (2009) tested the ability of highly tolerant pairs of chimpanzees to negotiate when the payoffs from mutual effort were potentially unequal. Pairs of tolerant chimpanzees were faced with the choice of retrieving an out-of-reach food platform with equal payoffs for each or a platform that had the same amount of food as the other but distributed unequally on either end of the platform. The more dominant in each pair was then released to make an initial "offer" by sitting in front of one of the four food dishes at the end of each platform while holding the rope. Shortly after, the subordinate of the pair was then released to either accept the offer by pulling the other end of the same rope or to make a "counteroffer" by sitting in front of another food dish on the other food platform. The dominants tended to offer the unequal split, and subordinates then refused to accept this offer in the majority of trials. However, if the chimpanzees could understand they needed each other to obtain the food, they should be able to "negotiate" a compromise. Perhaps remarkably, subjects were able to quickly come to an agreement and successfully cooperate in almost 90 percent of trials. Moreover, in the majority of trials where dominants initially offered the unequal split, the dyad ended up retrieving the food platform with equal payoffs. Crucially, if the dominant accepted the subordinates' counteroffer of the equal payoff, the subordinate did not then attempt to counteroffer the unequal platform in order to obtain the largest food payoff (i.e., subordinates were not guided by the simple rule of sitting in front of the largest pile of unoccupied food). Therefore, both individuals recognized that the other possessed leverage and recognized the limits of their own power in this situation.

Taken together, these two studies show that semicaptive chimpanzees are capable of spontaneously and flexibly collaborating (no experiments have been conducted with bonobos). Given that wild chimpanzees have decades to practice hunting and forming alliances, it seems that skills revealed in the experiments are likely used ubiquitously in the natural interactions of chimpanzees. This may account for novel cooperative behaviors that have been observed in wild populations, such as guarding during street crossings at Bossou (Hockings et al. 2006). However, there are limits to the collaborative abilities of chimpanzees. For example, while a pair of chimpanzees learned to solve the Hirata paradigm by recruiting the help of a human, they were unable to spontaneously recruit a human to help them pull a heavy object (Hirata and Fuwa 2007). Experiments will help us reveal more about the constraints as well as the potential chimpanzees have as collaborators. In turn we will gain insight about how our species' collaborative skills built on those observed in other apes, but now surpass them in terms of their flexible application.

Costly Cooperation

All the cooperation experiments described thus far resulted in a beneficial outcome for both the subjects tested. Bonobos and chimpanzees in their natural interactions also share prized food items and support each other in conflicts with conspecifics and predators. These types of behaviors can potentially be costly not just energetically but also toward an actor's inclusive fitness. One hypothesis suggests that this type of helping is egocentrically motivated, while others have argued that helping is more prosocial in nature (Silk et al. 2005; Gilby 2006; de Waal 2008).

In order to understand the psychology behind the seemingly costly behavior of chimpanzees and bonobos in their natural interactions, several experiments have been conducted to examine the prosocial tendency of chimpanzees and bonobos. Initial experiments examining chimpanzee food-sharing preferences support the view that much sharing that is observed in their natural interactions occurs in attempts to reduce the cost of harassment and does not result from a preference for sharing (Silk et al. 2005; Jensen et al. 2006; Vonk et al. 2008). In these studies chimpanzees did not prefer to retrieve a food tray that provided food both to them and to a different room with another chimpanzee. Instead, they chose randomly between the option providing food to two rooms over another option where they would receive the same amount of food but the other chimpanzee would receive nothing. These findings have, in part, led some to suggest that the prosocial tendency seen in humans is derived and potentially evolved together with another social trait such as our shift to cooperative breeding (Burkart et al. 2007). However, none of these initial studies successfully demonstrated in a pretest that the subjects understood the physical properties of the task.

It is necessary to demonstrate that when tested alone, subjects prefer the option where food is provided in two rooms if they themselves are allowed to retrieve all the food. Otherwise a null result in the test is likely due to a subject's failure to understand the physical setup. Unfortunately, these crucial controls were run in only one study, and when run they did not necessarily work (i.e., Jensen et al. 2006, experiment 1). Moreover, when human infants were tested in the same contexts they did not consistently show a prosocial tendency (Brownell et al. 2009). Therefore, until the appropriate controls are run, it is difficult to make strong conclusions from this test situation. (Note: in experiment 2 of Jensen et al. 2006, the controls worked and two subjects did show a prosocial tendency).

Because of the limitations of the food-tray tasks, Warneken et al. (2007) used a new helping paradigm to test the motivation and skills of chimpanzees, based on tasks previously validated with children (Warneken and Tomasello 2006). While a chimpanzee watched, a human played with a toy. Another experimenter then came and stole the toy and placed it out of the other person's reach but within reach of the chimpanzee. The question was whether the chimpanzee would return the stolen toy to the human who had been playing with it. Regardless of whether the chimpanzee was rewarded with food for returning the toy, the chimpanzees repeatedly helped the human. Across 10 trials there was no decrease in helping, even though the chimpanzees were repeatedly requested to help without receiving a food reward. They even helped when retrieving the toy required climbing several meters into a tunnel before bringing it back to the human. When the exact same procedure was run with young infants, there was no species difference in the level of helping. Thus, when tested in the same situation (which did not require understanding the physical properties of an apparatus), chimpanzees were as helpful as human children.

Although the previous experiment shows chimpanzees helping humans, it is possible they are not motivated to help other chimpanzees. Therefore, Warneken et al. (2007) designed another study in which a chimpanzee could potentially release another chimpanzee into a food room that the subject chimpanzee did not have access to (Figure 11.3a). The subjects had previously learned that if they pulled a key, they themselves could obtain food by passing from the room with the key into a second room. From this second room they could then open the door into the food room (if they didn't remove the key before entering this second room it remained locked). They also had learned that if they were locked into the room with the key they could not obtain the food even if they pulled the key that freed the door between the second room and food room. This means that Warneken et al. (2007) demonstrated that their subjects, before being tested, understood the physical properties of the task, making it easier to interpret a positive or negative result as a valid measure of helping. In the test, subjects were locked in the key room and could choose to open the door for a recipient in the second room or ignore the potential recipient. Subjects opened the door into the food room for the recipient in the majority of experimental trials, with no decrease in opening across trials. They also opened in the experimental condition more often than in the control, when opening did not help the recipient obtain food (i.e., food was not placed in the food room, Figure 11.3b). The subject could easily see the recipient obtain and eat the food (although in each trial it was placed out of the subject's view), and recipients never shared their food with the subject. Importantly, there was no way for recipients to physically harass subjects. Therefore, subjects repeatedly gave another chimpanzee access to food, even though they received no reward for doing so (see also Greenberg et al. 2010; Melis, Warneken, et al. 2010). Studies on a variety of other primates suggest that it is not nonhuman apes alone who help in such ways (Burkart et al. 2007; Lakshminarayanan and Santos 2008; Cronin and Snowdon 2007; de Waal, Leimgruber, and Greenberg 2008).

Although chimpanzees helped others in Warneken et al. (2007), subjects never had "to sacrifice their own rewards to provide rewards to others" (Silk 2008, 279). Subjects had to either retrieve an object or give access to food they could not obtain. While this shows a willingness to help another individual, it is low-level helping, in the sense that the costs are relatively low to the actor. Therefore, helping observed in this experiment likely does not simulate the costs associated with actively sharing meat or other high-quality food. Moreover, it is unlikely that chimpanzees would sacrifice their own food given the results of Melis et al. (2006b, 2009). In one experiment, chimpanzees avoided opening a door when they did not need the help of the partner (i.e., so they could eat all the food; Melis et al. 2006b). In another experiment, dominant chimpanzees almost never offered the larger amount of food on the unequal food tray to their subordinate partners (Melis et al. 2009). Meanwhile, humans often publicly distribute food they have collected to others in their groups (Boehm 1999; K. Hill 2002).

This work might suggest high-cost food sharing is uniquely human, but it is difficult to come to any conclusions about the evolution of human food sharing without experimentally examining bonobos. Therefore, B. Hare and Kwetuenda (2010) gave bonobos the opportunity to either monopolize food or to open a door and share part of their food with a conspecific (Figure 11.3c). We found that bonobos preferred to release a recipient from an adjacent room and feed together instead of eating all the food alone. They shared with another bonobo even though 1) they ate all the food alone in a control when given the same amount as used in the test, and 2) they could have eaten all the food in the test and then opened the door for the other bonobo (i.e., to have a social interaction). Over 10 trials, bonobos continued to voluntarily share their food,

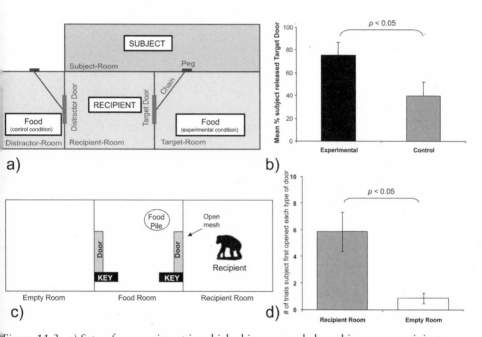

Figure 11.3. a) Setup for experiment in which chimpanzees help a chimpanzee recipient retrieve food (Warneken et al. 2007). Both a target and distracter door were held shut by chains. The recipient could not access either chain, but the subject could release the chain of the target door. In the experimental condition, food was placed in the target room, so that the subject could potentially help the recipient by releasing the target chain, while in the control condition food was placed in the distracter room and the subject was unable to help the recipient. b) shows in this experiment that subjects opened the target door for the recipient twice as often in the experimental condition than in the control. c) The setup for the experiment examining voluntary sharing in bonobos (B. Hare and Kwetuenda 2010). A food pile was placed in the center room, one-way keys were placed in adjacent doors, and a recipient was placed in an adjacent room. Subjects were released into the food room and could open one, both, or none of the keys into either of the adjacent rooms. In a control (not depicted), the recipient was replaced with a second food pile. d) Represents the mean number of trials that subjects either opened the recipient's door first or the empty room first out of the 10 test trials (subjects refrained from opening either door in 3.4 ± 1.26 trials out of 10). Error bars indicate standard error of the mean.

allowing the recipient to feed during 80 percent of the total feeding time needed to eat all the food provided. The sharing cannot be explained by 1) harassment, since recipients could not approach the food without the subject's help, 2) kinship, since none of the subjects were related, 3) attempts by subjects to reciprocate previous favors, since subjects even shared with recipients who were non-group-members, or 4) a desire for genital-genital contact, since they could (and did) obtain this without opening the door. Therefore, subjects preferred to voluntarily open the recipient's door to allow the other to share highly desirable food that they could have easily eaten alone—with no signs of aggression, frustration, or change in the speed or rate of sharing across trials.

This stable sharing pattern is particularly striking, since in other non-sharing contexts bonobos are averse to food loss and adjust to minimize such losses (e.g., Wobber, Wrangham, and Hare 2010). Motivationally it is possible, if not likely, that bonobos voluntarily share because they prefer to eat socially rather than alone. However, it may also be that bonobos share out of concern for the other bonobo. Regardless of the underlying motivation, it remains striking that bonobos would voluntarily prefer to share food with non-kin, non-group members in any context. More work is clearly needed to understand more precisely how human helping and sharing might differ from that seen in chimpanzees and bonobos. While humans and Panins will certainly differ in this respect, the differences are likely to be more nuanced than initially hypothesized.

Maintaining Cooperation

Whether a cooperative relationship is based on mutualism or a more altruistic exchange, there must be a mechanism by which cooperation is maintained. However, it has been suggested that reciprocal altruism will be rare in nature because of the high rate of discounting and a lack of inhibitory control in animals (Stevens and Hauser 2004). It has also been suggested that humans have a unique motivation to pay a cost in order to punish a cooperator who defects (Fehr and Gächter 2002). It may be that calculated reciprocity and active punishment play important roles in allowing for unique forms of human cooperation. Meanwhile, there is abundant evidence for reciprocal exchange of grooming, coalitionary support, and food sharing in chimpanzees and bonobos. Moreover, the reciprocal exchanges seen in philopatric wild male chimpanzees cannot be explained by kin relations (Langergraber et al. 2007). Given that Panins have reciprocal relationships, experimenters have begun exploring what cognitive mechanism might allow for and constrain chimpanzee reciprocity (no experiments have been conducted with bonobos).

First, Melis et al. (2006b) used the Hirata paradigm to test if chimpanzees can distinguish between a skilled and unskillful cooperative partner. Two potential partners were positioned in two separate rooms adjacent to the testing room. In pretests, one potential partner demonstrated skill with the Hirata apparatus, while the other always immediately caused a failure by pulling the rope without his pretest partner. Moreover, although all subjects were highly tolerant with both potential partners, no subject had ever been paired with either partner to solve the cooperation task. In the test, a subject was released into the testing room, where the rope from the out-of-reach food tray was positioned. Subjects could then choose which partner to solve the cooperation problem with by removing one of the one-way keys that locked the door between one of the adjacent rooms and the test room. Subjects initially did not have a preference between the two potential partners in the six-trial introduction. However, trial-by-trial analysis revealed that subjects did use a win-stay, lose-shift strategy in the introduction. Subjects preferred to open the door for a partner if the choice of that partner on the previous trial had led to success. Likewise, they switched partners on a trial following failure. Perhaps most remarkably, in the six-trial test session conducted as much as a week after the introduction, subjects almost exclusively chose the more skillful partner.

This study demonstrates that chimpanzees form and then remember a reputation about the cooperative skill level of different individuals for a significant period of time. Social memories of this kind could provide a foundation for reciprocal relationships. It also suggests that chimpanzees are capable of maintaining cooperative relationships through shunning (i.e., simply avoiding unreliable partners). This highly effective, low-cost form of punishment does not require physical contact that can lead to injury. Instead, in a fluid fission-fusion society, individuals can completely escape the dilemma of the second-order cost of punishment by simply avoiding problematic conspecifics in favor of beneficial social partners (Barrett et al. 1999).

The ability of chimpanzees to quickly assess the skill of another chimpanzee and to encode this information suggests their reciprocal relationships could be based on the memory of previous social interactions. Melis et al. (2008) used two experimental techniques to examine whether six dyads of chimpanzees were capable of calculated reciprocity. The first experiment was based on Melis et al. (2006b), in which subjects could choose to open a door and cooperate together with a partner who had chosen to cooperate with them, or a partner that had avoided choosing them as a cooperative partner. The second experiment was based on Warneken et al. (2007), in which subjects could choose between releasing into a room with food either another chimpanzee who had previously released them, or a chimpanzee who had refused to release them. In both experiments there was a significant but weak tendency for chimpanzees to choose the individual who had chosen them before.

These two experiments suggest that with a larger sample it may be possible to demonstrate calculated reciprocity (although note that a larger sample was not needed when the same subjects were tested in Melis et al. 2006a, b, and 2009, and Warneken et al. 2007). However, a number of other experiments have revealed little evidence for contingent reciprocity. In many of these studies, chimpanzees needed to exchange tokens with one another in order to obtain food, but in this context reciprocal exchanges either quickly break down or never develop (Brosnan and Beran 2009; Dufour et al. 2008; Pelé et al. 2009; Yamamoto and Tanaka 2009a, b; Yamamoto et al. 2009). Finally, in Melis et al. (2009), in which chimpanzees negotiated conflicting interests, a condition was run in which the food platform with unequal amounts of food had nearly twice as much food as the equal platform. In this situation the most advantageous strategy would be to reciprocally take turns receiving the largest food pile on the unequal platform; in this way both individuals receive more food than if they only are able to agree on the platform with equal payoff. The chimpanzees did not develop a reciprocal strategy in this situation, although they were still very successful at agreeing to retrieve one of the two platforms (i.e., when they agreed to pull the unequal tray, the dominant always received the larger payoff; see also Brosnan, Silk, et al. 2009; Yamamoto and Tanaka, in press).

Overall the picture regarding nonhuman ape reciprocity remains cloudy. While chimpanzees and bonobos clearly have reciprocal relationships, experimental methods have yet to be developed that provide the most powerful test possible. In the one context in which chimpanzees seem to potentially show contingent reciprocity, the sample size could be characterized as relatively small ($N = 6$ dyads). Moreover, bonobos have never been experimentally tested in a test of reciprocity. Therefore, it is still possible that future experiments will find evidence for contingent reciprocity (and perhaps even calculated reciprocity, since Panins have low discounting rates, are skilled at simple arithmetic, and have shown impressive inhibitory control in a number of cooperative tasks; Rosati et al. 2007; Hanus and Call 2007; Melis et al. 2006b, 2009). However, based on the current evidence, it seems most likely that symmetry within chimpanzee relationships is based on attitudinal reciprocity or long-term relationship factors that cut across currencies (Melis et al. 2008). Even if calculated reciprocity is demonstrated, it likely plays a minor role in maintaining cooperation within Panins. In addition, we predict that chimpanzees will avoid direct conflict when cooperation breaks down and simply shun unprofitable social partners. Humans, then, may have additional mechanisms for the maintenance of cooperation that are not found in the Panins, but it will be important to demonstrate why attitudinal reciprocity and shunning cannot explain the cooperation we see in our species.

Going Wild with Captive Cooperation

Anytime we observe a behavior, it can have multiple mechanistic explanations regarding the underlying psychology that produces it. Experiments allow us to probe these mechanisms and test between different hypotheses that cannot be tested using observational techniques alone. Based on the newest experiments reviewed above, we now know that a major constraint on cooperation is tolerance. Almost certainly during human evolution there was a major shift in tolerance levels that allowed for more forms of joint activity that could then develop into collaborative endeavors.

We also now know that chimpanzees are capable of spontaneously and actively recruiting a collaborative partner when faced with a novel problem that requires cooperation. They also can assess the leverage they have in deciding which cooperative problem they wish to solve. This suggest that wild chimpanzees are capable of recruiting alliance partners in social competitions and when hunting. It also suggests they have some understanding that cooperation changes the balance of power within the cooperative dyad itself and not just between those within the alliance and those outside it (i.e., they realize they need each other, which reduces the asymmetry in the relationship).

We have also observed that not all food sharing between apes is likely explained by harassment. Chimpanzees will help another retrieve an object or food even if no reward is given in return. Meanwhile, bonobos in some contexts prefer to share food rather than simply eat it all alone. This type of sharing is of relatively high cost, likely due to prosocial motivations, and may explain the increased gregariousness in this species.

Finally, studies of mechanisms maintaining cooperation have shown that chimpanzees can shun unskillful cooperators, but there still remains only weak evidence for contingent reciprocity. Therefore, we suspect that simpler mechanisms such as attitudinal reciprocity that can rely on low-cost punishment in the form of shunning maintain most cooperative relationships in apes.

Based on all the research just described, it would be easy to think that Crawford (1937) was completely misguided for ever thinking that cooperative behavior that was not already observed in the natural behavior of chimpanzees might emerge in laboratory experiments. However, this is exactly what we have found in studying the behavior of ape cooperation in the lab: apes can flexibly adapt their cooperative behavior in completely novel situations.

Recall that there is very little evidence that female chimpanzees actively cooperate in the wild (e.g., in hunting or in preventing male coercion; Muller and Mitani 2005). Although there is variance in general, it is male and not female chimpanzees who work together in seemingly complex ways. However, a significant number of females were extremely skilled in our tasks (Melis et al. 2006a,b, 2008, 2009). It seems that females living in a rich environment, when paired together with a highly tolerant partner, have skills of cooperation as sophisticated as those seen in male chimpanzees.

Even though bonobos have never been observed to cooperate in many of the ways observed in wild chimpanzees, they are more skilled than chimpanzees in the Hirata pulling task if the food reward in the experiment is highly monopolizable (B. Hare et al. 2007). It seems that the higher levels of food tolerance in bonobos allow them to outperform chimpanzees in some cooperative tasks in captivity, even though they never obtain food through joint effort in the wild (arguably supporting Nissen and Crawford's initial idea that a rich and predictable environment may not promote cooperative behavior that then might be expressed in the lab).

We highlight these examples because there has often been tension between those studying wild primates using observational methods and those studying primates in captivity using more-experimental methods. Both approaches have much to offer if we want to understand the full range of behavioral flexibility. Just as, without field studies, we would never have known that chimpanzees can hunt in groups, without experimental studies in captivity we would never have known how skilled female chimpanzees and bonobos are in collaborative activities. Ultimately we want to know the range of flexibility apes show when cooperating, and an accurate assessment requires observing these species in as many contexts as possible.

Even after considering ape cooperation across all contexts, what remains puzzling is the fact that there is substantial evidence that apes possess cognitive skills that allow for mutualistic endeavors, but this flexibility remains limited to a few contexts in comparison to what is observed in humans. We suspect that it is not a tendency to act altruistically that makes humans unique. Instead, it seems more likely our species is unusually cooperative because of our flexible ability to avoid high-cost helping (i.e., harmful to reproductive success) while recognizing the potential benefit of mutualistic endeavors—particularly low-cost acts that can have beneficial reputational effects (i.e., evolutionarily neutral or positive). After all, there is no puzzle to explain when mutualism or no-cost cooperation evolves—particularly when it can have beneficial effects for the actor. Instead, the puzzle is why we do not see more-complicated forms of it more often when we look outside human behavior. It also raises the question of whether human cooperation really can ever be characterized as altruistically motivated when even the strictest demonstrations cannot rule out reputation effects (i.e., even in "anonymous" experiments, subtle social cues suggesting that a subject's actions are being watched shape our cooperative tendencies, while subjects can always brag about how cooperative they were in experiments when they go home; Burnham and Hare 2007). Answering such questions will be the key to understanding both how much of our cooperation is human and how it evolved.

CHAPTER TWELVE

Fetal Testosterone in Mind: Human Sex Differences and Autism

Bonnie Auyeung and Simon Baron-Cohen

The Extreme Male Brain Theory of Autism

Autism, high-functioning autism, Asperger syndrome, and pervasive developmental disorder (not otherwise specified; PDD/NOS) are thought to lie on the same continuum and can be referred to as autism spectrum conditions (ASC). These conditions are characterized by impairments in reciprocal social interaction, as well as verbal and nonverbal communication, alongside strongly repetitive behaviors and unusually narrow interests (APA 1994). Recent epidemiological studies have shown that as many as 1 percent of people could have an ASC (Baird et al. 2006). The incidence of ASC is strongly biased toward males, with a male:female ratio of 4:1 for classic autism (Chakrabarti and Fombonne 2005) and as high as 10.8:1 in individuals with Asperger syndrome (Gillberg et al. 2006). The cause of the observed sex difference in ASC remains a topic of debate.

One method of investigating the increased male incidence of ASC is to examine how individuals with ASC perform in areas of behavior that have previously shown sex differences in the general population. Baron-Cohen (2002) examines these behaviors through an empathizing-systemizing (E-S) model of typical sex differences, which proposes that females on average have a stronger drive to empathize (to identify another person's emotions and thoughts and to respond to these with an appropriate emotion), while males tend to have a stronger drive to systemize (to analyze or construct rule-based systems—whether mechanical, abstract, or another type). The Empathy Quotient (EQ) and Systemizing Quotient (SQ) were developed to measure these dimensions in an individual. Using the difference between a person's EQ and SQ, individual "brain types" can be calculated where individuals who are equal in their empathizing (E) and systemizing (S) are said to have a balanced (B) brain type (E = S). The type S (S > E) brain type is more common in males, while the type E (E > S) is more common in females. Extreme types are also found, and the majority (61.6%) of adults with ASC fall in the extreme S (S >> E), compared with 1 percent of typical females (Wheelwright et al. 2006). See Figure 12.1.

A range of experiments have examined the possibility of a link between ASC and "extreme male" behavior. These can be variously considered as measures of "systemizing" and "empathizing" ability.

Individuals with ASC score higher on the SQ, where typical males also score higher than typical females in both adults (Baron-Cohen et al. 2003) and children (Auyeung, Baron-Cohen, Wheelwright, et al. 2009). Individuals with ASC are also superior to controls on the Embedded Figures Test (EFT), a task on which typical males perform better than typical females (Shah and Frith 1983; Jolliffe and Baron-Cohen 1997). The EFT requires good attention to detail, a prerequisite of systemizing. Individuals with ASC also have intact or superior functioning on tests of intuitive physics, a

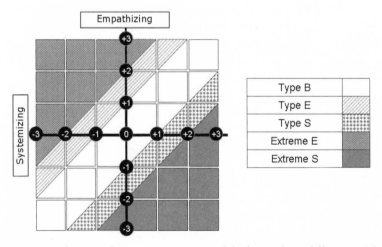

Figure 12.1. The empathizing-systemizing model of typical sex differences. The main brain types are illustrated on axes of empathizing (E) and systemizing (S) dimensions (numbers represent standard deviations from the mean). Balanced brain (type B); female brain (type E), male brain (type S); the extreme types E and S lie at the outer borders. According to the "extreme male brain" theory of autism, people with ASC will generally fall in the extreme type S region. Modified from Baron-Cohen et al. (2002).

domain that shows a sex difference in favor of males in adulthood (Lawson et al. 2004). Sex differences have been found on the block design subscale of the WISC-R intelligence test, with typical males performing better than females (Lynn et al. 2005), and children with autism demonstrating superior functioning on this test (Allen et al. 1991; Lincoln et al. 1988; Shah and Frith 1993).

Studies using the EQ report that individuals with ASC score lower on both the adult and child versions than the control groups, where typical females score higher than typical males (Auyeung, Baron-Cohen, Wheelwright, et al. 2009; Baron-Cohen and Wheelwright 2003). Individuals with ASC are also impaired on certain measures where women tend to score higher than men. For example, individuals with ASC score lower than control males in the "Reading the Mind in the Eyes" test (considered to be an advanced test of empathizing) (Baron-Cohen et al. 1997), on the Social Stories Questionnaire (Lawson et al. 2004), on the Friendship Questionnaire (which tests the importance of emotional intimacy and sharing in relationships) (Baron-Cohen and Wheelwright 2003), and in recognition of complex emotions from videos of facial expressions or audios of vocalizations (Golan et al. 2006).

In the typically developing population, the Childhood Autism Spectrum Test (CAST; formerly known as the Childhood Asperger Syndrome Test) has been used to screen for Asperger syndrome and ASC. Boys score higher than girls on this measure, and children with ASC score higher than typically developing children (Williams et al. 2008). On the Autism Spectrum Quotient (AQ), a measure developed to help quantify the number of autistic traits an individual displays, individuals with Asperger syndrome or high-functioning autism score higher than those without a diagnosis. Among controls, males again score higher than females, and these results are consistent in adults, adolescents, and children (Baron-Cohen et al. 2001; Baron-Cohen, Hoekstra, et al. 2006; Auyeung et al. 2008). Similar results have been found using the Social Responsiveness Scale (SRS), a 65-item rating scale designed to measure the severity of autistic symptoms, demonstrating that individuals with an ASC diagnosis score higher than typical males, who in turn score higher than typical females (Constantino and Todd 2003).

In addition to the evidence at the psychological level, it has been suggested that characteristics of neurodevelopment in autism, such as larger overall brain volumes and greater growth of the amygdala during childhood, may also represent an exaggeration of typical sex differences in brain development (Baron-Cohen et al. 2005). Studies using fMRI indicate that typical females show increased activity in the extrastriate cortex during the Embedded Figures Test and increased activity bilaterally in the inferior frontal cortex during the "Reading the Mind in the Eyes" test. Parents of children with ASC also tend to show hypermasculinization of brain activity (Baron-Cohen, Ring, et al. 2006), suggesting that hypermasculinization may be part of the broader autism phenotype.

It remains important to identify the biological mechanisms that cause such sexual dimorphism. One study has shown sexual dimorphism in looking preferences in 102 newborn infants who were approximately 37 hours old. Boys were found to exhibit a preference for mobiles, while girls tended to prefer looking at faces (Connellan et al. 2000). Whilst these simple experiments are not an indication of ASC, these early sex differences suggest a biological basis for sex differences in behavior, since there had been no opportunity for postnatal influence of social or cultural factors.

Other evidence for a biological mechanism comes from studies of exposure to fetal testosterone (fT). Recent studies have shown that exposure to this hormone is positively correlated with the number of autistic traits a typically developing child exhibits (Auyeung, Baron-Cohen, Chapman, et al. 2009; Auyeung et al. 2010). Other evidence that supports a role for hormones in the development of ASC are that androgen-related medical conditions such as polycystic ovary syndrome (PCOS), ovarian growths, and hirsutism occur with elevated rates in both women with Asperger syndrome and in mothers of children with autism (Ingudomnukul et al. 2007).

Hormones and Sexual Differentiation

Hormones are essential to reproduction, growth and development, maintenance of the internal environment, and the production, use, and storage of energy (Knickmeyer and Baron-Cohen 2006a). There are marked physical and behavioral consequences of exposure to hormones throughout life. Prenatally, the presence or absence of specific hormones (or their receptors) is known to be essential to the sexual differentiation of the fetus. In addition to affecting physical development, there is increasing evidence that prenatal hormones have a substantial effect on gender-typical aspects of behavior (Cohen-Bendahan et al. 2005; Hines 2004). The links between hormone levels, physical development, and behavior are complex and not fully understood. Hormone levels can be measured at particular points in time, but levels could vary on a daily basis (van de Beek et al. 2004). Prenatal measurements are particularly difficult, and correlations with behavioral measurements are always complicated by the need to determine the presence of a particular trait, without artificially inducing the behavior or creating bias in the result. A useful way of controlling some of these variables is to examine results from animals, where it has been possible to directly manipulate and monitor the levels of hormones throughout pregnancy and to control for environmental effects. As a result, we often look to confirm effects observed in animals with similar measurements in humans. Even then, the correlation between animals and humans is not always clear-cut, with the potential for quite different mechanisms.

Though genetic sex is determined at conception, it is the gonadal hormones (i.e., androgens, estrogens, and progestins) that are responsible for differentiation of the male and female phenotypes in the developing human fetus (Novy and Resko 1981; Fuchs and Klopper 1983; Tulchinsky et al. 1994; Hines 2004; Kimura 1999). Androgens such as testosterone and dihydrotestosterone (DHT—a hormone formed from testosterone) are of particular interest to the study of male-typical behavior because when these androgens and the appropriate receptors are present, the male genital phenotype will develop. If androgens (or their receptors) are not present, then the female genital phenotype will develop (such as in female fetuses or in individuals with androgen insensitivity syndrome) (Jost 1961, 1970, 1972; George and Wilson 1992). Another hormone that forms from prenatal testosterone is the estrogen hormone estradiol, which has been observed to promote male-typical behavior in rats and other rodents (Collaer and Hines 1995). The relative contributions of DHT and estradiol to development of male-typical human behaviors are less certain.

Behavioral studies in nonhuman mammals have shown that the same prenatal hormones that are involved in sexual differentiation of the body are also involved in sexual differentiation of behavior (Breedlove 1992; Goy and McEwen 1980). In animals, higher doses of hormones have been seen to masculinize behavior more than lower doses, though the effect of concentration is not uniform for different behaviors (Goy and McEwen 1980). Effects are also likely to be nonlinear and include both lower and upper threshold values, beyond which changes in concentration have no effect (Cohen-Bendahan et al. 2005). The interaction between hormones is also important to consider (Goy and McEwen 1980).

Atypical Fetal Hormone Environments

In humans, the manipulation or even direct measurement of hormone levels is considered unethical because of the potential dangers involved. However, some information is available from specific abnormalities that occur naturally. Such abnormalities can lead to considerable difficulties for the individual, and fortunately such instances are rare. However, some studies have obtained sufficient participation to render useful information about how abnormal environments influence behavior. A detailed review of many of the studies surrounding these conditions has been provided elsewhere (Baron-Cohen et al. 2004; Cohen-Bendahan et al. 2005; Hines 2004; Knickmeyer and Baron-Cohen 2006a), so we focus our discussion here on findings relevant to characteristics of ASC.

Congenital adrenal hyperplasia (CAH) is a genetic disorder that causes excess adrenal androgen production beginning prenatally (New 1998). CAH affects both males and females but is most clearly observed in females because of their typically low androgen levels. Female fetuses with CAH have similar androgen levels to those found in typical males (Hines 2004). Behavioral studies of females with CAH show a more masculinized profile compared to unaffected female siblings or matched controls.

In terms of specific behaviors, girls with CAH show masculinization of ability in activities typically dominated by males. These include spatial orientation, visualization, targeting, personality, cognitive abilities, and sexuality (Hampson et al. 1998; Hines, Fane, et al. 2003; Resnick et al. 1986). Females with CAH may also be more likely to be left-handed (Nass et al. 1987) and are more interested in male-typical activities and less interested in female-typical activities throughout life (Berenbaum and Hines 1992; Berenbaum and Snyder 1995; Berenbaum 1999; Ehrhardt and Baker 1974; Hines et al. 2004).

Studies relating CAH and autism are limited. Since the condition typically introduces masculinization, effects are more apparent in girls than boys. For these girls, behavior tends to become more aligned with expectations of behavior from typical males, and few cases of ASC with CAH are reported. Results from one study of girls with CAH suggest that they exhibit more autistic traits, measured by the Autism Spectrum Quotient, compared with their unaffected sisters (Knickmeyer, Baron-Cohen, Fane, et al. 2006). Individuals with CAH also demonstrate higher levels of language and learning difficulties compared with unaffected family members (Resnick et al. 1986), as do people with ASC. While CAH provides an interesting window on additional androgen exposure, the relatively rare occurrence of CAH in conjunction with ASC makes it difficult to obtain large enough sample sizes for generalization to the wider population. In addition, it has been suggested that CAH-related disease characteristics, rather than prenatal androgen exposure, could be responsible for atypical cognitive profiles (Fausto-Sterling 1992; Quadagno et al. 1977).

Complete androgen insensitivity syndrome (CAIS) occurs when there is a complete deficiency of androgen receptors and is more common in males, with incidence between 1 in 60,000 and 1 in 20,000 births. At birth, infants with CAIS are phenotypically female, despite an XY (male typical) complement and are usually raised as girls with no knowledge of the underlying disorder. Although breasts develop, diagnosis usually takes place when menarche fails to occur (Larsen et al. 2002; Nordenstrom et al. 2002).

Investigation of behavior such as gender identity, sexual orientation, gender role behavior in childhood and adulthood, personality traits that show sex differences and hand preferences have suggested that males with this condition do not significantly differ from same-sex controls (Hines, Ahmed, et al. 2003; Quadagno et al. 1977). However, other results suggest that individuals with CAIS tend to show feminized performance on tests of visuo-spatial ability (Money et al. 1984). If replicable, this finding lends support to the notion that androgens enhance male-typical behaviors. Specific evidence for ASC is not available due to the low incidence of this condition.

Idiopathic hypogonadotrophic hypogonadism (IHH) occurs when an individual's gonads lack sufficient stimulation to produce normal levels of hormones, and the disorder can occur congenitally or after puberty. These individuals have normal male genitalia at birth, so it can be assumed that their prenatal testosterone levels were normal (Knickmeyer and Baron-Cohen 2006a). Men with IHH perform worse on the Embedded Figures Test, the space relations subtests from the differential aptitude tests, and the block design subtest of the Weschler Adult Intelligence Scale, when compared with control males and males with acquired hypogonadotrophic hypogonadism after puberty (Hier et al. 1982). However, another study found that males with IHH do not show deficits on the same scale (Cappa et al. 1988). More research needs to be conducted in order to resolve these findings and relate the effects to ASC.

Hormonal Effects: Indirect Studies in Typical Populations

A steady body of evidence indicates that fetal hormone levels influence certain physical characteristics that can be observed after birth. These "proxy" measurements have been used to indicate the levels of prenatal androgen exposure and have been examined extensively in relation to behavioral traits. Several reviews of these measurements exist (e.g., Cohen-Bendahan et al. 2005; Kimura 1999), and we focus the discussion here on studies most relevant to behaviors associated with ASC.

The ratio between the length of the second and fourth digit (2D:4D) has been found to be sexually dimorphic, being lower in males than in females. Significant sex differences have been observed by week 14 of fetal life (Malas et al. 2006; Galis et al. 2010), and it has been hypothesized that it might reflect fetal exposure to prenatal sex hormones in early gestation (Manning et al. 2002).

Measurements indicate an association between fT levels and 2D:4D ratio for the right hand after controlling for sex (Lutchmaya et al. 2004). For individuals with CAH, females show lower (more masculinized) 2D:4D on the right hand compared to unaffected females, and men with CAH have lower 2D:4D on the left hand compared to unaffected males (W. Brown et al. 2002). Results in this sample are consistent with the notion that prenatal androgen exposure masculinizes 2D:4D ratio. This measure has been widely used as a proxy for prenatal testosterone exposure because of the ease and simplicity of measurement. However, it is likely that 2D:4D ratio is affected by multiple factors (Cohen-Bendahan et al. 2005).

The findings in studies with 2D:4D ratio tend to support the suggestion that higher fT levels are a risk factor for ASC. Lower (i.e., hypermasculinized) digit ratios have been found in children with autism compared with typically developing children, and this was also found in the siblings and parents of children with autism, suggesting genetically based elevated fT levels in autism (Manning et al. 2001; Milne et al. 2006).

It has been proposed that some observable sex differences in human behavior and cognition may be accounted for by differences in cerebral lateralization (Hines and Shipley 1984). Prenatal exposure to testosterone has also been implicated in left-handedness and asymmetrical lateralization (Fein et al. 1985; McManus et al. 1992; Satz et al. 1985; Soper et al. 1986). Left-handedness and ambidexterity are more common in typical males (Peters 1991) as well as in individuals with autism (Gillberg 1983).

Pubertal onset has been used to investigate variations in prenatal hormones. Females typically enter puberty earlier than males (Cohen-Bendahan et al. 2005). Research examining the physical indicators of hormone exposure and autism have found that a subset of male adolescents with autism show hyperandrogeny, or elevated levels of androgens, and precocious puberty (Tordjman et al. 1997). These findings suggest that individuals with autism have atypical hormonal activity around this time. Other research has also shown that androgen-related medical conditions such as polycystic ovary syndrome (PCOS), ovarian growths, and hirsutism (Ingudomnukul et al. 2007) occur with elevated rates in both women with Asperger syndrome and in mothers of children with autism (Ingudomnukul et al. 2007). Delayed menarche has also been observed in females with Asperger syndrome (Ingudomnukul et al. 2007; Knickmeyer, Wheelwright, et al. 2006). These may reflect early abnormalities in level of fT, though this would require testing in a longitudinal study.

Hormonal Effects: Measurements of Fetal Testosterone

While many convenient methods have been recommended, the ability to infer prenatal hormone exposure through abnormal environments or proxy measures has obvious limitations. Although evidence for the influence of androgens can be obtained, there is (as yet) little direct support for the predictors discussed in the previous section as a way of studying prenatal hormone influence.

Ideally, we would like to make direct measurements of testosterone at regular intervals throughout gestation and into postnatal life. However, it would be extremely hazardous to attempt direct measurements from the fetus itself. Maternal samples are readily available and may provide some indication of fetal exposure and have been linked to increased male-typical behavior in later life (Udry et al. 1995). However, there is little evidence to suggest that these correlate well with the fetal environment, which is protected by the placenta (Hines 2004).

The timing of hormonal effects is also crucial when studying lasting effects on development. There are thought to be two general types of hormonal effects: organizational and activational (Phoenix et al. 1959). Organizational effects are most likely to occur during early development when most neural structures are becoming established and produce permanent changes in the brain (Phoenix et al. 1959), whereas activational effects are short term and are dependent on current hormone levels. Since ASC are typically persistent with an early onset, any hormonal influence on the development of ASC is likely to be organizational in nature.

It is widely thought that organizational effects are maximal during sensitive periods, which are hypothetical windows of time in which a tissue can be formed (Hines 2004). Outside the sensitive period, the effect of the hormone will be limited, protecting the animal from disruptive influences. This means, for example, that circulating sex hormones necessary for adult sexual functioning do not cause unwanted alterations to tissues, even though the same hormones might have been essential to the initial development of those tissues. Different behaviors may also have different sensitive periods for development. The importance of sensitivity to organizational effects was seen by researchers who showed that androgens masculinize different behaviors at different times during gestation for rhesus macaques (Goy et al. 1988).

For typical human males, there is believed to be a surge in fetal testosterone at around weeks 8–24 of gestation (Baron-Cohen et al. 2004; Collaer and Hines 1995; Hines 2004), with a decline to barely detectable levels from the end of this period until birth. As a result, any effects of fetal testosterone on development are most likely to be determined in this period. For typical human females, levels are generally very low throughout pregnancy and childhood (Hines 2004).

In addition to the fetal surge, two other periods of elevated testosterone have been observed in typical males. The first takes place shortly after birth and lasts for approximately 3–4 months (Smail et al. 1981), after which levels return to very low levels until puberty. Results show that neonatal testosterone (measured shortly after birth) is important for genital development (Brown et al. 1999), but the evidence for its role in behavioral development is unclear. Early pubertal effects are the first visible signs of rising androgen levels in childhood and occur in both boys and girls. Due to the early onset of ASC, the pubertal surge in testosterone is of less interest in determining etiology of these conditions.

Few studies have been conducted on the effects of neonatal testosterone; however, an increasing body of evidence suggests that prenatal androgens may be involved in determining sexually dimorphic traits. In the remainder of this chapter, we discuss direct measurements of testosterone and our ability to correlate this with the development of ASC.

Some studies have examined relationships between umbilical cord (perinatal) hormones and later behavior such as temperament and mood. High perinatal testosterone and estradiol levels were significantly related to low timidity in boys (Jacklin et al. 1983; Jacklin et al. 1988; Marcus et al. 1985). In girls, no relationships were observed. Other studies of umbilical cord hormones have shown inconsistent results (Abramovich and Rowe 1973; Forest et al. 1974; Pang et al. 1979).

A recent study using umbilical cord blood testosterone measures examined pragmatic language ability in girls followed up at 10 years of age. Results showed that the higher a girl's free testosterone level at birth, the higher the scores on a pragmatic language difficulties questionnaire (Whitehouse et al., in press). However, levels of fT are typically at very low levels from about week 24 of gestation, and umbilical cord samples can contain blood from the mother as well as the fetus (and hormone levels may vary due to labor itself) (Jacklin et al. 1988), so umbilical cord blood testosterone does not allow one to test if outcomes reflect prenatal testosterone exposure per se.

One of the most promising methods for obtaining information about the fetal exposure to androgens appears to be the direct sampling of fT levels in amniotic fluid, obtained from routine diagnostic amniocentesis. This is performed for clinical reasons in order to detect genetic abnormalities in the fetus. As a result, it is typically performed in a relatively narrow time window, which is thought to coincide with the peak in fetal testosterone for male fetuses. This peak is also apparent in amniotic fluid, and several studies have documented a large sex difference in amniotic androgens. There are significant risks associated with the procedure itself, so that it cannot be performed solely for research. However, the process itself does not appear to have any negative effects on later development (Judd et al. 1976).

The origins of androgens in amniotic fluid are not fully understood, but the main source seems to be the fetus itself (Cohen-Bendahan et al. 2005). Hormones enter the amniotic fluid in two ways: via diffusion through the fetal skin in early pregnancy, and via fetal urine in later pregnancy (Judd et al. 1976; Schindler 1982). Given the risk entailed in obtaining blood from the fetus, there are very limited data directly comparing testosterone in amniotic fluid to that in fetal blood. Androgens in amniotic fluid are unrelated to androgens measured in maternal blood in the same period, as shown in studies in early and mid-gestation (Rodeck et al. 1985; van de Beek et al. 2004). Based on these findings, testosterone obtained in amniotic fluid appears to be a good reflection of the levels in the fetus and represents an alternative to direct assay of the more risky process of collecting fetal serum (Cohen-Bendahan et al. 2005).

Finegan et al. (1992) conducted the first study that explored the relationship between prenatal hormone levels in amniotic fluid and later behavior on a broad range of cognitive functions at age four. The findings are difficult to interpret, since the authors used measures that did not show sex differences. However, the same children were followed up at seven years of age, and associations between spatial ability and fT were examined (Grimshaw, Sitarenios, and Finegan 1995). A significant positive association between fT levels and faster performance on a mental rotation task was observed in a small subgroup of girls, but not boys. At 10 years of age, prenatal testosterone levels were found to relate to handedness and dichotic listening tasks (Grimshaw, Bryden, and Finegan 1995), and the results were interpreted as providing support for the hypothesis that higher levels of prenatal sex hormones are related to lateralization in boys and girls (Witelson 1991).

The Cambridge Fetal Testosterone Project is an ongoing longitudinal study investigating the relationship between fT levels and the development of behaviors relating to ASC (Baron-Cohen et al. 2004; Knickmeyer and Baron-Cohen 2006b).

The first study aimed to measure fT in relation to eye contact at 12 months after birth (Lutchmaya et al. 2002b). Reduced eye contact is a characteristic common in children with autism (Swettenham et al. 1998). Sex differences were found, with girls making significantly more eye contact than boys. The amount of eye contact varied quadratically with fT levels when the sexes were combined. Within the sexes, a relationship was only found for boys (Lutchmaya et al. 2002b).

Another study focused on the relationship between vocabulary size in relation to fT levels from amniocentesis. In some subgroups within ASC, such as classic autism, vocabulary development is also delayed (APA 1994). Girls were found to have a significantly larger vocabulary than boys. Results showed that levels of fT inversely predicted the rate of vocabulary development in typically developing children between the ages of 18 and 24 months (Lutchmaya et al. 2002a).

These children were next followed up at four years of age and completed a "moving geometric shapes" task where they were asked to describe cartoons with two moving triangles whose interaction with each other suggested social relationships and psychological motivations (Knickmeyer, Baron-Cohen, et al. 2006). Sex differences were observed, with girls using more mental and affective-state terms to describe the cartoons compared with boys; however, no relationships between fT levels and mental or affective-state terms were observed. Girls were found to use more intentional propositions than males, and a negative relationship between fT levels and frequency of intentional propositions was observed when the sexes were combined and in boys. Boys used more neutral propositions than females, and fT was related with the frequency of neutral propositions when the sexes were combined. However, no significant relationships were observed when boys and girls were examined separately.

A second follow-up at four years of age examining quality of social relationships demonstrated higher fT levels to be associated with poorer quality of social relationships for both sexes combined. Fetal testosterone levels were also associated with more-narrow interests when the sexes were combined and in boys only (Knickmeyer et al. 2005). Sex differences are reported, with males scoring higher (i.e., having more-narrow interests) than females (Knickmeyer et al. 2005). Individuals with ASC demonstrate more-restricted interests as well as difficulties with social relationships (Knickmeyer et al. 2005).

When the children were six to eight years old, the parents were asked to complete children's versions of the Systemizing Quotient (SQ-C) and Empathy Quotient (EQ-C). Boys scored higher than girls on the SQ-C, and levels of fT positively predicted SQ-C scores in boys and girls individually (Auyeung et al. 2006). Sex differences were observed in EQ-C scores, with girls scoring higher than boys. A significant negative correlation between fT levels and EQ-C was observed when the sexes were combined and within boys. For the EQ-C, a main effect of sex was found, but no main effect of fT. However, the effect of fT cannot be disregarded, since sex and fT are strongly correlated (Chapman et al. 2006). A subset of these children completed the children's version of the Reading the Mind in the Eyes test (Eyes-C). No significant differences were found between sexes, though a significant relationship between fT levels and Eyes-C was observed for both boys and girls (Chapman et al. 2006).

The lack of ability to empathize and drive to systemize appear to be characteristic of ASC (Baron-Cohen 2002). While these behaviors do not confirm a clear link between fT and ASC, the results are broadly consistent with a role for fT in shaping sexually dimorphic behavior.

In light of some of the above results, a more direct approach of evaluating the links between autistic traits and fT was implemented. In this study, effects of fT were directly evaluated against autistic traits as measured by the Childhood Autism Spectrum Test (CAST) (Scott et al. 2002; Williams et al. 2005) and the Autism Spectrum Quotient—Child Version (AQ-C) (Auyeung et al. 2008). The CAST was used because it has shown good test-retest reliability, good positive predictive value (50%), and high specificity (97%) and sensitivity (100%) for ASC (Williams et al. 2005). The AQ-C has also shown good test-retest reliability, high sensitivity (95%), and high specificity (95%) (Auyeung et al. 2008).

Fetal testosterone levels were positively associated with higher scores (indicating greater number of autistic traits) on the CAST as well as on the AQ-C. For the AQ-C, this relationship was seen within sex as well as when the sexes were combined, suggesting this is an effect of fT rather than an effect of sex. The relationship between CAST scores and fT was also seen within males, but not within females (Auyeung, Baron-Cohen, Chapman, et al. 2009). This has recently been replicated using the Quantitative Checklist for Autism in Toddlers, a measure of autistic traits in toddlerhood showing a positive relationship between fT levels and autistic traits from as young as 18 to 24 months of age (Auyeung et al. 2010). These findings, from three measures of autistic traits, are consistent with the notion that higher levels of fT may be associated with the development of autistic traits.

Research suggests that amniotic fluid provides the best direct measurement of fetal hormones compared with maternal serum and is therefore probably the best choice for studying the behavioral effects of variations in prenatal androgen exposure. Using this method, research has shown that fT levels are significantly associated with behaviors associated with ASC, providing strong evidence for a role for prenatal hormones on typical development. However, it is worth remembering that studies that use amniocentesis to measure fT are restricted to individuals where this procedure is used for clinical diagnostic purposes. This means that individual participants are selected in several ways that may influence the generalizability of the results. In addition, these studies only utilize measurements of total extractable testosterone, which may not be directly responsible for the interactions that masculinize behavior (Hines 2004). A more detailed understanding of the chemistry of masculinization would be useful in extrapolating the effects of testosterone in the development of conditions such as ASC. Against these limitations should be weighed the strengths of amniocentesis, which mainly concern its timing and measurement of the fetal environment while avoiding unnecessary additional risk.

Conclusions

Autism spectrum conditions (ASC) are characterized by social impairments, restricted and repetitive interests, accompanied by language delay. ASC are believed to lie on a spectrum, reflecting the range of individual ability in each of these areas.

There is a significant body of evidence connecting the characteristic behaviors of ASC to extremes of certain male-typical behaviors. Evidence includes superior performance on a range of tasks where males typically outperform females but impairment compared to typical males on tasks showing a female advantage. This observation has led to the development of the "extreme male brain" theory of autism. Support for this theory can also be found very early in life and also in some primate studies, suggesting the development of sex-typical behaviors is at least partly biological.

Additional evidence linking ASC to an extreme form of the male brain comes from measurements of physical characteristics where males and females typically differ. In general, physical sexual differentiation is attributed to the gonadal hormones and in particular testosterone and its derivatives. Animal studies have suggested that the same hormones might also control the development of sex-typical behaviors.

In humans, the direct manipulation of hormones to study their effects in early development is not possible or desirable. While studies of naturally occurring atypical prenatal hormone environments yield some information about the role of testosterone in behavioral development, sample sizes are very limited, particularly for individuals who also have ASC. Physical indicators have been used to infer information about levels of exposure to prenatal hormones. These include finger-length ratio and lateralization. Evidence obtained from these proxy measurements shows some correlation with behavior in later life, though the link to actual hormone levels is not clear.

Direct study of the effects of testosterone is difficult because levels rise and fall in the fetal environment over the course of gestation. In addition, there is limited evidence suggesting that maternal hormone levels during pregnancy are representative of fetal levels. The optimal way to directly measure fetal testosterone appears to be via amniotic fluid obtained during clinical amniocentesis. This method is not ideal because it limits the sample available. However, some results have been obtained linking behavior and fT levels in typically developing children.

In this sample, findings suggest that behaviors known to be affected by ASC also tend to be related to elevated fT levels. These behaviors include measures of social and communicative development, empathy and systemizing. In addition, more recent studies using the Childhood Autism Spectrum Test, the Autism Spectrum Quotient—Children's Version, and the Quantitative Checklist for Autism in Toddlers have demonstrated a relationship between the number of autistic traits a child exhibits and fT levels measured via amniocentesis.

Although the findings presented in this chapter lend support to the "extreme male brain" theory of ASC and its link to fT, a thorough evaluation of this theory will require testing not just for associations between fT and autistic traits, but between fT and clinically diagnosed ASC. This remains an active area of research.

Acknowledgments

Bonnie Auyeung was supported by a scholarship from Trinity College, and Simon Baron-Cohen by grants from the Nancy Lurie Marks Family Foundation and the MRC during the period of this work. We are grateful to Emma Chapman, Svetlana Lutchmaya, Bhismadev Chakrabarti, Liliana Ruta, Mike Lombardo, and Rebecca Knickmeyer for valuable discussions. This chapter is based on our article "A Role for Fetal Testosterone in Human Sex Differences," which appeared in Andrew Zimmerman et al., eds. (2008), *Autism: Current Theories and Evidence*, Humana Press.

Memory, Emotions, and Communication

CHAPTER THIRTEEN

The Role of Broca's Area in Socio-Communicative Processes of Chimpanzees

William D. Hopkins and Jared P. Taglialatela

NE OF THE MOST basic and fundamental milestones in human language development is the occurrence of pointing in a social context (Bates et al. 1979; Bates et al. 1975). Beginning around 12–15 months of age, developing preverbal human children begin to point to objects in their environment in the presence of another social agent. In addition to pointing to objects within a social context, children at this age will also alternate their gaze between the referent and the social agent. This alternating in gaze is thought to reflect the child's understanding that the signaling is intentional and thus communicative in function, as opposed to simply a frustrated attempt to reach toward the object. Not only do children alternate their gaze between the referent and social agent—there is also evidence that they will "repair" and elaborate their communicative signaling when previous signals fail. For example, if they point to an out-of-reach object and the social agent does not respond, infants will produce the point response again, point for a longer period of time, or use a different type of signal, such as a vocalization. Finally, many of the early pointing and other gestures produced by children are accompanied by prelinguistic sounds, and in some cases there is significant synchrony in the production of gestures and sounds (Iverson and Thelen 1999; Lock 2001). Eventually, by approximately 18 to 24 months of age, when motor control of the peripheral speech organs becomes increasingly developed, the gestures are supplanted or replaced by spoken words. Thus, one can account for the emergence and transition from an early preverbal socio-communicative system to a simple and effective linguistic system in the developing child.

From an evolutionary standpoint, though historically many have argued for discontinuity of language and speech in modern humans compared with apes, it is increasingly clear that some parallels can be found in the linguistic system of human language and speech and the communicative systems of other species, particularly other hominids, including the great apes. It has been known for more than 40 years now that human-raised apes can acquire and use simple nonspeech symbolic communication systems such as American Sign Language (ASL) or visual-graphic systems for interspecies communication (Gardner and Gardner 1971; Gardner and Gardner 1969; Premack 1971b, a; Rumbaugh 1977; Rumbaugh et al. 1973; Savage-Rumbaugh 1986; Savage-Rumbaugh and Lewin 1994; Savage-Rumbaugh et al. 1993; Savage-Rumbaugh et al. 1978). Similarly, in the past 15–20 years, considerable evidence has demonstrated significant parallels in the cognitive foundation underlying gestural communication in great apes and the preverbal pointing behavior of developing human children (Pika 2008). One aim of this chapter is to summarize some basic facts discovered in the past 15 years regarding ape gestural communication and to discuss how children and apes may differ with respect to their motivation for communication.

A second aim of this chapter is to discuss the neural correlates of sociocommunicative competencies of chimpanzees, particularly as they relate to lateralization in function. In adult humans, clinical, neuropsychological, and more-recent functional imaging studies have clearly implicated the left inferior frontal gyrus (Broca's area) and posterior temporal lobe (Wernicke's area) as key brain regions involved in the production and comprehension of language and speech (Frost et al. 1999; Binder et al. 1996; Sommer et al. 2001; Josse and Tzouio-Mazoyer 2004; A. Damasio and Geschwind 1984). For example, lesions to Broca's and Wernicke's areas result in linguistic deficits in the production and perception of language and speech. PET and fMRI studies have shown that during certain linguistic tasks, such as naming, verbal fluency, and receptive comprehension of meaningful words, activation of the left Broca's and Wernicke's areas is often found, particularly among right-handed individuals. Lastly, structural MRI and postmortem studies have shown that the left posterior temporal lobe (PT) and, less consistently, the left inferior frontal gyrus (IFG) are larger in the left hemisphere, which some believe may account for the functional asymmetries observed in language and speech processing (Beaton 1997; S. Keller, Crow, et al. 2009; S. Keller et al. 2007, Shapleske et al. 1999).

Over the past 10 years, our laboratory has accumulated a number of structural MRI scans in chimpanzees and other great apes. We have been specifically interested in assessing whether chimpanzees show population-level asymmetries in brain regions considered the homologues to Broca's and Wernicke's area in the human brain, including the IFG and PT. Moreover, we have been interested in assessing individual differences in asymmetries in the IFG and PT with measures of handedness, particularly for gestural communication. In the second part of this chapter, we summarize the findings from the neuroanatomical and functional imaging studies we have conducted in chimpanzees and how they relate to lateralization in socio-communicative processes.

Seven Observations about Socio-Communicative Processes in Chimpanzees and Other Great Apes

1. Ape communication is sensitive to an audience. A number of different laboratories have investigated different cognitive and communicative processes underlying intra- and interspecies communication paradigms primarily in captive great apes. One of the most robust findings across laboratories and settings with respect to social communication in apes is the existence of the so-called audience effect (Leavens et al. 2008; Leavens et al. 2009). In the audience effect, food or some other type of object is placed outside the ape's enclosure, and a human is either present or absent. A number of studies have shown that the majority of apes will attempt to communicate, usually in the form of a gesture, *only* when humans are present and not when they are absent. This indicates that the apes use these signals communicatively and that the signals do not reflect frustrated emotional states.

2. Apes alternate their gaze between a referent and social agent. Like developing children, when an object is placed out of their reach (such as outside their cage), a significant majority of apes will gesture to the objects and alternate their gaze between the referent and the social agent (Leavens and Hopkins 1998; Leavens et al. 1996; Leavens, Hopkins, and Thomas 2004). These findings clearly demonstrate that the apes' gestural responses are intentional and referential.

3. In addition to alternating their gaze between the referent and social agent, it has also been reported that chimpanzees and orangutans will alter the modality or type of communicative signal they use in response to the attentional state of the human (Cartmill and Byrne 2007; Hostetter et al. 2001; Hostetter et al. 2007). For instance, Leavens, Hostetter, et al. (2004) recorded the communicative

behaviors of subject chimpanzees under three experimental conditions, including (a) offering food and looking at the subject (b) offering food and looking at another chimpanzee living in the same group, and (c) offering food and looking at a chimpanzee living in an enclosure adjacent to the subject's. When the experimenter offered food and looked at the subject chimpanzee, the chimpanzee attempted to communicate with the experimenter using primarily visual cues such as manual gesture or lip pouts. In contrast, when the experimenter was offering the food and looking at a chimpanzee with or adjacent to the subject, the subject chimpanzee attempted to communicate with the human using auditory cues such as attention-getting vocalizations or tactile cues including spitting, cage banging, or clapping. The significance of these findings is twofold. First, the results suggest that ape communication is tactical in that apes deploy different types of signals depending on the attentional status of the human, rather than just produce a litany of behaviors in a random manner with the aim that they will eventually get rewarded. Second, indirectly, the results may suggest some basic theory-of-mind abilities in the apes. For example, when a chimpanzee selectively uses an attention-getting sound when a human is facing away, it might be suggested that the ape "knows" the human will not see a visual communication signal and therefore knows it needs to deploy a different, nonvisual signal.

4. One of the indicators of intentionality in social communication in developing children is that they will repair or elaborate when their attempted communicative behaviors fail. For instance, a child might point to an object to request it from a caretaker, but if the caretaker fails to respond, the child may point again (in this case "repair") or point to the object and vocalize until the caretaker acknowledges or provides it to the infant. At least one study has reported that chimpanzees will similarly repair and elaborate failed communicative events. Leavens, Russell, and Hopkins (2005) designed a study in which two foods were presented to chimpanzees, one highly desirable (bananas) and one less desirable (monkey chow). A human experimenter would approach the subject's home cage after the foods were baited, and inevitably the chimpanzees would point to the bananas to request that the human give them some. When this occurred, the experimenter did one of three things: (a) give the banana to the chimpanzee, (b) give the chimpanzee chow (appear to misinterpret the gesture), or (c) break the banana in half and give one portion to the ape and place the

remaining half in his or her lab coat pocket. Following this manipulation, the subsequent communicative behaviors of the chimpanzees were recorded for 30 seconds. When the human experimenter misinterpreted the apes' gesture (i.e., gave chow) or partially misinterpreted (gave only half the banana), the chimpanzees produced significantly more posttest communicative behaviors compared to when the human correctly interpreted the signaling (i.e., gave the banana). Similar results using a slightly different paradigm have been reported in orangutans (Cartmill and Byrne 2007).

5. The use of so-called attention-getting sounds that have been described for chimpanzees (the "raspberry" and extended food grunt) and orangutans (the kiss squeak) indicate that apes have voluntary control over the use of these sounds. It has been a long-standing claim that nonhuman primates do not have volitional control of their vocal apparatus or facial expressions (Aitken and Wilson 1979; Premack 2004, 2007) and that all these communicative signals are tied exclusively to emotional contexts. The evidence demonstrating that chimpanzees and orangutans can selectively use attention-getting sounds and produce them in relatively emotionally benign contexts clearly challenges the historical and contemporary view that they do not have voluntary control of these sounds (Hopkins et al. 2007).

6. In unbiased testing environments, a significant majority of chimpanzees prefer to gesture with their right hand. For example, Hopkins et al. (2005) measured hand preference in 227 chimpanzees when producing food beg gestures and found population-level right-handedness. In a follow-up study, Meguerditchian, Vauclair, and Hopkins (2010) recorded gestural signals during inter- and intraspecies communication and similarly found population-level right-handedness. Interestingly, the degree of right-handedness found for manual gestures is significantly higher in chimpanzees when compared with hand preferences for manual actions that are not communicative in function, such as simple reaching or coordinated bimanual actions. This indicates that manual gestures, in comparison to hand preferences for other kinds of tasks, may be more indicative of hemispheric specialization. Similar results have recently been reported in developing human children (Couchet and Vauclair 2010; Couchet and Vauclair, in press) and in baboons for a hand-slapping gesture they produce during agonistic interactions (Meguerditchian and Vauclair 2006).

7. Finally, in chimpanzees, it has been observed that attention-getting sounds are often accompanied by a manual gesture, and these signals are synchronous (Hopkins and Cantero 2003). In other words, in some but not all chimpanzees, when they produce an attention-getting sound, they often also produce a manual gesture. The co-activation of hand and sound production has similarly been described in developing children (Iverson and Wozniak 2007; Iverson and Thelen 1999; Lock 2001) and is considered an important milestone in language development.

Broca's Area from a Comparative Perspective

Broca's area in the human brain is found within the frontal operculum and comprises the pars opercularis and pars triangularis (S. Keller, Roberts, and Hopkins 2009; Keller et al. 2007). Cytoarchitectonically, Broca's area is composed of Brodmann's areas 44 (Ar44) and 45 (Ar45), which are generally found within the border of the pars opercularis and pars triangularis, respectively (Amunts et al. 1999). In chimpanzees, anatomically, Broca's area corresponds to the inferior frontal gyrus (IFG), which is bordered anteriorly by the fronto-orbital sulcus and posteriorly by the precentral inferior sulcus (S. Keller, Roberts, and Hopkins 2009) (see Figure 13.1). Basically, the IFG of chimpanzees corresponds to the pars opercularis of the human brain, whereas the pars triangularis seems to be a unique sulcal pattern in human brains. Recently, in chimpanzees, cytoarchitectonic studies have shown that Ar44 is principally found in the dorsal portion of the IFG, whereas Ar45 is located in the

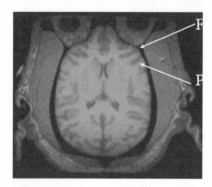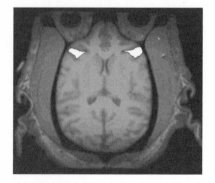

Figure 13.1. Axial view of the inferior frontal gyrus (IFG) in a chimpanzee brain. FO = fronto-orbital sulcus, PCI = precentral inferior sulcus. The IFG lies between FO and PCI (indicated in white in right image).

gyrus immediately anterior to and within the fronto-orbital sulcus (Schenker et al. 2010).

One of the more controversial topics in comparative neuroscience is the question of neuroanatomical asymmetries in Broca's area in both human and nonhuman primates when measured at the gross anatomical level of analysis (Sherwood et al. 2003). In humans, the pars opercularis and pars triangularis have been quantified, and the results are complex, with some reporting evidence of leftward asymmetries, and others, rightward (S. Keller, Crow, et al. 2009). Moreover, variation in lateralization in the pars opercularis and pars triangularis can differ between right- and left-handers and between individuals with different linguistic skills (Foundas et al. 1995; Foundas et al. 1998). More recently, rather than quantify the entire IFG, several investigators have examined lateralization in the IFG for gray matter alone using either region-of-interest or voxel-based morphometry, and these findings have failed to report any consistent evidence of population-level asymmetries (Good et al. 2001; Knaus et al. 2004, 2006; Watkins et al. 2001; Tomaiuolo et al. 1999).

The findings on asymmetries in the IFG in chimpanzees are, in many respects, like the reports in humans. In one of the initial studies in great apes, Cantalupo and Hopkins (2001) measured the volume of the IFG in the parasagittal plane in a sample of 20 great apes and reported a population-level leftward asymmetry for this region. Since this original publication, my colleagues and I have also measured the volume of IFG in the axial plane in a much larger sample of chimpanzees and failed to find population-level asymmetries when considering the entire volume of the gyrus (Hopkins et al. 2008).

The previous work on asymmetries in the IFG has centered on measuring the entire volume of the gyrus (gray + white matter). More recently, in our laboratory, we have turned our attention to quantifying the gray-matter tissue comprising the IFG in chimpanzees using voxel-based morphometry (Hopkins et al. 2008) or with probabilistic maps. For the analysis of the IFG using probabilistic maps, prior to tracing the IFG on each individual brain, the skulls were removed from the raw MRI scans and the brains were realigned in the axial planes and subsequently virtually cut into consecutive 1 mm slices (Hopkins et al. 1998). Each subject's brain volume was then co-registered to a template of a chimpanzee brain using three-dimensional voxel registration with a linear transformation (Analyze 8.1, Mayo Clinic) (Rilling et al. 2007). The IFG was then quantified separately for the left and right hemispheres, in the axial (transverse) plane on each spatially aligned brain (see Figure 13.1) and the object maps saved for each individual subject. The individual IFG object maps were then summed across all subjects and subsequently thresholded at three different criterion levels, 30 percent, 50 percent, and 70 percent. The different thresholded object maps represented the voxels within the individual IFG tracings that were present in 30 percent of the chimpanzee sample (common voxels in at least 14 individuals), 50 percent of the sample (common voxels in at least 24 individuals), and 70 percent of the sample (common voxels in at least 34 individuals). For each of the thresholded volumes, the resulting object map within the left and right hemisphere was quantified and saved for subsequent use with the segmented gray-matter volumes. Specifically, the IFG volumetric object maps for the left and right hemispheres that were derived for the 30 percent, 50 percent, and 70 percent probabilistic maps were then applied to the co-registered GM volumes. The volume of GM voxels found within the left and right hemispheres for the three different probabilistic maps (30%, 50%, 70%) were quantified across subjects. Asymmetry quotients were then created for the GM following the formula $[AQ = (R - L) / ((R + L) *.5)]$ where R and L reflect the percentage of GM found for the right and left hemispheres.

From the GM probabilistic maps of the IFG, several interesting results have emerged. First, population-level asymmetries are found, but they are specific to the highest threshold level (70%) and are absent at the 50 percent and 30 percent threshold levels. This suggests that leftward GM asymmetries are present only within voxels of the IFG regions that are most consistently present across a majority of the chimpanzees. Second, lateralization in gestural and vocal communication is associated with variation in GM asymmetries in the IFG. Specifically, recall that some chimpanzees reliably produce attention-getting sounds and some do not. Moreover, chimpanzees that produce the attention-getting sounds are much more right-handed for gestures than those that do not. To test whether differences in asymmetries within the IFG may be associated with gesture handedness and oro-facial motor control, we compared the AQ values from the three threshold levels between right- and non-right-handed chimpanzees who do (VOCAL+) or do not (VOCAL−) produce attention-getting sounds. This analysis revealed significant two-way interactions between threshold level and vocal group $[F(2, 146) = 4.05, p < .02]$ as well as between gesture hand and vocal group $[F(1, 73) = 5.14, p < .03]$. For the vocal group by threshold interaction, no differences were found at 30 percent and 50 percent, but at 70 percent the VOCAL+ chimpanzees showed a significantly greater leftward asymmetry than the VOCAL− group (see Figure 13.2). For the gesture hand by vocal group interaction, left-handed VOCAL− chimpanzees showed significantly higher AQ values (indicative of right

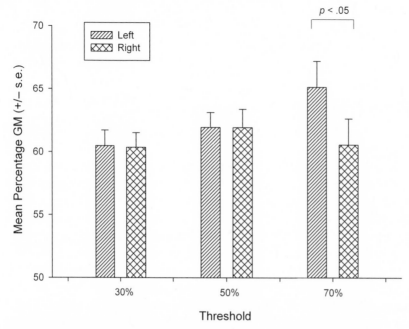

Figure 13.2. Mean percentage of GM (+/− s.e.) for the left and right hemispheres at each threshold level.

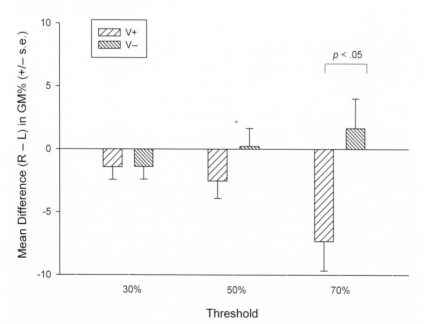

Figure 13.3. Mean difference between the right and left hemisphere in the percentage of gray matter (+/− s.e.) in chimpanzees that do (V+) or do not (V−) consistently produce attention-getting sounds at the three different threshold levels.

hemisphere asymmetries) than left-handed VOCAL+, right-handed VO-
CAL–, and right-handed VOCAL+ chimpanzees (see Figure 13.3).

Neurofunctional Correlates of Gestural Communication in Chimpanzees

The previously described work from our laboratory has focused on correlating behavior with structural asymmetries in the IFG. These studies do
not speak to the question of whether the IFG is functionally involved in
communication in chimpanzees. To address this issue, my colleagues and I
have recently employed positron emission tomography (PET) to examine
the neurofunctional correlates of manual and vocal communication in
chimpanzees (Taglialatela et al. 2008). In this PET study, in two separate
tests, we orally dosed chimpanzee with fluorodeoxyglucose (FDG) and
subsequently had the subjects engage in a behavioral task. During one PET
imaging session, the chimpanzee engaged in gestural and vocal signaling
behavior with a human. In the second PET session, as a comparison condition, the chimpanzees picked up small objects in their enclosure and
handed them to an experimenter during the uptake period. We then compared the patterns of neural activation in the communication and grasping
conditions using a voxel-by-voxel t-test analysis. A number of significant
clusters were found in this analysis, but germane to this paper was the evidence for activation in the dorsal portion of the left IFG. Thus, we have at
least some evidence that the IFG (among a number of regions) is functionally involved in socio-communicative processes in chimpanzees.

Conclusions

With respect to socio-communicative processes, the collective results on
great apes, and particularly chimpanzees, clearly show that intentional, referential communication is present in this taxonomic family and that the
behavior is homologous to what has been described in developing human
children. Though some have suggested that the cognitive processes underlying gestural communication in human children and chimpanzees are
different (Povinelli et al. 2000, 2003), my colleagues and I have argued that
this claim seems unjustifiable and that some rather simplistic noncognitive
mechanisms can explain the emergence of imperative pointing in both species (Leavens, Hopkins, and Bard 2005). More crucial to the discussion of
ape and human pointing are recent claims that the function of pointing is
different between humans and chimpanzees. Specifically, some have argued
that human infants engage in both imperative and declarative pointing,
whereas apes only engage in imperative pointing (Tomasello 1999; Moll

and Tomasello 2007). Imperative pointing is instrumental communication that is motivated solely to "request" something from another individual. In contrast, declarative pointing is informative—that is to say, the infant points to objects with the motivation being solely communicative or to share in attention about the object with another social agent.

One body of work cited to support this view is the evidence that chimpanzees cannot solve the "object choice" (OC) task (Míklósi and Soproni 2006). In the OC task, a human experimenter points to one of two containers that were baited with food out of sight from the ape subject. The apes can then point to one of the containers to request the food. A number of studies have reported that apes perform very poorly on this task, and some suggest that one reason for their failure is the apes do not understand the "informative" nature of gestural signals.

Recently, my colleagues and I have demonstrated that chimpanzees and bonobos reared in human socio-linguistic environments perform significantly better on the object choice task than apes raised in standard laboratory or zoo settings (Lyn, Russell, and Hopkins 2010). In light of the fact that all the apes previously tested on the OC task had been raised in standard laboratory or zoo settings, our findings suggests that performance on the OC task is strongly influenced by early rearing experiences. Thus, apes raised by humans in an environment that emphasizes comprehension of pointing and other forms of communication have little trouble with the OC task. I believe these findings raise some questions regarding the claims that poor performance on the OC task reflects an inherent limitation in comprehending "declarative" information by great apes. Indeed, I would argue that the ability to engage in joint attention must be a foundational skill for the abilities of apes to acquire and use symbols of the type described in many of the ape language studies. Consider what must be necessary for an ape to acquire the meaning of symbols; the referent must be paired with the symbol via pointing. For example, an apple might be shown to the ape, and the human may point to the symbol for "apple" while verbally labeling the food. For the apes to acquire the symbolic meaning, they must engage in joint attention with the human when pointing to the symbol for "apple" in order to associate the symbol with the food. I am not suggesting here that real differences in the motivation to communicate with other social beings may differ between human children and chimpanzees (or other great apes). My main concern with the claims regarding the absence of declarative pointing in apes is that a) there are almost no data from apes that have employed paradigms similar to those used with human children, and b) the evidence we have showing that early social rearing experiences can influence the comprehension of declarative signals raises some doubts about the robustness of these claims. It is likely not a matter of kind but rather amount when considering imperative and declarative pointing.

With respect to the role of the IFG in socio-communicative processes, though there are some issues with quantification of the IFG in chimpanzees, our results clearly implicate this brain region in gestural communication in chimpanzees, notably in terms of variation in gray matter lateralization. It should be noted that asymmetries in the IFG are not related to handedness for manual actions that are not communicative in function, save tool use. Thus, the behavioral correlates of asymmetries in the IFG appear to be specific to manual actions involved in communication. The PET results extend the neuroanatomical data and suggest that the IFG is functionally involved in gestural communication and possibly oro-facial motor control in chimpanzees. In short, though more functional imaging studies are needed, there appears to be convergent evidence that the IFG of chimpanzees is involved in communication, and this neural substrate may have been the foundation from which language and speech evolved in humans, after the split from chimpanzees. One of the most striking aspects of the human IFG from a comparative perspective is the sheer size. From the extant cytoarchitectonic data on Ar44 and Ar45 in humans and chimpanzees, the data suggest that the human Ar44 and Ar45 are nearly seven times larger than the chimpanzee's (note that overall brain size differences are only threefold) (Schenker et al. 2010). The relatively large size of Broca's area in humans compared with chimpanzees clearly suggests that there has been intense selection for expansion of this region after the split between chimpanzees and humans from the common ancestor. The expansion of the IFG in humans likely reflects, in part, the increasing demands for motor control of the hand and oro-facial musculature associated with the emergence of language and speech. However, it should not be ignored that the IFG of humans is also involved in other complex socio-communicative processes, such as imitation and theory of mind. The extent to which the IFG plays a role in complex social learning, such as imitation, remains an empirical question, but future research should focus on this question. The evidence of mirror neurons in regions corresponding to Broca's area in monkeys and the fact that these neurons respond to the initiation and perception of actions suggest that perhaps this forerunner to Broca's area played a prominent role in social learning in the common ancestor of humans and chimpanzees and that this area was subsequently co-opted for use in language and speech, as has been suggested by some (Arbib et al. 2008). The development of in vivo structural and functional imaging with nonhuman primates, and particularly chimpanzees, provides an important platform for addressing these important evolutionary questions.

Acknowledgments

This research was supported in part by NIH grants NS-36605, NS-42867, HD-38051, and HD-56232. American Psychological Association guidelines for the ethical treatment of animals were adhered to during all aspects of this study. We are grateful to the entire veterinary staff at the Yerkes Center for their helpful assistance in collection of the MRI and PET scans. I am thankful for the helpful assistance of Ms. Jamie Russell, Ms. Jennifer Schaeffer, and Dr. Heidi Lyn for their assistance in this work. Correspondence concerning this chapter should be addressed to Dr. William Hopkins, Department of Psychology and Neuroscience, Agnes Scott College, 141 E. College Avenue, Decatur, GA 30030. E-mail: whopkins@ agnesscott.edu or whopkin@emory.edu.

CHAPTER FOURTEEN

Emotional Engagement: How Chimpanzee Minds Develop

Kim A. Bard

COMPARATIVE DEVELOPMENTAL psychology is a perspective with which we can view the similarities and differences in developmental processes that occur across primate species. My developmental area of specialty is infancy, and I find topics involving emotion particularly interesting. As a developmental psychologist, I ask questions about how the emotional system of chimpanzees develops, and in particular I am interested in how developmental outcomes change as a function of social-cultural environments (e.g., Bard 2005). Comparative psychologists are interested in similarities and differences across species in order to address questions of species-unique characteristics or characteristics shared by species based on evolutionary history. The last common ancestor (LCA) of humans, chimpanzees, and monkeys lived approximately 30 million years ago, and the LCA of humans and chimpanzees lived approximately 7 million years ago (e.g., Steiper and Young 2006), which means that humans and chimpanzees have 23 million years of shared evolutionary history. In fact, I am interested in how developmental processes of chimpanzees compare with developmental process of humans. To determine those characteristics that are uniquely human, those that are uniquely chimpanzee, or those shared by humans and chimpanzees, a comparative approach must be combined with the study of development across species (Bullowa 1979; Johnson-Pynn et al. 2003).

The most common question I am asked is how does the development of chimpanzees compare to that of humans. The desired answer is typically a cross-species comparison on a single dimension; for example, "the chimpanzee" is like a two-year-old human. The answer, however, is quite a bit more complex, because it depends upon which dimension is chosen, both for control and for comparison. Chimpanzees, like humans, show developmental changes in motor skills, in emotionality, in cognition, and in social skills. The outcomes within these domains vary as a function of "lived experiences" (sometimes called rearing, or learning, or eco-cultural experiences, or simply environment). Moreover, outcomes in infancy form the foundational skills for adult competencies. In this chapter, I provide the broad strokes of a comparative developmental answer to this question of how the development of chimpanzees compares to that of humans, with a special focus on the role of engagement.

The study of young infants makes somewhat easier the task of comparing development across species and of understanding common developmental processes (which is my main goal). To understand developmental processes, we need to understand the influences of maturation, social-cultural environments, and of evolutionary history on outcomes (e.g., Jablonka and Lamb 2007). Combining a comparative perspective with truly developmental study allows us to address the flexibility and plasticity in outcomes. In this chapter, I review a number of studies highlighting the relevance of this developmental approach to topics that are of interest to comparative psychologists, primatologists, evolutionary biologists, etc. The conclusion is that social cognition has an important developmental history of lived experiences.

The idea that one can pinpoint an age for "the chimpanzee" to equate them to "the human" of a certain age is perpetuated (repeatedly) in the comparative literature (Leavens and Bard 2011). Part of the problem is that outcomes from "chimpanzees," usually adults, are compared to outcomes of "humans," typically two-year-old infants (e.g., Herrmann, Hernandez-Lloreda, et al. 2010). When authors find that chimpanzees performed less well on a particular task than 2.5-year-old humans, the assumption is that chimpanzees are not as developed as 2.5-year-old humans. Of course, to make conclusions of species differences (if one could control the additional confounds of rearing, for example), the addition of adult humans and chimpanzee infants is required (to have an appropriate experimental design to test for main effects of species and age at testing, and to test for the interaction of species by age; Leavens, Hopkins, and Bard 2005, 2008). I resolve this problem by testing chimpanzees and humans at the same developmental age, choosing to equate the groups on duration of postnatal lived experiences. In this review, I discuss the effects of early experiences (i.e., engagement) on numerous milestones in development: 1) newborn neurobehavioral integrity; 2) early social cognitive processes such as primary intersubjectivity, neonatal imitation, and mutual gaze; 3) emotional expressions and social referencing; 4) attachment; and 5) self-awareness. With this database, I highlight the relevance of rearing histories (of lived experiences) to the development of social cognition in chimpanzees, and hence the importance of developmental studies to comparative psychology (see also Boesch 2007; Bjorklund 2006; Burghardt 2009; Suomi 2006).

Characteristics of Newborns

In this section I address the question of how chimpanzees and humans compare at birth. First, we should compare gestational age. Full-term age in humans is approximately 266 days, or 38 weeks, when measured from conception to birth, and average full-term age for chimpanzees is approximately 225 days, or 32 weeks from the day of ovulation to birth (e.g., Wallis 1997). This is corrected for the fact that typically, for humans, the length of pregnancy is measured from the last menstrual period until birth (adding 2 weeks = 40 weeks of pregnancy), whereas for chimpanzees (as in most nonhuman animals), pregnancy is measured from ovulation to birth. So, when compared from conception to birth, humans have a longer period of in utero development than chimpanzees, approximately six weeks.

Second, we should compare the sizes of infants at birth. Newborn chimpanzees weigh 1.5 kilograms on average, or 3 percent of their adult weight (Fragaszy and Bard 1997). Newborn humans weigh 3.25 kilograms (7 lbs, 4 oz) on average, or 4 percent of their adult weight (approximated for U.S. males and females combined). It is interesting to note that DeSilva and Lesnik (2006) estimate that chimpanzees' brains have reached 35–40 percent of their adult growth at birth, and that the brain size of human newborns is 25–30 percent of their adult size. Thus, chimpanzees have a shorter in utero period, weigh less at birth, but have relatively more brain growth than human newborns.

Chimpanzees have inborn capacities for emotional engagement with caregivers and possess developmental processes, such as neonatal imitation and primary intersubjectivity, by which emotional experiences become integrated in communication and cognition. The remainder of this section reviews these inborn capacities in chimpanzees and comes to the surprising conclusion that from birth through the first month, the neurobehavioral integrity and other expressions of inborn capacity of newborn chimpanzees are not easily distinguished from newborn humans (see review in Bard, Brent, Lester, Worobey, and Suomi, 2011).

newborns, and arrived at different conclusions (Bard et al. 2011). There were three nursery-raised chimpanzee groups—two from the Yerkes Research Center nursery at Emory University (standard care, or ST, and responsive care, RC), and one group raised at the Southwest Foundation for Biomedical Research (SW). The three nursery settings (for chimpanzee infants whose mothers did not have sufficient maternal skills) differed in terms of their rearing practices and socialization goals. Briefly, the ST nursery consisted of groups of four to six infants of the same age together constantly (24/7), with intermittent human contact with technicians who fed and cleaned the infants (scheduled once every two hours in the first weeks after placement in the nursery and every four hours in later weeks). In the RC nursery, there was additional contact with specially trained research assistants (for four continuous hours per day, five days a week), who nurtured the socioemotional and communicative development of each infant (Bard 1996). The SW nursery was similar to the Yerkes ST nursery, with infrequent contact, but animal care technicians consistently wore biosafety masks and gloves and had very limited cradling contact with the chimpanzees. An additional group of chimpanzees was reared by their biological mothers (Mo), with constant cradling contact (24/7), except for the two times that they were temporarily "borrowed" from their mothers and given NBAS tests at 2 and 28 days (Hallock et al. 1989). The chimpanzee groups were compared with one human group, reared by their biological mothers, in a middle- to upper-middle-class urban part of Providence, Rhode Island (see Lester et al. 1989 for full details). All infants were assessed using the same tool, the NBAS, tested by certified examiners, at the same ages, 2 days and 28 days.

This cross-species analysis with multiple chimpanzee groups revealed very different findings from the original comparison between a single group of chimpanzees and of humans. On the second postnatal day, the human group was significantly different from all the chimpanzee groups in only 3 of 25 scores, specifically in alertness, in muscle tone, and pull-to-sit (from eight scores in orientation, six scores in motor performance, five scores in range of state, five scores in state regulation, and the single score of smiling). On day 30, there was only a single difference that distinguished the human group from all the chimpanzee groups, and it was muscle tone, with the chimpanzees significantly more hypertonic than the human newborns. On the remaining 24 of 25 NBAS scores at 30 days, the human group was not distinct. Moreover, there was not a simple explanation for the pattern of group differences. This range in chimpanzee neonatal behavioral outcomes encompassed the neonatal behavioral outcomes found in the human group. "In terms of describing chimpanzee newborns, we conclude that how they performed depended to a large degree on what they have experienced" (Bard et al. 2011, p. 88).

Therefore, in comparing chimpanzee newborns to human newborns, with data from multiple chimpanzee groups, I would summarize by saying that it is very difficult to find species differences. In fact, in the analyses that I report above, only a single human group was used to compare with the multiple chimpanzee groups. There are significant cross-group differences in humans as well (e.g., Nugent et al. 1989), especially in range of state, state regulation, and some motor performance. Therefore, the degree to which the neonatal system matures in interaction with specific features of the postnatal environment illustrates the flexibility inherent in the genome of chimpanzees and humans to respond to particular types of emotional engagements.

Neonatal Imitation

Imitative behaviors differ across primate species: there are a few examples of imitation in newborn rhesus monkeys (see Chapter 2 by Ferrari and Fogassi) but quite a number of instances of imitation in chimpanzees from infancy through adulthood (see review by Myowa-Yamakoshi 2006). Additionally, it's clear that imitative abilities differ across development: newborns don't copy their parent banging a hammer, for example, but two-year-olds just might. Imitative abilities in nonhuman primates need to be viewed from both comparative and developmental perspectives. Here I present an overview of research on neonatal imitation in chimpanzees (from Bard 2007).

We used two procedures to assess imitation in healthy full-term chimpanzee newborns: one was very rigid and standardized, the other was both communicative and interactive. In the structured procedure, the chimpanzee infant was placed facing the adult human demonstrator. The adult human followed a predetermined order of demonstration: either mouth-opening (MO) first and tongue protrusion (TP) second, or TP first and MO second. There were four presentations of the first demonstrated action within a 20-second period, followed by a passive face for 20 seconds, and this lasted for the first four minutes. Immediately thereafter, the second demonstrated action was given in the same cycle of four demonstrations within a 20-second period, followed by 20 seconds of a passive face. From the beginning through the entire eight-minute session, the actions of the adult human were completely determined by the structure of the test and did not change regardless of the actions of the infant. Using the structured procedure, all five chimpan-

Table 14.1   Imitative performance of newborn chimpanzees in two test
paradigms (adapted from Bard 2007)

	First test when chimpanzees were 7–15 days old		
Structured paradigm			
	No response or no match	*Match 1 model*	*Match both models*
		Lindsey	
		Rosemary	
		Wilson	
		Claus	
		Nugget	
Interactive paradigm			
	No match	*Match 1 model*	*Match 2 models*	*Match all 3 models*
		Wilson	Rosemary	Lindsey
			Claus	
			Nugget	

zees matched one of the modeled actions, but none matched both actions (see Table 14.1).

In the Interactive procedure, the infant was either held or laid comfort-
ably within arm's reach. The order of modeled actions was determined in
advance, just as in the structured procedure. However, the number of presentations depended on the infant's response. The first demonstrated action was presented, and a five-second pause allowed the infant to respond.
If the infant emitted any of the target actions, then this was considered the
infant's response, and the demonstration concluded. If the infant did not
respond, up to five repetitions of the demonstrated action were given. After
the infant responded, or after five demonstrations, there was a 10-second
pause, before the next action was demonstrated. The series of modeled actions was provided twice, so that the infants had two opportunities to exhibit imitation of each action.

In the interactive procedure, we used three modeled actions: MO, TP,
and Tongue Clicks (TC). The MO and TP models were identical to those
in the structured procedure. One demonstration of a TC included three
rhythmic clicks, a rather complex model of both sound production and a
series of three actions. Using the interactive procedure, one subject
matched one modeled action, three subjects matched two modeled actions, and one subject matched all three modeled actions (Table 14.1).

The best performance in neonatal imitation was given by Lindsey, when she was seven days old. During the first presentation of TC, she listened and occasionally looked at the model during the first four demonstrations, and then with the fifth demonstration, after a pause, she emitted a single tongue click. During the second pass, and again after the fourth demonstration of TC, she responded. But this time she imitated the series of three actions (two MOs) and matched a TC.

When chimpanzees are 7–15 days of age, imitation was exhibited in both the structured and interactive procedures by all five chimpanzees (Bard 2007, table 14.1). Imitation of facial expressions by chimpanzee neonates appears to represent a capacity to engage in intersubjectivity that is similar to that of human infants. Furthermore, four of five subjects performed better in the interactive paradigm, indicating that chimpanzees share with humans a developmental mechanism (i.e., neonatal imitation) that may allow outcomes to be highly influenced by early interactive experiences.

Emotional Expressions

How do chimpanzees and humans compare in their development of emotional expressions? In the first 30 days of life, chimpanzees, like humans, express happiness with movements of the mouth. We call this smiling, if matching the emotion across species. We call the facial expressions something different, however, if we are matching on facial movement morphology (i.e., with a standardized coding of facial action units, such as with ChimpFACS; Vick et al. 2007). Chimpanzee infants express happiness with an open-mouth expression, a play face, whereas human infants express happiness most often with lip corners being raised, although humans also have open-mouth smiles (e.g., Messinger 2002), and chimpanzees also raise their lip corners (Thorsteinsson and Bard 2009; Vick et al. 2007). Given the same positive context, say of a caregiver vocalizing to an infant while tickling, the prototypic facial movements of human and chimpanzee infant might look very much the same (compare the large smile of the human infant in Oster 2005, 281, with the big play face of the chimpanzee infant in peekaboo in Bard 2005, 47). But note that these are three-month-old infants, although newborns also smile (see Table 14.2). Of course, chimpanzee newborns and human newborns cry and fuss with distinctive facial expressions and vocalizations (see Bard 2000 for details of crying in chimpanzees). We found that species-typical chimpanzee vocalizations of greeting, threat bark, and alarm call occurred in some chimpanzees during the newborn period, although not all occurred in appropriate contexts (see Bard 2003, 1996 for more details).

Table 14.2   Emotional expressions found in neonatal chimpanzees (Bard 1998, 2003)

Interest	Joy	Upset/Distress	Anger	Fear
Brightening	Greet	Pout	Mad face	Alarm
Effort grunt	Smile	Fuss	Threat	
Tongue click	Laughter	Cry		
		Cry face		

We know that the chimpanzee emotional system, like the human emotional system, develops in interaction with the social environment (Bard 2005). Beginning in the neonatal period, chimpanzee infants, like human infants, smile in response to familiar caregivers in face-to-face contexts (see also Tomonaga 2006). By 30 days of age, chimpanzees raised in a nursery by humans smile significantly more often than do mother-raised chimpanzees, when tested by a human examiner (Bard 2005). Perhaps most interesting is that human infants (from middle-class America) smile more than mother-raised chimpanzees but less than nursery-raised chimpanzees at 30 days of age (Bard et al. 2011).

Socioemotional experiences can impact the development of negative emotional expressions. The amount of crying and fussing in chimpanzees varies as a function of rearing experiences: two groups of nursery-reared chimpanzees differed significantly in the amount of fussing and crying evident in the second month of life (Bard 2000); and extended periods of crying occurred much more rarely in nursery-raised chimpanzees compared with mother-raised chimpanzees, when given the NBAS standardized test (e.g., Bard et al. 2011). Interestingly, by 21 days of age, on average, some chimpanzee newborns displayed angry faces, although there was a significant difference in occurrence as a function of the early rearing environment (Bard 2003). Therefore, it is clear that a variety of emotional expressions, both facial and vocal, occur in young chimpanzees, but the prevalence or even the age of onset may vary widely as a function of the socioemotional environment in which chimpanzees are developing.

Characteristics of Three-Month-Olds

"Intersubjectivity" is Trevarthen's (1979) term for human infants' ability to adapt their own purposeful acts to the subjectivity of others. Primary intersubjectivity is most often seen in face-to-face interactions of parents with their very young infants. One question that has motivated much of my research relates to whether chimpanzee infants, like human infants, are born with this capacity for intersubjectivity. Four strands of evidence are required to conclude that humans, or chimpanzees, have primary intersubjectivity, including (1) neonatal imitation; (2) mutual gaze; (3) the extent to which emotional behaviors can acquire communicative meaning; and (4) the degree of flexibility in emotional and/or communicative meanings (i.e., based on differences in early rearing). If we find evidence of each in chimpanzees, then we might conclude that primary intersubjectivity was a capacity common to the great ape–human lineage (for further discussion of this topic see Bard 2009).

In order to enter into a communicative system, an individual must have behaviors that can assume communicative meaning. These meaningful behaviors can develop from the most basic elements of neonatal behavior, such as crying and alert attention. I observed chimpanzee mothers with their newborn infants and found that the mothers spend time engaged in face-to-face interactions with their infants. They spend lots of time in communicative, tactile games (Bard 1994, 2002). Primary intersubjectivity appears to be expressed in social games in chimpanzees. For example, Sheena, when she was three months of age, was assessed with a NBAS exam, the item where a cloth is placed lightly covering the face. She pulled the cloth down and smiled at me, initiating a game of peekaboo. I replaced the cloth over her face, and when she pulled it down the second time, her smile was even broader. Nursery chimpanzees are active partners in social games (see figure 2.8, page 47 in Bard 2005).

So there is evidence that young chimpanzees have neonatal imitation and behaviors that assume communicative meaning (i.e., emotional expressions), and there is some evidence already from the newborn period that socialization exerts a powerful influence on the development of emotional expressions in chimpanzees. What about one other marker of primary intersubjectivity—that is, mutual gaze?

In chimpanzee mother-infant interactions at the Yerkes Research Center, there were between 8 and 10 instances of mutual gaze in an hour. That level was steady as the infant grew from 2–4 weeks, to 6–8 weeks, and 10–12 weeks of age. This is in marked contrast to my observation of the skills of infant chimpanzees raised in the Yerkes nurseries—that is, interacting with human partners. In the nursery, I was impressed with infant chimpanzees' ability to engage in emotionally meaningful, face-to-face interactions, with extended periods of mutual gaze.

By using an approach that combined a developmental perspective (describing changes in mutual gaze in the first three months of life) with comparative data (documenting mutual gaze and intuitive parenting behaviors across different groups and across species), my colleagues and I explored developmental prerequisites and correlates of mutual gaze in chimpanzees (Bard et al. 2005). This study confirmed previous evidence that mutual gaze occurs in mother-raised chimpanzees (16.9 times per hour). Moreover, mutual gaze occurred in chimpanzees at all ages sampled, and at both centers; but we found different developmental patterns of mutual gaze in the two groups of chimpanzee mother-infants pairs. There was a significantly higher rate of mutual gaze in the mother-infant pairs residing at the Primate Research Institute of Kyoto University (PRI) compared with mother-infant pairs residing at the Yerkes Research Center of Emory University (an average of 22 mutual gaze bouts per hour at PRI, and an average of 12 mutual gaze bouts per hour at Yerkes).

It appeared that mothers at PRI were actively encouraging mutual gaze by tilting or holding up the infant's chin with a finger while looking into the infant's eyes. This behavior occurred in each of the PRI mothers with their three-month-old infant but was not observed to occur in any Yerkes chimpanzee mother. We found that the rate of mutual gaze was significantly and inversely correlated with percent of time spent cradling three-month-old infants. Reduced face-to-face interactions and reduced amounts of mutual gaze are reported in some human cultures that have increased physical contact with infants, compared with Western norms (reviewed in Bard et al. 2005; H. Keller 2007). In an experimental study (Lavelli and Fogel 2002) with human infants, mutual gaze increased when two- and three-month-old infants were out of contact compared with being held by their mothers. Thus, the enculturation process of intuitive parenting behavior in chimpanzees appears strikingly similar to the process used by humans.

We proposed the Interchangeability Hypothesis (Bard et al. 2005) to explain variations of mutual engagement based in physical contact without mutual gaze (the basic primate system, also found in traditional human societies; H. Keller 2007), or on mutual gaze without high levels of physical contact (an evolutionary derived system, found in middle-class human societies). My observations of nursery chimpanzees fits in nicely, as those chimpanzees raised by human caregivers are most often out of physical contact, and they have the highest levels of mutual gaze among three-month-old chimpanzees studied to date. There is a range of outcomes in mutual engagement, explainable by the same development mechanism, active in chimpanzees and humans (e.g., Bard et al. 2005; H. Keller 2007; Plooij 1984). Furthermore, by identifying a similar developmental process that accounts for higher (and lower) amounts of mutual gaze, we confirm further evidence for primary intersubjectivity in chimpanzees.

Early Cognition and Social Cognition (Six to Nine Months of Age)

The most widely used assessment of early cognition in human infants is the Bayley Scales of Infant Development (BSID) (Bayley 1969, reviewed in Bard and Gardner 1996). My research team administered the Bayley test to chimpanzee infants raised in the three nursery settings of the Yerkes Research Center of Emory University (Bard and Gardner 1996; van IJzendoorn et al. 2009). At three to five months of age, chimpanzees oriented well to sights and sounds, and grasped and mouthed object, scoring significantly above human infants of the same age (humans score an average of 100 whereas chimpanzees scored an average of over 130 on the Mental Development Index of the Bayley test). By six to seven months, the chimpanzees from both standard-care (ST) and responsive-care (RC) nurseries were still scoring significantly above humans of the same age. At eight to nine months the chimpanzees overall were not different from eight- to nine-month human infants. But by 10–12 months, when turn taking, pointing to the parts of a doll, and completing puzzle boards were required, the chimpanzees scored significantly lower than 10- to 12-month-old humans.

Once again, we found differences among the chimpanzee nursery groups. At nine months, for example, the RC chimpanzees scored higher than nine-month-old human infants, but the ST chimpanzees score lower than the humans (van IJzendoorn et al. 2009). Moreover, the chimpanzees that experienced early stress performed significantly less well on the BSID throughout the first year of life, with a spike of fearful anxiety during testing at six to seven months. It is evident that the institutionalized rearing was responsible for many of the poor cognitive outcomes in year-old chimpanzees, as their rearing was impoverished (see, e.g., van IJzendoorn et al. 2009; Racine et al. 2008), especially compared with that of typical middle-class urban human infants, for whom the Bayley tests were devised.

It is interesting that the extent to which parents use objects in their interactions with young human infants varies widely across eco-cultural contexts. H. Keller and colleagues (2007) have demonstrated that for three-month-old infants living in rural subsistence eco-cultures (for instance, the Nso community in Cameroon) objects are much less often the focus of interactions (less than 1 percent of waking time compared with 2.5 percent in urban settings), and body contact is much more often experienced (~80 percent in rural Nso compared with 43 percent in urban German infants; p. 258). Bakeman et al. (1990) found !Kung (a foraging human culture) caregivers most often gave objects to their infants as a distraction, so that the caregivers can carry on with their work undisturbed by infants. E.

Menzel (1964a, b) found that chimpanzees raised in isolation exhibited extreme fear when they first encountered novel objects, and early rearing had an impact on tool use (E. Menzel et al. 1970). Thus, it is clear that (the lack of) previous experience with objects has a long-term effect on chimpanzee emotional engagement with objects. The need to habituate captive apes to novel objects is a constant reminder of this fact in experimental contexts.

Caregivers in biomedical nursery environments do not typically nurture enhanced object manipulation in chimpanzees (e.g., Vauclair and Bard 1983); however, inanimate objects are usually provided to stimulate well-being in captive chimpanzees from a very young age (e.g., Bard and Nadler 1983). In the Bayley (cognitive) tests, we scored the degree to which young chimpanzees manipulated objects with their hands. There were significant improvements from 3–5 months through 10–12 months in manipulations (Figure 14.1), and significant differences among the different nursery rearing groups, beginning at 6 months and continuing through the first year of life (Bard and Gardner 1996). The RC infants performed significantly bet-

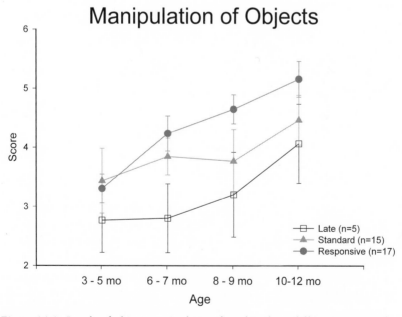

Figure 14.1. Levels of object manipulation found in three different groups of nursery chimpanzees during their first year of life. Group differences indicate that object manipulation varies as a function of early socioemotional experiences. Object manipulation also varies as a function of age. Adapted from Bard and Gardner 1996.

ter than the ST infants. Even in relatively basic skills, such as object manipulation, the early lived experiences of chimpanzees have a significant
impact.

Characteristics of 12-Month-Olds

Locomotor Ability

Chimpanzee infants begin to walk on (hands and) feet around five to seven
months of age. This early locomotor ability may have an important influence on the chimpanzees developing social cognition (Vauclair and Bard
1983). Human infants begin to walk (crawling then taking first steps)
around 12 months of age. This means when humans first begin to coordinate their engagement with objects and with social partners (i.e., coordinated joint attention; Adamson and Bakeman 1991) at nine months, they
are not yet able to independently locomote. Therefore they engage in a lot
of referential communication (e.g., asking caregivers to give them out-of-
reach objects). Chimpanzees at nine months appear also to be coordinat-
ing their engagement with objects and with social partners (along with
other tertiary circular reactions; see overview in Bard 1992), but since
chimpanzees are able to independently locomote, they appear to bypass
the Referential Problem Space (see diagram on page 188 in Leavens, Hopkins, and Bard 2005 for further details) and typically engage in less referential communication (Leavens, Hopkins, and Bard 2005; see also Vauclair
and Bard 1983). Therefore, understanding developmental processes—here
the interplay between locomotor (in)ability and the lack of need for using
referential communication, for example, pointing—helps us to make sense
of the findings of comparative studies.

Attachment

A major factor of early rearing, with long-term consequences for social
cognition, is the quality of attachment bonds. Van IJzendoorn and colleagues (2009) found that one-year-old chimpanzees exhibit the balance
of exploration (play with toys when the caregiver is present) and security
seeking (distress when separated from their caregiver) that is found in
one-year-old humans when tested in the Strange Situation Procedure. The
chimpanzee infants from Yerkes were primarily secure (54 percent), but
33 percent were classified with insecure-ambivalent, and 7 percent with
insecure-avoidant, when classifications were conducted. This compares
favorably to Ainsworth et al.'s (1978) original sample of U.S. middle-class
human infants, who were 57 percent secure (and 42 percent insecure), but

the relatively higher ambivalent classifications relative to avoidant classifications in the chimpanzees compares more favorably to Japanese samples with limited experiences in being alone and encountering strangers
(e.g., Miyake et al. 1985).

Children with disorganized attachment systems have serious problems
in childhood and can continue to have "poor emotional health" later in
life. One cause of disorganized attachments is a developmental history of
neglect or abuse (van IJzendoorn et al. 2009). Whereas 15–20 percent of
nonclinical and day-care human samples have disorganized attachments,
the percentage rises to over 60 percent for children living in Romanian,
Hungarian, or Greek orphanages (see Figure 14.2). By one year of age, the
attachment system of nursery-raised chimpanzees was more like that of
human infants raised in Romanian orphanages than like that of human
infants raised in middle-class families (van IJzendoorn et al. 2009). At one
year of age, chimpanzees with dysfunctional attachments were more likely
to exhibit stereotyped rocking and tended not to contact the caregivers
during reunion episodes but rather clutched towels (van IJzendoorn et al.

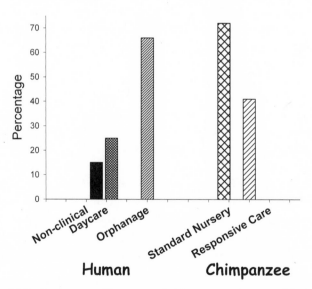

Disorganized attachment in SSP

Figure 14.2. Percentage of human and chimpanzee one-year-olds that have
disorganized attachments with favorite caregivers (measured in the Strange
Situation Procedure, SSP) as a function of early rearing environments. Adapted
from van IJzendoorn et al. 2009.

2009). Raising chimpanzees in biomedical centers significantly and negatively impacts their emotional and cognitive systems, with potentially long-lasting effects (e.g., Brune et al. 2006; Kalcher et al. 2008).

It is important to note explicitly that the attachment system is highly influenced by early rearing. Chimpanzees raised in the enriched RC nursery, compared with those raised in the standard care nursery at Yerkes, had significantly fewer dysfunctional attachments, had significantly higher nine-month cognitive scores, and were significantly less likely at nine months to exhibit an attachment to a towel. Early experiences with caregivers during the first year of life have important and long-lasting effects on developmental outcomes in chimpanzees.

Social Referencing and Cooperative Motivations

In social referencing experiments, when the infant looks to the caregiver, seeking information about the novel object, the caregiver is instructed to give an emotional message about the object. Joint Attention is demonstrated when the infant (chimpanzee in this case) is influenced in his or her behavior toward the novel object as a result of the caregiver's emotional message (e.g., Bard and Leavens 2009, 2011). In social referencing, the key is whether the chimpanzee associates the caregiver's message with his or her actions on the object. Russell et al. (1997) reported that (1) all the chimpanzees, from as young as 14 months, looked to their favorite caregiver and engaged in alternating gaze between the caregiver and a novel object. This indicates that the young chimpanzees, like human infants, actively seek information about the novel object from their caregiver; (2) moreover, their behavior is regulated by the emotional message that they receive. Young chimpanzees withdraw from a novel object when the caregiver gives a distressed facial and/or vocal expression, and chimpanzees look longer at the toy when the caregiver gives a happy message about the toy. Note that early rearing does not appear to influence the basic ability to follow another's gaze (Pitman and Shumaker 2009).

Social referencing is a form of secondary intersubjectivity, in which the infant demonstrates the ability to coordinate engagement with a social partner (attending to the emotional message given by the caregiver) with engagement with an object (to regulate interactions with the novel toy; Russell et al. 1997). Clearly there can be wide individual differences in caregivers' natural emotions toward objects in the world, and individual differences in infants' sensitivity to emotional messages. For example, in general, the nursery-raised chimpanzees of the Russell et al. study were more wary of novel objects than were middle-class human infants. Therefore, this developmental system of social referencing, common to chimpanzees and humans, can account for individual differences, as well as rearing-group differences in outcomes for human and chimpanzee infants.

Characteristics of 2- to 2.5-Year-Olds

Mirror Self-Recognition (Bard et al. 2006)

The identification of the self is a cognitive ability found in humans and great apes. Evolutionarily, mirror self-recognition is a cognitive capacity shared by chimpanzees and humans—but is not present in any species of monkey, and so it is thought to have evolved in the past 30 million years (for reviews see Bard et al. 2006; Inoue-Nakamura 2001). Developmentally, mirror self-recognition appears to develop around two years of age in humans and in chimpanzees, although there has been some debate concerning the age of acquisition for chimpanzees (see Bard et al. 2006).

An objective methodology to index self-awareness in nonverbal or pre-verbal primates, the mark-and-mirror test was proposed by Gallup (1970) for comparative psychologists, and independently by Amsterdam (1972) for developmental psychologists. The methods involve application of makeup (as used by Amsterdam for human children) or dye (as used by Gallup for nonhuman primates) to the face and placement of the marked individual in front of the reflective surface of a mirror. If the individual used the mirror image to touch the marked area on his or her own face, then this constituted behavioral evidence of understanding that the image in the mirror was the self, distinct from any other being.

In order to be more precise about age of onset, Bard et al. (2006) conducted a replication of an earlier study, extending the age range to include more young chimpanzees. In the design of this experiment, we were faced with a comparative/development dilemma: do we treat infant chimpanzees as if they were infant humans (and use the methods developed by developmental psychology), or do we treat infant chimpanzees as if they were adult chimpanzees (and use the methods developed by the comparative psychologists)? We decided to take a primarily developmental approach, choosing to minimize distress and not to render them unconscious during the marking. In fact, we marked the chimpanzees with Halloween makeup and tested them while they remained with their normal social partners, in their social groups and in their familiar rooms. The question was whether the infants could be exhibiting mark-directed behavior when they in fact did not have the cognitive ability, labeled "false positives."

Specifically, we proposed that the method of overt marking with makeup, while awake and in a social group, did not cause false positives. Two groups of human infants were tested—one group of 15-month-olds (none expected to pass) and a group of 24-month-olds (all expected to pass). Did this method create self-awareness in 15-month-olds? No, it did not! Even though the mark was placed on each child's face with the child's full awareness, not a single 15-month-old touched the mark while looking the in the mirror, or passed the mark test by any criteria, whether mark directed or self labeling. Even when the infants were prompted to respond, when their mother asked "Who's that in the mirror?" none of the 15-month-olds said their name or gave any other self label. Yet four of the five 15-month-old children demonstrated that they understood the task, by wiping the mark on the doll when asked to do so by the mother.

In contrast, half of the 24-month-olds passed the mark test, all of them labeled the self, and all understood the task. Viewing the data on mirror-guided mark-directed behavior separately from verbal self-labels presents another interesting finding. Only half the 24-month-olds actually touched the mark, when the literature suggests that most, if not all infants of this age possess self-awareness. The amount of mark touching was comparable in the 24-month children and the 28-month chimpanzees. This study raises questions about the age of onset of mark touching distinct from self-labeling in the human literature, and about the influence of the social-cultural environment on the development of self-naming in the mirror (e.g., does playing "name-the-self games" during the first 18 months of life facilitate passing the test?).

With a comparative developmental perspective applied to studies of self-recognition, we learn that self-awareness is a cognitive characteristic shared by humans and chimpanzees. Moreover, there are similarities in the developmental patterns, age of onset, mirror behavior, and probably socio-communicative variables that culminate in self-awareness for both human and chimpanzee infants (Bard et al. 2006). However, it appears that mirror self-recognition is a relatively robust phenomenon in humans, as it may well be found across all human cultures (but see H. Keller et al. 2004), and little evidence exists to date that rearing environment influences age of onset of self-recognition in chimpanzees.

Advanced Cognitive Skills (For Example, Language)

The literature on ape language and ape cross-fostered projects presents overwhelming evidence of the impact of early experiences on social and cognitive outcomes (see review in Leavens and Bard, 2011). Young chimpanzees have learned symbol systems comparable to those learned by two- to four-year-old human infants (e.g., Fouts and Mills 1997; Savage-Rumbaugh 1986). Young chimpanzees show self-recognition comparable to two-year-old human infants (reviewed in Bard et al. 2006). There are additional reports of triadic engagement in chimpanzees who do not learn symbol systems. For example, at Yerkes in the responsive care program (Bard 1996), young chimpanzees often worked jointly in activities such as opening a padlock with a key. Kellogg and Kellogg (1933) reported that their son, Donald, at 18.5 months, and their chimpanzee, Gua, at 16 months, often engaged in cooperative games, such as rolling a ball back and forth. These examples illustrate commonly observed instances of mutual triadic engagements when chimpanzees are raised with warm relationships and industrially manufactured objects. In contrast, when young chimpanzees are tested in conditions in which there are few triadic engagement opportunities with human testers, outcomes can be relatively poor (e.g., Tomonaga et al. 2004).

Summary

Emotional engagement plays a critically important role in the development of social cognition skills, by providing foundational motivation for the expression of competencies, or by actually co-constituting social cognition. Chimpanzees have inborn capacities for emotional engagement with caregivers and possess developmental processes, such as neonatal imitation and intersubjectivity, by which emotion becomes integrated in communication and cognition. In making cross-species comparisons, it is important to note the effects of early experiences and socialization practices, as they have a major impact on the nature and developmental trajectories of emotional engagements in primates, with potentially long-lasting effects. Socialization effects on expressions of emotion and on regulation of emotion are evident within the first month of life in chimpanzees (Bard et al. 2011); and by three months, socialization has an impact on many social, emotional, and interactional outcomes, such as mutual gaze (Bard et al. 2005). By six months, the effects of early experiences become evident in cognitive outcomes in chimpanzees (e.g., early stress compromises attention, leading to poor cognitive performance, whereas positive emotional experiences enhance cooperation, leading to enhanced cognitive outcomes; Bard and Gardner 1996). At 8–10 months, chimpanzees spontaneously exhibit joint engagement with caregivers and objects, and by one year, emotional attachments with specific caregivers have developed. When caregivers are emotionally responsive, chimpanzees' cognitive and emotional development is positively stimulated (van IJzen-doorn et al. 2009). Many chimpanzees raised in institutional settings, however, do not develop secure attachments, impairing their ability to emotionally engage, which can seriously compromise performance on tests of social cognition later in life. The study of development in primates is important for understanding cross-species comparisons in communication and cognition, especially social cognition.

Primary intersubjectivity, moreover, provides the foundation for the development of secondary intersubjectivity. Secondary intersubjectivity is when an infant can coordinate attention and action on an object with the attention and action with a social partner (reviewed in Trevarthen and Aitken 2001). We have evidence that chimpanzees engage in at least two different forms of secondary intersubjectivity: intentional communication and social referencing. Intentional communication is evident when chimpanzees point to an object while alternating gaze between the object and the person (Leavens, Hopkins, and Bard 2005, 2008). Although Leavens and colleagues have not conducted developmental studies of intentional communication, fieldwork strongly suggests that referential behavior is evident as early as 9–12 months in wild chimpanzees, the same age as in human infants (van Lawick-Goodall 1968).

I use the comparative developmental perspective to compare results from my own studies with those reported in the literature to arrive at an informed view of how development milestones in chimpanzees compare with those in humans, and how developmental history of engagements might impact outcomes. Overall, these studies show that young chimpanzees compare favorably to humans as newborns through to 2.5 years olds. Chimpanzees' earlier development of motor skills, independently locomoting by 5–6 months, in contrast to the 11–12 months for humans, has consequences in their developmental entry into the Referential Problem Space (Leavens, Hopkins, and Bard 2005, 2008), and their advanced manipulative ability has consequences in assessments of early cognition. However, even early developmental milestones in chimpanzees, as in humans, are significantly influenced by early experiences.

Much of our scientific knowledge of "universal" psychological outcomes of humans is based on mono-cultural data (e.g., Henrich et al. 2010). Across many domains of adult cognition, Western adults were found to be extreme outliers when compared with humans raised in non-Western cultures (Henrich et al. 2010). Leavens, Bard, and Hopkins (2010) suggest that conclusions drawn from studies using laboratory chimpanzees similarly be viewed with caution. As reviewed in this chapter, emotional and cognitive outcomes of chimpanzees are sensitive to socioemotional engagement experiences. The search for universal or species-unique social cognition in chimpanzees must be conducted with consideration of groups raised across many different "lived experiences."

Acknowledgments

Research reported in this chapter was funded in part by NIH grants RR-00165, RR-03591, RR-06158, and HD-07105, a Max-Planck Society stipend in collaboration with H. Papousek, and by European Commission FP6-IST-045169. I extend grateful thanks to the many researchers and colleagues who have assisted with the studies reported here. In particular, I thank Duane Rumbaugh for recent discussions of lived experiences, and especially thank David Leavens for constructive and extensive discussions about many of the developmental comparative topics presented in this chapter.

CHAPTER FIFTEEN

Distress Alleviation in Monkeys and Apes: A Window into the Primate Mind?

Filippo Aureli and Orlaith N. Fraser

Introduction

A large body of research has dealt with stress and its negative consequences. Much less is known about how animals deal with its alleviation. Stress is a multifaceted phenomenon, and various terms have been used to capture the various aspects. Although the terminology has overall been useful, it has also created some confusion. For example, one of the definitions of stress in the *Webster's Encyclopedic Unabridged Dictionary* (1989) is "physical, mental or emotional strain or tension," whereas distress is defined as "great pain, anxiety or sorrow," and anxiety as "distress or uneasiness of mind caused by apprehension of danger or misfortune." There is obviously a lot of overlap between the definitions of these concepts.

The term "stress" has been used to describe the biological response elicited when an individual perceives a threat to its homeostasis (Moberg 2000). In this respect, stress is an adaptive response. However, "stress" has also been used to refer to the stimulus that provokes a response, in addition to the resulting changes induced by the stimulus. Therefore, it has been suggested that the term "stress" should be replaced with the terms "stressor" for the stimulus and "stress response" for the resulting physiological changes (Creel 2001). In addition, not all stressors are detrimental to an individual, and the stress response can be beneficial (e.g., Engh et al. 2006b). This "good stress" has been labeled "eustress" (Selye 1974). By contrast, when a stressor has a negative impact on the individual's welfare, "distress" may be a more appropriate label (Moberg 2000). Given this distinction and the apparent overlap of definitions shown above, we decided to use the term "distress" in the rest of the chapter to capture the negative consequences of stressors.

In this chapter we focused on distress-alleviating mechanisms as a window into the primate mind. As distress alleviation is a composite area of research, it was beyond the scope of this chapter to review the entire body of literature encompassing social buffers (e.g., Kikusui et al. 2006), coping styles (e.g., Koolhaas et al. 2007), and the role of environmental enrichment (Carlstead and Shepherdson 2000). Instead, we mainly focused on recent findings showing that nonhuman primates cope with distress by self-alleviation or alleviation provided by other group members. We highlighted short- and long-term distress alleviation effects and discussed their implications for empathy and social networks. We believe that several of the mechanisms we covered are not restricted to primates. Thus, we used a comparative perspective and included relevant examples from other taxa. Distress can be detected and measured in a variety of ways, and the type of measure used may have implications for the conclusions drawn. Here we focus primarily on studies that have used noninvasive physiological measures (e.g., fecal glucocorticoids, or GC) or behavioral measures (rates of self-directed behavior, or SDB) that have been shown to be reliable indicators of distress in primates. (See Appendix to this chapter for a review of all measures and their implications.)

Distress Alleviation

Many mechanisms for distress alleviation have been suggested and demonstrated in human and nonhuman animals. For example, coping mechanisms range from active/passive styles (Koolhaas et al. 2007) to avoidance, inhibition of social activity, and increased affiliation, depending on the duration of crowding (Judge 2000). Another example is social support, in which the behavior, or the mere presence, of a familiar individual functions in decreasing distress (reviewed in Cheney and Seyfarth 2009). Sapolski's studies provide a well-integrated example: male olive baboons who had social support, social control, and were better at predicting the outcome of their interactions coped best with distress (Sapolsky 1998). A detailed review of these mechanisms can be found in Cheney and Seyfarth (2009). Here we review recent developments in how grooming other individuals, signaling benign intent, and using post-conflict mechanisms are effective in distress alleviation.

Grooming and Grunting

Grooming other individuals is probably the most common social behavior among primates and, possibly, among other mammals and birds (Goosen 1987; Kutsukake and Clutton-Brock 2006; Radford and Du Plessis 2006). It has clearly a hygienic function in terms of removal of ectoparasites (Zamma 2002). It is also generally considered a mechanism to establish and maintain social bonds between group members, being exchanged for further grooming or other services (reviewed in Schino 2007; Schino and Aureli 2008). This social function may be mediated by the release of brain opioids such as b-endorphins (Keverne et al. 1989).

The usually relaxed posture of the groomee and the massage-like actions of the groomer are probably what inspired Terry (1970) to propose a tension-reduction function for grooming. A distress alleviating function has recently been supported by studies using physiological and behavioral measures. In studies of pigtail macaques *(M. nemestrina)* and rhesus macaques *(M. mulatta),* individuals receiving grooming showed a reduction in heart rate when compared with appropriate control conditions (Boccia et al. 1989; Aureli et al. 1999). This physiological evidence is further supported by a study using SDBs. Long-tail macaques were found to display lower levels of SDBs after they received grooming (Schino et al. 1988).

Approaching other group members and being in close proximity to others are potentially stressful events given the higher risk of aggression. In support of this view, macaques, baboons, capuchin monkeys, and chimpanzees display more SDBs and increased heart rate when in proximity with group members, depending on their relative dominance rank and degree of familiarity and affiliation (e.g., Aureli et al. 1999; Castles et al. 1999; Kutsukake 2003). Thus, grooming may function as a mechanism to alleviate or even prevent distress under these circumstances. However, whereas there is evidence that grooming provides relief to the groomee (see above), until recently there was no indication that grooming is beneficial to the groomer in terms of distress alleviation. The first evidence for short-term distress reduction in the groomer was found in a study of female grooming in crested black macaques *(M. nigra)* (Aureli and Yates 2009). Seven adult females displayed lower levels of self-scratching and self-grooming soon after they stopped grooming a group member than they did in other conditions (Figure 15.1). In addition, the former groomee was more likely to be the nearest neighbor of the former groomer in the 10 minutes after the end of grooming, compared to control conditions. Such an increase in tolerance may mediate the short-term distress-alleviating effect of grooming others (Aureli and Yates 2009). Another short-term effect may explain the pattern of highest intragroup allopreening rates in green woodhoopoe *(Phoeniculus purpureus)* groups that have the greatest involvement in intergroup conflicts (Radford 2008). This effect was mainly due to the post-conflict increase in allopreening. Group members allopreened more following long conflicts and those they lost compared with short conflicts and those they won, perhaps because after the former there was more need for distress alleviation (Radford 2008).

There is also evidence that grooming others may have a longer effect. In female Barbary macaques *(M. sylvanus),* giving rather than receiving grooming was associated with lower fecal GC levels. Both the duration of grooming given and the number of grooming partners were negatively associated with fecal GC levels (Shutt et al. 2007). Because fecal GC levels reflect the cumulative result of stressors, this effect is likely linked to the distribution of grooming over time, that is, grooming networks. Studies on female chacma baboons found that grooming networks, which focus on a few predictable partners, were associated with lower GC levels (Crockford et al. 2008). During periods of instability, all females contracted

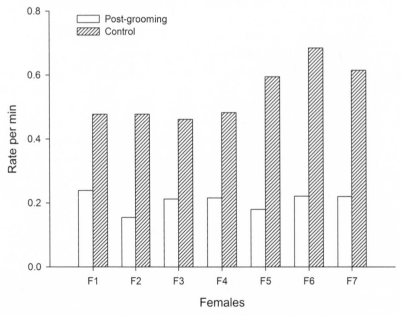

Figure 15.1. Rates of self-scratching by seven adult female crested black macaques during post-grooming and matched-control 10-minute periods. In the post-grooming periods the seven females had just terminated to groom another group member. Data from Aureli and Yates (2009).

their grooming network even further, and females whose grooming network had been more focused experienced smaller increases in GC levels (Wittig et al. 2008). Given that concentrated grooming networks are critical for distress alleviation, female baboons experienced high GC levels when this network was damaged by the death of a close grooming partner (Engh et al. 2006a). Distress was transitory, however, as females sought out social contact and increased the number of grooming partners following this loss. Whereas it is clear that receiving and giving grooming alleviate distress, it is uncertain whether the mere social contact is sufficient or whether the identity of the partners receiving and giving grooming is an essential issue (e.g., the concentrated grooming network with a few select companions; Cheney and Seyfarth 2009).

Vocalizations, such as grunts in baboons and certain macaques, are signals of benign intent and are associated with a low probability of aggression from the vocalizing individual (Cheney et al. 1995). For example, mothers of young infants receive grunts by a higher-ranking female during approaches to handle their infants and are then less likely to receive aggression (Silk et al. 2003). Such vocalizations can then play a role in distress alleviation. There is evidence that pregnant and cycling female baboons who receive grunts at a high frequency from approaching dominant females have lower GC levels than those who receive grunts at a lower frequency (Crockford et al. 2008). Thus, grunts may alleviate distress associated with aggression in part because they permit females to assess and predict the behavior of higher-ranking animals (Cheney and Seyfarth 2009).

In this section, we reviewed evidence emphasizing that both initiating and receiving social interactions are important in providing distress alleviation. In the next section we separately treat self-alleviation and alleviation provided by others. We first focus on how distress spreads to bystanders (i.e., individuals witnessing, but not participating in, a social interaction), as this effect can be relevant to how actions aimed at self-alleviation may result in providing alleviation to others.

Effect of Distress on Bystanders

In recent years there has been an increased interest in the impact of social exchanges on bystanders. Witnessing one individual experiencing an event can affect the behavior and emotion of other individuals. For example, in an experimental setting, a Japanese macaque (the observer) could watch a conspecific target monitoring an unfamiliar third monkey who was out of the observer's sight. The observer scratched itself more often when the target self-scratched while monitoring the third monkey than when the target monitored without scratching (Nakayama 2004). The finding indicates a contagion of self-scratching from the target to the observer, suggesting a possible contagion of distress.

The mere observation of events does not only affect bystanders' behavior, but also elicits physiological changes. For example, greylag geese *(Anser anser)* experience a higher increase in their heart rates when watching social events, such as departing or landing geese and aggressive interactions, than when witnessing nonsocial events, such as vehicles passing by and loud noise (Wascher et al. 2008). The response to social events differs depending on the quality of the relationship (see below) that the bystander has with the individuals involved in the events. The heart-rate increase was higher when the bystander was watching aggressive interactions involving close associates than when watching those involving less-close individuals (Wascher et al. 2008). Similarly, in a series of elegant but invasive experiments, Miller (1967) documented the behavioral and physiological responses of rhesus macaques when observing a conspecific companion facing a positive or negative stimulus. The facial expression of the companion conveyed the nature of the stimulus, and the observer monkey's heart rate changed accordingly. Most interestingly, when the stimulus was negative, the acceleration of the observer monkey's heart rate quickly followed the companion's increase in heart rate.

These findings suggest that bystanders experience emotional changes that may be at the basis of simple forms of empathy and serve in promoting distress-alleviation mechanisms. This view can somewhat be supported by negative results. In an attempt to find evidence for bystanders' involvement in post-conflict interactions in monkeys (see below), Schino et al. (2004) monitored the behavioral responses of the mother of the recipient of aggression in Japanese macaques. They did not find evidence for an increase in the mother's SDBs after the aggression compared to control periods, whereas the offspring recipient of aggression increased its SDBs. The lack of the mother's emotional response was accompanied by a lack of initiating affiliative contact with her distressed offspring (Schino et al. 2004).

Post-Conflict Mechanisms of Distress Alleviation

Former opponents may experience distress after aggressive conflict as a result of the damage to their relationship, and thus the loss of benefits

Table 15.1 The distress-alleviating function of post-conflict interactions: Grooming given and grooming received in published studies of primate species

Interaction	Studied Species	Distress-alleviation function demonstrated?	Measure of distress	Source
Reconciliation	Macaca fascicularis	Yes	SDB	Aureli, van Schaik, & van Hooff 1989
		Yes	SDB	Das et al. 1998
	Macaca fuscata	Yes	SDB	Kutsukake & Castles 2001
	Macaca mulatta	Yes	Heart rate	Smucny et al. 1997
	Macaca radiata	Yes	SDB	Cooper et al. 2007
	Papio hamadryas anubis	Yes	SDB	Castles & Whiten 1998
	Papio hamadryas hamadryas	Yes	SDB	Romero, Colmenares, & Aureli 2009
	Pan troglodytes	Yes	SDB	Fraser, Stahl, & Aurelli 2010
		No	SDB	Koski, Koops, & Sterck 2007
	Homo sapiens	Yes	SDB	Fujisawa et al. 2005
		Yes	sGC	Butovskaya 2008
Redirected aggression	Macaca fascicularis	Yes	SDB	Aureli & van Schaik 1991b

Table 15.1 (continued)

Interaction	Studied Species	Distress-alleviation function demonstrated?	Measure of distress	Source
Quadratic affiliation	*Papio hamadryas hamadryas*	Yes	SDB	Judge & Mullen 2005
Bystander affiliation toward victim (consolation?)	*Pan troglodytes*	No Yes	SDB SDB	Koski & Sterck 2007 Fraser, Stahl, & Aureli 2008
Affiliation between bystander and aggressor	*Macaca fascicularis* *Papio hamadryas hamadryas*	No No	SDB SDB	Das et al. 1998 Romero et al. 2009
Grooming received	*Macaca nemestrina*	Yes	Heart rate	Boccia, Reite, & Laudenslager 1989
	Macaca mulatta	Yes	Heart rate	Aureli, Preston, & de Waal 1999
	Macaca fascicularis	Yes	SDB	Schino, Scucchi, et al. 1988
Grooming given	*Macaca sylvanus*	Yes	fGC	Shutt et al. 2007
	Macaca nigra	Yes	SDB	Aureli & Yates 2009

Notes: For post-conflict interactions, only studies that demonstrated an increase in distress following aggressive conflict have been included.
SDB = self-directed behavior; fGC = fecal glucocorticoids; sGC = salivary glucocorticoids.

afforded by the relationship, as well as the risk of renewed aggression (reviewed in Arnold et al. 2010). A stress response may enable the conflict participants to cope with the uncertainty of the post-conflict situation, but it could also lead to distress (Aureli and van Schaik 1991b). Accordingly, in many primate species, elevated post-conflict SDB levels (and saliva cortisol levels in humans; Butovskaya 2008) have been reported for both the recipient of aggression and, often to a lesser extent, the aggressor (Arnold et al. 2010). In Japanese macaques, former opponents experienced negative effects in terms of increased hostility and reduced approach, proximity, and grooming rates for at least 10 days following the initial aggressive conflict (Koyama 2001). Despite these and other possible costs, aggressive conflict is prevalent in the lives of all group-living animals, possibly because some of these costs may be mitigated through post-conflict interactions. In particular, some such interactions may function to alleviate post-conflict distress. As the majority of studies have focused on the distress experienced by the recipient of aggression, we focus here on distress-alleviating post-conflict mechanisms for recipients of aggression, although similar mechanisms may also be applicable to aggressors (Table 15.1).

Self-Alleviation Mechanisms

RECONCILIATION

After aggressive conflicts, rather than dispersing, former opponents may in fact approach and affiliate with each other. Such post-conflict reunions are termed "reconciliation" (de Waal and van Roosmalen 1979), implying a relationship repair function, and indeed such reunions have been shown to restore the opponents' relationship by reducing the risk of renewed aggression and restoring affiliation and tolerance around resources to baseline levels (reviewed in Arnold et al. 2010). Moreover, reconciliation alleviates post-conflict distress. SDB levels in the recipient of aggression were lower following reconciliation than during post-conflict observations in which reconciliation does not occur (Table 15.1; Figure 15.2; reviewed in Aureli et al. 2002; Arnold et al. 2010). Similar findings in school-age children (Fujisawa et al. 2005) have been supported using cortisol levels (Butovskaya 2008). Heart rate has also been shown to follow a similar pattern in rhesus macaques (Aureli and Smucny 2000). By repairing the opponents' relationship, restoring its benefits, and reducing the risk of renewed aggression, reconciliation may reduce the uncertainty about the post-conflict situation and thus alleviate post-conflict distress experienced by the recipient of aggression (Aureli and van Schaik 1991b).

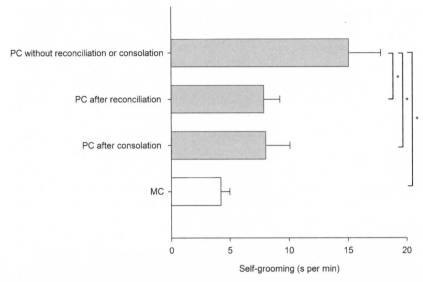

Figure 15.2. Mean (±s.e.) self-grooming levels in chimpanzees for PCs without reconciliation or consolation, for MCs, and for minutes 2–10 of PCs where either only reconciliation or only consolation has occurred in the first minute. Data from Fraser et al. (2008, 2010).

REDIRECTED AGGRESSION

Attempting reconciliation is likely to entail a risk of renewed aggression in itself, and thus if the benefits of relationship repair upon reconciliation do not outweigh the costs, alternative post-conflict distress-alleviating interactions may occur. Recipients of aggression may redirect aggression to other group members, thus reversing the negative effects of losing the previous conflict and reducing the likelihood of becoming targets of renewed aggression from the aggressor or other group members (Aureli and van Schaik 1991a; Kazem and Aureli 2005). Reducing the likelihood of renewed aggression may alleviate post-conflict distress in the original recipient of aggression. For example, among wild olive baboons, males who frequently respond to losing fights by redirecting aggression against others have significantly lower basal GC levels and a better response to acute challenges than similarly ranked individuals who do not exhibit this behavior pattern as frequently (Virgin and Sapolsky 1997). However, the extent to which this can be attributed to beneficial consequences of redirection per se is unclear, as redirection was only one of a suite of traits for coping with male-male competition that characterized these individuals. An example of a more direct distress-alleviating function is the reduction of post-conflict SDB levels in long-tailed macaques following redirected aggression compared to when neither reconciliation nor redirected aggression took place (Aureli and van Schaik 1991b). By reducing the distress of the original recipient of aggression, redirecting aggression may even facilitate reconciliation, as the former was less likely to occur after the latter took place (as it was no longer necessary), but reconciliation was more likely to occur after redirected aggression occurred (Aureli and van Schaik 1991a). The reduction of distress through redirected aggression may also be achieved by signaling to the former aggressor and bystanders that the original recipient of aggression has not been unduly compromised by its preceding defeat and thus is not an easy target for further aggression (Kazem and Aureli 2005).

QUADRATIC AFFILIATION

Recipients of aggression and aggressors are not the only group members to become distressed after an aggressive conflict. Bystanders may also become distressed, possibly as a result of the risk of further aggression, either redirected by the victim or directed by another group member. This may be particularly relevant to the opponents' kin, as they are more likely to become involved in an aggressive conflict after a conflict between their relatives (Cheney and Seyfarth 1989; Aureli et al. 1992). However, such groups of kin may also be more likely to affiliate with each other after their relatives have been involved in an aggressive conflict (Cheney and Seyfarth 1989; Judge 1991). As the conflict between two individuals subsequently affects the behavior of two other individuals, this interaction was termed "quadratic affiliation" (Judge and Mullen 2005). In hamadryas baboons *(Papio hamadryas hamadryas)*, bystanders increased affiliative contact with other non-kin bystanders, who were preferred social partners, following an aggressive conflict between other group members. Bystanders' SDB levels were elevated above baseline levels during the post-conflict period but were reduced following affiliation with other bystanders, suggesting that such quadratic affiliation has a distress-alleviating function (Judge and Mullen 2005). Bystanders may therefore seek to alleviate distress experienced by witnessing a conflict between other group members by affiliating with partners with whom they share a good relationship. While quadratic affiliation has also recently been demonstrated in two groups of Tonkean macaques *(Macaca tonkeana)*, bystanders did not show elevated post-conflict levels of SDB in this species (De Marco et al. 2010). Tonkean macaques are characterized by a more relaxed dominance style than hamadryas baboons, which may lead to lower risks of aggression for Tonkean macaque bystanders and thus less pronounced post-conflict distress. Alternatively, quadratic affiliation may play a more important role in maintaining social cohesion than in distress alleviation for some species (De Marco et al. 2010).

Alleviation Provided by Others

CONSOLATION

As well as affiliating with each other after conflicts, bystanders may affiliate with the conflict opponents (known as "post-conflict bystander affiliation"; Fraser et al. 2009). In chimpanzees, for example, bystanders may embrace or kiss the recipient of aggression after a conflict (de Waal and van Roosmalen 1979; Fraser and Aureli 2008). The function of such contacts may vary according to the quality of the relationships between those involved and the role of the opponent in the conflict (Das 2000; Watts et al. 2000; Fraser et al. 2009). Post-conflict affiliation directed from a

Figure 15.3. Example of affiliative behavior typical of post-conflict consolation in chimpanzees. Photo by Orlaith N. Fraser.

bystander to a recipient of aggression is likely to alleviate the recipient's distress and thus function as "consolation" (Figure 15.3; de Waal and van Roosmalen 1979). In chimpanzees, recipients of aggression showed lower post-conflict SDB levels following affiliation from a bystander than either before bystander affiliation or if such affiliation did not occur (Figure 15.2; Fraser et al. 2008), providing support for a distress-alleviating function and thus the label "consolation." Unlike redirected aggression, however, consolation does not appear to facilitate reconciliation, as former opponents may actually be less likely to reconcile after consolation occurs (Fraser et al. 2009). Consolation may thus function as an alternative distress-alleviating mechanism to reconciliation when the risks of renewed aggression upon reconciliation are too high (Wittig and Boesch 2005; Fraser et al. 2009). Whereas the term "consolation" in a post-conflict context refers to affiliation from a bystander to the recipient of aggression, affiliation from a bystander to the aggressor may serve a similar distress-alleviating function, although evidence for such a function is lacking (Das et al. 1998; S. Koski and Sterck 2007). Equally, it is possible that in reducing the recipient's distress, the bystander simultaneously alleviates its own distress, either as a direct result of the affiliative contact or because the bystander's distress was contingent upon the distress of the recipient of aggression (see below).

Relationship Quality and Emotional Mediation

Distress alleviation is not necessarily a simple consequence of affiliative contact with any individual, as the identity of the affiliating partners with the distressed party may play a pivotal role in the effectiveness of such contact. Long-tail macaque aggressors experienced distress after aggressive conflict that was not alleviated by affiliation from bystanders, but was alleviated by affiliation with the former opponent (Das et al. 1998). Similarly, hamadryas baboon aggressors sought post-conflict affiliation with bystanders, but only affiliation with the former recipient of aggression alleviated their distress (Romero et al. 2009). The quality of the relationship between the opponents may also affect the degree of post-conflict distress experienced by the opponents. For partners who share a more valuable relationship, the costs of losing the benefits afforded by that relationship are higher, and thus they are likely to experience a higher degree of post-conflict distress than opponents who share a less valuable relationship. The relatively high level of post-conflict distress experienced by opponents who share a valuable relationship may in fact serve as a facilitator or "emotional mediator" for the occurrence of reconciliation by motivating both opponents to take steps to repair their relationship (Aureli 1997). In support of this view, individuals displayed higher SDB levels after conflicts with higher-quality partners than after conflicts with lower-quality partners (Aureli 1997; Kutsukake and Castles 2001; Cooper et al. 2007; S. Koski et al. 2007; but see Schino et al. 2007). If reconciliation is mediated by the level of post-conflict distress, reconciliation is expected to be more likely after conflicts between valuable partners, who would be most distressed and for whom relationship repair would afford the greatest benefits (Aureli 1997). Accordingly, evidence supporting increased conciliatory tendencies between partners who share a valuable relationship is abundant and consistent across species (Aureli et al. 2002; Watts 2006; Arnold et al. 2010).

The distress-alleviating function of post-conflict affiliation between opponents and bystanders may be equally contingent upon the quality of the relationships between those involved. As reviewed above, post-conflict bystander affiliation may not reduce distress at all (Das et al. 1998; S. Koski and Sterck 2007; Romero et al. 2009), and only one study has thus far found support for this function (Fraser et al. 2008). Bystanders may affiliate with conflict participants for a number of reasons, including self-protection from redirected aggression, kin-mediated or triadic reconciliation, reinforcement of an alliance, appeasement, or consolation (Das 2000; Watts et al. 2000; Fraser et al. 2009). Whether such contacts reduce distress may depend upon the relationship between the conflict participant and the bystander and whether the participant is distressed, which is likely to be dictated primarily by the quality of the opponents' relationship and conflict characteristics. As expected, when post-conflict affiliation by bystanders with the recipient of aggression reduced behavioral indicators of distress (thus functioning as consolation) in chimpanzees, the bystanders (consolers) were found to be more likely to share a valuable relationship with the recipient of aggression (Fraser et al. 2008). Although, among primates, bystander affiliation provided by valuable or close partners has so far been found only in chimpanzees (Fraser et al. 2008; Romero and de Waal 2010), such affiliation was provided by pair mates in rooks (*Corvus frugilegus;* Seed et al. 2007) and by those sharing a good relationship with the recipient in wolves (*Canis lupus;* Palagi and Cordoni 2009). Additionally, valuable partners in ravens *(Corvus corax)* affiliated with recipients of aggression after more-intense conflicts, when such recipients were likely to be more distressed, and thus in need of consolation (Fraser and Bugnyar 2010).

Valuable partners may be more responsive to each other's distress as empathic responses are likely to be modulated by the quality of the partners' relationship (Aureli and Schaffner 2002), and social closeness or similarity has been shown to promote empathy (Preston and de Waal 2002). The degree to which chimpanzees and other primates are capable of empathy, however, is debated (Preston and de Waal 2002; Silk et al. 2005; Penn and Povinelli 2007b; Warneken et al. 2007; de Waal 2008), and a potential incapacity for the degree of empathy necessary for providing consolation has been suggested as a possible reason for why consolation does not appear to occur in any monkey species (de Waal and Aureli 1996; Schino et al. 2004; de Waal 2008). Consolation is thought to reflect a degree of empathy known as "sympathetic concern," which differs from simpler forms of empathy and personal distress in that the emotions of the two parties are distinct, which suggests that the consoler acts out of concern for the distressed party, rather than out of a selfish motive to alleviate its own arousal as a result of the partner's distress (de Waal 2008). The degree to which consolers themselves are distressed, however, is unknown, although hamadryas baboon bystanders, unlike Japanese macaque victims' mothers (Schino et al. 2004), showed distress following conflicts between other group members (Judge and Mullen 2005). Furthermore, while relationship quality may be a mediating factor in the occurrence of consolation, evidence that heart rate in geese increases more when observing their pair mate involved in aggressive conflict (Wascher et al. 2008) suggests that the quality of the relationship between a conflict opponent and a bystander may also mediate the bystander's distress.

Conclusions

Distress may be alleviated in many different ways, and the type of distress-alleviation mechanism is likely to be linked to the context of distress and, in particular, whether short-term or long-term alleviation is required. Although the negative effects of aggressive conflict on the relationship between former opponents can last at least 10 days (Koyama 2001), a single brief affiliative interaction between them may repair their relationship and alleviate distress (Aureli et al. 2002; Arnold et al. 2010). It is unlikely, however, that such an interaction on its own would be effective in alleviating chronic distress, such as that experienced by lactating baboons when the risk of infanticide is high, for which a focused grooming network of close partners appears to be most effective (Wittig et al. 2008). The difference between behaviors involved in short-term and long-term distress alleviation, however, is subtle, and indeed a clear distinction may not exist. Female baboons who received more grunts from dominants had lower GC levels over an eight-month period, suggesting that grunts may alleviate chronic distress associated with a high risk of aggression (Cheney and Seyfarth 2009). However, such grunts also function as reconciliation after aggressive conflicts (Cheney et al. 1995), thus acting as both a short-term and long-term distress alleviation mechanism. Similarly, grooming has been associated with low fecal GC levels (Shutt et al. 2007; Wittig et al. 2008), but has also immediate effects in reducing SDB levels (Schino et al. 1988; Aureli and Yates 2009) and heart rate (Aureli et al. 1999) and is one of the behaviors commonly used for reconciliation (Arnold et al. 2010). Thus, the difference between short- and long-term distress alleviation may not be related to the behaviors used but rather to the quality of the interactions. Accordingly, receiving grooming has a short-term distress-alleviating effect (Schino et al. 1988; Aureli et al. 1999), and thus an increase in the amount of grooming over time may be expected to have a long-term distress-alleviating effect. However, it is the selectivity of grooming partners, and thus the size and quality of grooming networks, as opposed to grooming rates, that is correlated with long-term distress alleviation (Crockford et al. 2008; Wittig et al. 2008).

Although this review has focused on grooming, grunting, and post-conflict interactions as distress-alleviation mechanisms, many other such mechanisms exist. We know, for example, that the rates of brief, affiliative behaviors (e.g., kisses, embraces, or gentle touches) and play increase during periods of tension, such as prior to scheduled feeding in provisioned groups (de Waal 1992; Palagi et al. 2006), suggesting that these behaviors may also play a distress-alleviating role. When subgroups of spider monkeys *(Ateles geoffroyi)* fuse, some members of each subgroup embrace one another (Aureli and Schaffner 2007). Under such circumstances, embraces may help to alleviate distress by reducing the risk of post-fusion aggression (Aureli and Schaffner 2007). Similar "greetings" occur when group members meet in many species—for example, *Papio hamadryas* (Colmenares 1990); *Colobus guereza* (Kutsukake et al. 2006); *Alouatta palliata* (Dias et al. 2008)—but whether they alleviate distress is as yet unknown.

Providing distress alleviation to others may be viewed as one of the best examples of expression of empathy in nonhuman primates (see above). In this respect distress-alleviation mechanisms may offer a window into the primate mind. Whereas there is increasing support for these mechanisms to be emotionally mediated, evidence is still missing about the ability to distinguish one's own emotions from the distress of others. Findings that bystanders change their behavior and physiology when witnessing others in distress are a first step in demonstrating "sympathetic concern" (de Waal 2008). The key aspect for future research is to examine whether the actions that lead to distress alleviation of others are based on a genuine concern for the distressed party, instead of a selfish way to alleviate one's own distress caused by witnessing the partner in distress.

If distress alleviation is mediated by empathy, then it can be viewed as an altruistic behavior. De Waal (2008) eloquently argued that empathy may motivate altruism, with individuals providing potentially costly behavior, such as supporting others in aggressive conflict, sharing food, and rescuing others from risky situations, out of concern for distressed group members. Self-alleviation of distress can also play a role. For example, the groomer's distress alleviation (Aureli and Yates 2009) fits the perspective that under certain conditions delivering benefits to others can be self-rewarding (de Waal, Leimgruber, and Greenberg 2008). Hence, the long-term benefit of social networks (Shutt et al. 2007; Crockford et al. 2008) may be subserved by a proximate mechanism based on the immediately rewarding consequences of giving grooming (Aureli and Yates 2009). This reasoning may be applicable to other distress-alleviation mechanisms, such as consolation, in which the bystander may be immediately rewarded with the reduction of its own distress for alleviating the partner's distress. Offering distress alleviation can then be reciprocated in kind or with other beneficial behaviors at a later stage (Romero et al. 2010), providing a reinforcing mechanism for the maintenance of strong social networks.

Appendix: Indicators of Distress

In order for distress to be detected, or for the effects of potential distress-alleviating mechanisms to be ascertained, we need reliable methods for quantifying individual levels of distress (for reviews see Aureli and Whiten 2003; Honess and Marin 2006). As invasive techniques—involving, for example, physiological assessment of distress from blood glucocorticoid levels—are likely to directly influence the level of distress experienced by the individual, noninvasive methods are preferable. Noninvasive quantitative measures of distress can be split into two broad categories: those that measure physiological correlates of distress, and those that use behavioral indicators as an indirect measure of distress.

The most common application of behavioral indicators of distress is the measurement of self-directed behaviors (SDBs) in primates, such as self-grooming, self-scratching, yawning, or body shaking. These behaviors have been reported to occur at above-baseline rates during periods of tension or uncertainty, such as when making decisions over the direction of travel (Kummer 1968) or whom to groom (Smuts 1985). The above-baseline rates of SDBs are displayed without any apparent contextual significance and appear to be related to autonomic arousal (Maestripieri et al. 1992; Troisi 2002). Direct evidence that SDBs reflect distress-related physiological changes in primates is lacking. There is, however, a growing body of circumstantial evidence that suggests that SDBs are reliable indicators of distress in primates.

Pharmacological studies have shown that rates of self-scratching in long-tailed macaques *(Macaca fascicularis)* and black tufted-ear marmosets *(Callithrix pencillata)* increased on administration of anxiogenic drugs, while anxiolytic drugs decreased rates of self-scratching (Schino et al. 1996; Barros and Tomaz 2002), providing strong support for the link between SDBs and anxiety. Rates of self-scratching have been shown to increase following aggression (Aureli et al. 1989), when heart rates are also elevated (Boccia et al. 1989). Heart rates and scratching rates have also both been shown to increase with risk of aggression, such as when in proximity to a dominant or nonaffiliative individual (Aureli et al. 1999; Castles et al. 1999; Kutsukake 2003). SDBs may be particularly useful in assessing distress in primates, as they are sensitive to the degree of distress experienced by the individual. Using an experimental approach, rates of self-scratching by a chimpanzee were found to increase with the difficulty of a task and the number of incorrect responses given (Leavens, Aureli, and Hopkins 2004). Levels of SDBs have also been used in primates to assess relationship security (Castles et al. 1999; Kutsukake 2003), the effects of crowding (e.g., van Wolkenten et al. 2006), and maternal style (Maestripieri 1993).

Direct, noninvasive, physiological measures of distress are commonly obtained by measuring glucocorticoid (GC) levels in feces or urine, or from saliva if the animals are trained. GCs are secreted by the adrenal glands in response to a physical or psychological stressor, forming part of the initial "alarm reaction" provoked by the hypothalamic pituitary adrenal system that diverts resources away from noncritical functions toward those enabling a return to homeostasis (Sapolsky 1998; Moberg and Mench 2000). Fluctuations in GC levels as a result of natural circadian variation, considerable variation in individual GC levels (Moberg and Mench 2000), and the delay from the time GCs are secreted by the adrenal glands to the time when they can be detected and variation therein can place major constraints on experimental design (Honess and Marin 2006). Nevertheless, GC levels have been successfully associated with increased competition during mating seasons, predation risk, dominance rank, and infanticide (reviewed in Cheney and Seyfarth 2009).

Animals may respond to stressors through behavioral, neuroendocrine, autonomic, and/or immunological means (Honess and Marin 2006).

Variation in the type of response elicited and the type of distress quantified by each measure may render comparisons between studies using different measures of distress difficult and may limit the interpretation of results obtained using a single measure. Accordingly, although salivary cortisol and SDB levels in school-age boys were highly correlated in post-conflict context (Butovskaya 2008), a recent investigation into stress levels in female wild olive baboons *(Papio hamadryas anubis)* found no correlation between day-to-day or long-term SDB levels and levels of fecal glucocorticoids (Higham et al. 2009). The difference in the findings between the two studies may be because SDB levels are related to acute stress responses in the short term to which fecal GC levels are not adequately sensitive. As there is a delay of from one to three days between hormone secretion and its excretion in feces (Palme et al. 2005), changes in fecal GC levels are likely to reflect long-term changes in distress experienced by the animal, rather than a response to specific single events (Cheney and Seyfarth 2009). However, differences in GC and SDB measures of distress may also stem from differences in the "type" of distress quantified by the two measures. SDBs may be linked specifically to anxiety or internal conflict related to uncertainty about events or decisions (Maestripieri et al. 1992) rather than a high-impact chronic stressor resulting in prolonged elevation of GC levels (Higham et al. 2009). It has also been suggested that rather than correlating with GC levels, SDBs represent a coping mechanism for dealing with short-term distress, which may in fact reduce the physiological response. In support of this view, women who displayed higher levels of SDB during a stressful interview showed lower heart rates and less pronounced vagal suppression during the post-distress recovery period (Pico-Alfonso et al. 2007).

Although more invasive procedures, such as measuring heart rate, are likely to be distressing themselves and thus potentially including confounds (Honess and Marin 2006), they have nevertheless been used successfully in a number of studies using implantable telemetry devices (e.g., Boccia et al. 1995; Aureli et al. 1999). Smith et al. (2000), for example, found that dominant male baboons showed an anticipatory increase in heart rate in the seconds preceding their locomotion with aggressive posture toward a group member when compared with locomotion without such a posture.

CHAPTER SIXTEEN

Enquiries Concerning Chimpanzee Understanding

Charles R. Menzel and Emil W. Menzel Jr.

When we entertain therefore, any suspicion that a philosophical term is employed without any meaning or idea (as is but too frequent), we need but enquire, from what impression is that supposed idea derived?

David Hume, *An Enquiry Concerning Human Understanding*

If man had not been his own classifier, he would never have thought of founding a separate order for his own reception.

Charles Darwin, *The Descent of Man*

ASKED WHAT he had learned about the mind of God from a study of nature, biologist J. B. S. Haldane is reported to have replied that God must have "an inordinate fondness for beetles." By the same token, our own studies of human and animal behavior have taught us at least one thing about the collective minds of our colleagues: they have an inordinate fondness for neologisms, for they continue to create more and more of them, ad infinitum. We have consulted the ghost of Emil W. Menzel Sr., who was an avid amateur naturalist and linguist as well as a Christian educational missionary, and we assure you that he would endorse both statements. After all—he used to say—the first human, Adam, was assigned the job of naming all the animals, if not also everything else, in his appointed garden. That has turned out to be an enormous, endless task. The more you look at just about anything, the more you will see, unless of course you lose your inborn, childhood zest for exploring, learning, naming, and classifying. As to precisely how Adam distinguished between animals and inanimate objects, and between people and other animals, and between various species of nonhumans, that is a puzzler—especially since he lived in an age when even serpents and rocks could sometimes talk.

That brings us to Premack and Woodruff's (1978b) famous neologism: "theory of mind" (in chimpanzees). The first time we heard of it we guessed, correctly, that their paper would attract more attention in psychology and be cited more frequently than other articles (e.g., Premack and Woodruff 1978a) that presented the same data, that appeared in equally or more prestigious journals, but that used less colorful language. We also were reminded of another choice quote from a Haldane book review:

> I suppose the process of acceptance will pass through the usual four stages:
>
> 1. This is worthless nonsense,
> 2. This is an interesting, but perverse, point of view,
> 3. This is true, but quite unimportant,
> 4. I always said so.
>
> (Haldane 1963, 464)

We confess that we have read only a small fraction of the by-now voluminous literature on theory of mind. Nor are we entirely sure which "stage of acceptance" we have reached. Let us therefore simply call it stage 2 and, as soon as possible, turn our attention to some of the more relevant "impressions" of the natural phenomena of animal behavior that have engaged our attention over the last 50 years. For a more standard, mentalistically oriented scholarly review of the last 30 years, see Call and Tomasello (2008). According to these authors, in some ways chimpanzees do have theory of mind, and in some ways they do not.

The principal concern of the present chapter is to address the question of how a chimpanzee knows what others know or don't know, and to place ongoing work on ape memory and communication abilities in the context of earlier primate research. The first part is a personal account of E. Menzel's work, for those who are interested in the history of primatology prior to 1978, when Premack and Woodruff's landmark paper introduced its instantly popular and influential neologism. The second part reviews C. Menzel's work, with emphasis on some new and previously unpublished experiments on a lexigram-trained chimpanzee named Panzee that add memory to the picture. Results from these and other experiments suggest that the process of initial adaptation to social beings in infant chimpanzees is, to an important degree, the same process that underlies adaptation to inanimate novel objects; that group-living chimpanzees come to learn what others know by moving with them and trying to see what they see; and that a chimpanzee's ability to extrapolate others' travel movements and directional gestures in space and time is enhanced by recall memory capabilities.

Part 1: E. Menzel's Research prior to 1978 (by E. W. Menzel Jr.)

Early Rearing Experiments

He or she who does not comprehend the uniqueness of individuals does not understand the workings of natural selection, let alone the workings of anyone's minds (Mayr 1982; Nissen 1956). I would say that for any animal I have ever studied. But it is literally my mantra about the first chimpanzees I really got to know as individuals—particularly No. 157.

No. 157 was the first chimpanzee assigned to Henry Nissen's final Infant Project (of many developmental studies that Nissen and his co-workers conducted at the Yale University—subsequently named Yerkes—Laboratories of Primate Biology starting in the early 1930s). The project was on the effects of "deprived" early experience upon the social and emotional behavior and learning abilities of chimpanzees. I in turn was Nissen's last doctoral-level research assistant. The project continued for many years after Nissen's death in 1958, with Richard Davenport as the leader; I myself worked on it, and with No. 157, for about six years. One of my main jobs, initially, was to "shape" No. 157 and then, subsequently, the other infants, as soon as possible, to obtain all their food in liquid form through a tube, from a standardized automated operant learning apparatus, without any human intervention. I assure you that this was a difficult, tedious, and intellectually demanding task, even though I had had plenty of prior experience with nonprimate animals and was sympathetic to and well versed in the writings of Skinner (1953). No single experience has so shaped my own opinions about learning theory and test apparatuses. Indeed, in retrospect I find it amazing that I and my fellow PhDs felt elated when we were eventually able to train a very young nursery infant on our apparatus in a month or less. After all, nature's standardized apparatus for this job (a lactating adult female chimpanzee with a modicum of social experience) ordinarily enables newborn infants to nurse on their own by at least day two. How much the adult female really knows about what its infant knows, and vice versa, is an open question. Nevertheless, neither one of them is, by any stretch of the imagination, a tabula rasa or an entirely independent agent in their joint achievements.

No. 157 did have another name ("Pix"), but we were not supposed to use it, on the presumption that it was not scientific or objective. He was raised from birth in our "maximally restricted condition," which amounted to a bare gray, enclosed cubicle (pictured in E. Menzel et al.

1963a). According to folklore, chimpanzees raised in this fashion might be expected to turn out like peas in a pod, not to mention "vegetables" more generally, intellectually speaking. Nothing could be farther from the truth. If anything, there was more to be learned from Pix than from more-enriched chimpanzees, especially from his reactions to novel objects and situations when practically everything was new for him. Most of the tests that I devised for other animals in the next decade, whether they were conducted in the laboratory or in the field, or on groups of animals or single individuals, were inspired by Pix. So what if it took him seemingly forever to adapt to almost any formal learning test apparatus of the day, or even to discriminate black from white, for food reinforcement? There is far more to the primate mind than that.

What does responsiveness to novel objects have to do with theories of mind? A good bit, actually. The prime question in the present scenario of ours is, how does a newborn chimpanzee, or a somewhat older one that has been raised with highly controlled and limited social and object experience, know, or come to learn, just by touching or looking at an object, whether it is living or dead, animal or not, or which "animating principle" might govern its reactions, especially toward other animals? This question reared its head most pointedly in the first tests we conducted with actual living beings—namely with human beings, including ourselves. My firm conclusion from many hours of serving as a novel test object is that the borderlines between social and nonsocial behavior in young chimpanzees—especially infant chimpanzees that are raised in a laboratory nursery by human beings—are almost totally nebulous. Anything that the animal can do with one human-defined class of object it can, and probably will at some time do, with almost any other class of object. As Freud might have put it, chimpanzees are polymorphous perverse. As Konrad Lorenz put it, the precise motor patterns that an animal performs and the precise objects toward which it directs its actions are two different, and complementary, problems. What internal emotions and mental representations are involved is yet another problem.

To habituate Pix and his cohorts to myself (or, if you prefer, to explore his Kaspar Hauser Mind), I found it not just a waste of time but also a hindrance to think of my own appearance and actions in anthropomorphic or even presumed "chimpo-morphic" terms. Who cares how friendly, comforting, or nurturing I thought I was being? It was more pertinent to ask how big and visually complex the novel object in question looked from Pix's perspective, whether it was moving or stationary, how abruptly and rapidly it moved (especially if moving toward rather than away from Pix), what volume and types of noise it generated, how predictable its appearance and actions were from moment to moment and from day to day, and above all, what might be done to modify these aspects of the "stimulus object" and correlate them, interactively, with Pix's actions in such a way as to facilitate, rather than retard, the desired effect (E. Menzel 1964a; Mason et al. 1968). The process of adaptation to inanimate novel objects in young chimpanzees is, to an important degree, the same process that underlies initial adaptation to social beings. In both realms, adaptation does not proceed in an "automatic" fashion but can depend critically on small details, such as the timing of sudden movements or sounds in relation to an animal's initial approach. From a broad perspective, the mechanism that leads an animal after one or a few trials to strongly avoid a mere block of wood that begins moving suddenly upon approach (without ever being struck or touched by the object), or to avoid a tiny object that produces a loud "click" at the right moment in the animal's cautious approach, also may underlie rapid establishment of dominance relationships between young chimpanzees. This is knowledge, of sorts, of the potential reactivity of another object or being (see also Tinbergen 1973; Grandin and Johnson 2005).

Did Pix ever show any signs that suggested he might attribute a mind, of sorts, to people or to other chimpanzees? I believe that he did, even within the first days that they met face to face, if not even before that time (as, for example, in his reactions to his own mirror image and to dolls and stuffed toy animals). I am not going to argue the case as Premack and Woodruff, and the host of other investigators who follow that same philosophical tradition, might. For one thing, I do not profess to know, formally, precisely, and generically, what "mind" is, let alone the exact criteria by which to distinguish its "higher," more mentalistic reaches from its presumably lower reaches or its accoutrements such as "mere" instinct, classical conditioning, operant conditioning, etc., etc. For another thing, after perhaps seven years of full-time primate research and some exposure to free-ranging primates, I came to be more inclined to simply assume cognitive concepts (at least for some purposes and for some animals) than to defend these concepts or to agonize about them (E. Menzel 2010).

Menzel and Johnson's Hypothetical "Lone Feral Monkey in the Woods" Scenario

Actually, I addressed the central problem of the present paper quite directly many years ago (E. Menzel and Johnson 1976, 1978; E. Menzel 1983, 1988). To quote from the 1978 paper: "Can one animal take into account the behavioral and cognitive capacities of another?"

CAN ONE ANIMAL TAKE INTO ACCOUNT THE BEHAVIORAL
AND COGNITIVE CAPACITIES OF ANOTHER?

It is hard for us to imagine how any animal could survive otherwise. We therefore concur with PW [Premack and Woodruff 1978b] that this question should be answered in the affirmative. We have developed our own argument in some detail elsewhere (E. Menzel and Johnson 1976). It should, however, be noted that even nonmentalistic accounts of evolutionary epistemology would have no quarrel with such a position. In the evolutionary sense, animals and even plants are problem solvers. The conjectures and tentative solutions that they incorporate into their anatomy and their behavior are (if one chooses to use such terminology) biological analogs of theories, and vice versa (D. Campbell 1974; Lorenz 1977; Popper 1972). Just where one is justified in attributing to animals phenomenological understanding of others, or theories of mind in the literal, anthropocentric sense, is an interesting and open question; and there are not many animal researchers who have addressed themselves to this question as directly and explicitly as have PW. However . . . the important issue is not whether animals have "theories of mind," but what these "theories" are, how they came to be "formulated," and so on. We would emphasize that certain concepts, seen in the overall context of the animal's natural behavior, including its economic situation and biological and social history, do not seem so strange and/or mystical as they might when proposed in the abstract.

The contrast between "the theory of mind approach" and "the Menzel approach" can be stated even more sharply by recounting a historical anecdote about Robert Yerkes. According to Dewsbury (2006, 145), Yerkes was characterized by several grant reviewers in 1937 as primarily an "old-fashioned naturalist" who "works on chimpanzees, not on problems." The reviewers (including the renowned psychologists Lashley and Tolman) presumably viewed this as a strongly negative criticism, and so too, we think, would Dewsbury, Premack, and many other experimental psychologists. I, on the other hand, am proud to stand as a latter-day part of the tradition of Yerkes (Yerkes and Yerkes 1929; Yerkes 1943), Nissen (1931, 1951), and C. Carpenter (1964).

As far as "problems" go, I certainly have no complaints with psychologists focusing on theory of mind or whatever else they might choose to study. But for any true or aspiring primatologist (or ethologist, anthropologist, ornithologist, ichthyologist, etc.), one's chosen organism always comes first, and almost nothing about it, from any academic discipline, should be dismissed as totally irrelevant or useless. (For that matter, mothers, caregivers, amateur naturalists, farmers, and hunters are often a better source of good ideas for research than most PhDs.) Yerkes was, quite simply, and obviously, a major early proponent of primatology and primate field research, and more modern in these respects than his critics. As Yerkes might well have put it for himself, and as Darwin did put it, "To study Metaphysics, as they always have been studied, appears to me like puzzling at astronomy without mechanics.—Experience shows that the problem of the mind cannot be solved by attacking the citadel itself.— the mind is function of body.—We must bring some stabile function to argue from" (Notebook M as quoted by Sulloway 1979, 241).

Similarly, Hölldobler and Wilson (1994, 67–68) distinguish between two groups of field biologists—theoreticians and naturalists—based on differences in approach to the organisms selected for study. Theoreticians conceive of an interesting biological problem and search for an organism that might be ideally suited to its solution. Naturalists, including Hölldobler and Wilson, proceed in exactly the opposite direction: they select an organism that they find interesting for intrinsic reasons, seek to learn all that they can about it, and look for biological problems which that organism is ideally suited to solve.

What we have said thus far is rather abstract and nonspecific (see also E. Menzel 1988). Every species, every individual organism, and every individual ecosystem is unique. What we, therefore, strongly prefer to do is to take as our fundamental "problem" particular individuals in particular environments, with particular past histories. Tolman (1938) said (and even his archrival Hull concurred), let us consider a lone, naïve albino rat at the choice point of a maze; Guthrie and Horton (1946) said, make that instead a cat in a puzzle box; primate field workers say, we'll take the howler monkeys of Barro Colorado or the chimpanzees of Gombe. I myself have chosen to study intensively, over long periods of time, individual chimpanzee infants in bare, gray nursery cubicles, especially when they are confronted with novel objects or environments (E. Menzel et al. 1963a, 1963b; E. Menzel 1963, 1964a), a group of young wild-born chimpanzees in a one-acre field (E. Menzel 1973, 1974, 1978, 1988, 2010), and family groups of tamarins in a Long Island greenhouse (E. Menzel and C. Menzel 1979; Menzel and Juno 1985). C. Menzel, after considerable experience with other species (e.g., Menzel 1980, 1991), has chosen instead to focus on a human-reared, language-trained adult female chimpanzee who once had fairly extensive experience with the outdoors. She is no longer allowed to get around as free-ranging animals do, but can exploit her caregivers into doing her outdoor foraging for her (C. Menzel 1999, 2005). Anyone—including experimentalists and naturalists, cognitivists and behaviorists, and those who, like Robert Yerkes, see no need to jump on any of these potential bandwagons—can create their own variation on this approach (E. Menzel 1969).

E. Menzel and Johnson's (1976) variation was a "thought experiment" rather than a real one: to examine how animals go about collecting and transmitting information, they said "consider a 'simple' interaction between an animal and an unknown object or being. Suppose that a normal, free-ranging monkey is strolling through the woods and suddenly comes into contact with an object. How would he go about sizing up the situation? How would he know whether he was dealing with another living being, and if so, how would he determine its intent? If it were a conspecific, how would he determine whether it knows better than he does the lay of the land, the whereabouts of food, predators, and the rest of the social group?" (p. 134).

IS THERE ANY SURE MARK AND CRITERION OF MENTALITY IN A PHENOMENON?

Heyes (1998) has argued that the problem of "seeing" might be as good a starting point as any in such analyses, and better than theory of mind. I concur, and submit as evidence two of my favorite old macaque photos—Figures 16.1 and 16.2 (also C. Menzel 1980). Johnson and I surmised from more-general ethological studies, and also from my observations of infant chimpanzees and of free-ranging macaques, that of all static visual cues our generic, hypothetical monkey would use for identifying an object as a living thing, one of the most important is anything that looks like eyes. Two buttons attached to a rectangular piece of fur, for example, will produce a much greater reaction than the fur alone. To be really effective the buttons should be located on one end of the rectangle. A monkey will characteristically circle the object and then make its first close approach from the "tail end," opposite to the "eyes." Using a doll as the test object, one can in fact probably cause the monkey to vary its direction of approach according to whether the doll's eyes are open or closed. Had the doll in Figure 16.2 been radio-controlled, and had we caused it to open its eyes and rotate its head to face the monkey, precisely as the monkey made its first cautious and tentative approach, the effect would, I expect, be striking. The larger the doll and the more rapid its motion and the more "dominant-acting" it could be made to act and sound, the greater the monkey's caution and the wider the berth it would be given (E. Menzel 1966). Object motion is probably the most salient of all visual cues, because it is so biologically significant. As M. F. Washburn (1923, 387) said, "a moving stimulus is a vitally important stimulus; it means life and hence may mean either food or danger." Still further, some specific types of movement are obviously more significant, biologically speaking, than others, and they deserve especially close study (Lorenz 1977; E. Menzel 1964b; Johansson 1973; Dasser et al. 1989).

Now then, how can we as observers tell whether our hypothetical primate "sees" the hypothetical object in the first place? As Wolfgang Köhler said at the opening of his classical monograph, which might well qualify as the first major, significant inquiry into chimpanzee understanding, or comprehension: "When any of those higher animals, which make use of vision, notice food (or any other objective) somewhere in their field of

Figure 16.1. Katsuyama, Japan, 1963. A juvenile Japanese macaque "tries to see what it can see." This bucket usually contained soybeans, but the last time any group member looked into it there was a doll or a raccoon tail as well, which produced marked caution. The monkeys do not merely imitate each other's overt behavior; they orient appropriately and move about as if trying to see what and where others see.

Figure 16.2. Cayo Santiago, Puerto Rico, 1966. This figure was published in
E. Menzel 1969; the caption read, "A juvenile [rhesus macaque] eases up toward
a stranger from the rear, after circling around the rock. Similar behavior is
displayed toward people; several rhesus followed up by jabbing me in the back
with a finger and racing up a tree."

vision, they tend—so long as no complications arise—to go after it in a
straight line. We may assume that this conduct is determined without any
previous experience, providing only that their nerves and muscles are
mature enough to carry it out" (1925, 11). Hull (1952) would have ques-
tioned the latter assumption, but otherwise in his own competing formulation of "behavior in relation to objects in space" he said about the same
thing on p. 227, and he formalized it as a fundamental theorem, citing
Euclid as his authority. Both authors would add that the larger the quantity of food and the higher its preference value for the animal, the faster
and more directly the animal is likely to move and the more time and
effort it will expend, if necessary, if complications or obstacles arise. Thus
we surmise that, everything else being equal, whichever animal in a group
takes off first, runs the fastest and in the straightest line, and works the
hardest, when necessary, not only probably either "directly sees" or otherwise somehow "knows" what sort of object is out there in the field and
where it is located, but also is better informed than any other animal in
its group. Alternatively, any molar, purposive behavior (as contrasted

with the so-called fixed action patterns of classical ethology, which are supposed to be described without regard to their goal or directionality) is a sufficient basis for objective communication about the environment (E. Menzel and Halperin 1975; Chomsky 2006, 60). And the better that every group member knows the group's common environment and each other member's idiosyncratic interests and habits and preferred goals, the easier and more reliable the process of communication. Perhaps, like well-trained human infantrymen on patrol in a combat zone in the jungle, chimpanzees chatter and gesticulate only insofar as it is really necessary to do so.

Needless to say, there are many cases in which complications do arise, and in which the optimal path for the animal to take is not a straight line (e.g., James 1890, 6–8; Köhler 1925; Hull 1952; Lorenz 1977; E. Menzel 1978, 1988; E. Menzel and C. Menzel 2007; C. Menzel et al. 2002). Individual primates often are able to detect, at the outset of trial 1 of a new problem on which all relevant details are visible from the outset, the need for executing a detour or roundabout route. Furthermore, in many instances other primates who observe them are equally able to detect, from the subject's movements relative to the various objects in the environment, almost precisely what the problem is and what the subject is up to, and how they might beat the subject to its goal (E. Menzel 1974; see also B. Hare et al. 2000, 2001). Premack and Woodruff describe their 1978 tests of a single chimpanzee, Sarah, as direct adaptations and extensions of Köhler's tests of chimpanzee comprehension. My observations and tests of group-living chimpanzees may be thought of in similar terms. Indeed, they were even simpler, more straightforward, and more obvious adaptations.

A Group of Young Wild-Born Chimpanzees in a One-Acre Field in Louisiana

My basic question here was simply how several animals rather than just one locate themselves and change their locations relative to various objects in the environment (including, obviously, one another). The basic experimental procedure was to initially have all animals in one or more of the sleeping cages that were attached to the periphery of the one-acre field; on some occasions perform some simple manipulation, such as placing one or more novel objects or pieces of food somewhere in the field, or showing them only to one animal; and then, after leaving the enclosure ourselves, turn everybody loose, usually but not always simultaneously. The hard-core data were analogous to snapshot photography: on ordinary scaled maps of the one-acre field, my invaluable assistant, Palmer Midgett, recorded the location of each and every animal, simultaneously, every X seconds. My job was to figure out where the animals were, really—a problem that harks back not just to "James, Kohler, and Tolman versus Thorndike, Guthrie, and Hull," and to "Descartes versus Newton," but all the way back to Aristotle's accounts of motion, both animate and inanimate. Where would the chimpanzees go next, and under what conditions would they all move together or split? Can any single animal lead or control the group's movement? If so, how does he or she do it?

No doubt the best-known aspects of this research, at least for students of theory of mind, are the experiments that used delayed response procedures in which we carried one or more animals (hereby operationally defined as "leaders") over to spots where novel objects or some food had been hidden, and all the other animals were uninformed, except insofar as they could get their cues from the "leader" (reviews: E. Menzel 1974, 2010). We did not have to carry the group "leaders" all the way over to the hidden objects. Merely taking a step or two in the appropriate direction or pointing manually sufficed as a cue of direction. Numerous recent papers (e.g., B. Hare 2007; Tomasello 2007; Lyn et al. 2010) have overlooked this finding, but C. Menzel greatly extends it in part 2 of this paper, with several new experiments on Panzee.

How, then, do young wild-born, group-living chimpanzees or, I presume, any wild chimpanzees come to know what others know? First and foremost, and especially at the outset, I believe, by going where others go, and by trying to see whatever they might be trying to see, especially if there is anything novel or unusual in the scenario. At first they literally ride on their mother's or mother surrogate's back or belly; as they mature and learn more about the environment and about each other's ways, they seem to operate more independently of others. Do not be fooled by appearances. No live primate is ever truly alone or entirely independent of all others. Group-living chimpanzees are always, in some sense, cooperating, and taking one another's actions if not intentions into account, and extrapolating farther in space and in time than any professedly hard-nosed scientist would ever have imagined 50 years ago. In many cases, this is just about as easily and reliably detectable by simple, direct observation as by elaborate formal experiments. A case in point is a qualitative summary note (E. Menzel 1973, 216–218) that I wrote two years before conducting my first delayed-response experiment. Young wild-born chimpanzees progressed from ventro-ventral clinging, to tandem walking, to ever more smoothly coordinated initiation of travel, to pack running, to learning the goal-directed character of others' behavior (quantitative data from the same test were presented in E. Menzel 1974). At the time the note was written, my procedures were not much different from those that Davenport, Rogers, and I used to study Pix and his cohorts—except, of course, that I tested several animals at a time, and they were all wild-born and thoroughly socialized, and the test area was outdoors. Still, the relevance of the data to what might readily be seen in the wild, without any obvious human intervention, should also be quite apparent: see, for example, C. Carpenter's (1964), Goodall's (1968) and van de Rijt-Plooij and Plooij's (1987) accounts of the development of mother-infant interactions. It would amaze me if the interactions between Panzee and her caregivers were in a totally different ballpark (see also Chomsky 2006, 60).

Part 2: New Research (1996–Current)

One of the prime ways in which chimpanzees come to read the goal-directed character of others' behavior is, we suggest, through personal experience, particularly early experience moving with others. The upper limits of a chimpanzee's ability to extrapolate the movements of others in space and time have seldom been studied adequately and, we suggest, have been greatly underestimated. As a case in point, we will discuss how the ability of a female captive chimpanzee to read her caregiver's directional pointing entails recall of distant objects and events.

A Chimpanzee in a Watchtower on the Edge of a Small Forest at the Language Research Center

C. Menzel's (1999) basic procedure for studying memory and communication in chimpanzees differs in several respects from that used in E. Menzel's field enclosure studies of juvenile chimpanzees. The subject of the study, Panzee, is a captive-born adult chimpanzee rather than a wild-born juvenile, and she directs humans to hidden goals in a small wood. Panzee's experimental social group consists of her caregivers and the experimenters who put objects out for her, not other chimpanzees (although she is not housed alone otherwise). The time elapsed between seeing items hidden and the recovery of those items can be hours or days, not just minutes. Reporting involves the use of lexigrams, and the tests entail recall of objects in areas that are not visible. Despite these differences in procedure, the approach used with Panzee addresses many of the issues discussed earlier. It extends the temporal scale, the range of spatial conditions, and the variety of different food and object types that the subject may report.

Panzee is a female *Pan troglodytes,* currently 25 years old. She was reared in a highly enriched environment, and as an infant she traveled almost daily with her human caregivers in the 20-hectare forest of the Language Research Center. Her early travel experiences involved riding on the shoulders of her caregivers and directing them with an extended arm. At a later age, she walked on her own but did not simply head off independently; she would look back to confirm that her caregivers were following. Currently her travels are restricted to a single building (the Lanson building) and its attached outdoor enclosures. Panzee's early rearing and her exposure to lexigrams were conducted by Karen Brakke, Sue Savage-Rumbaugh, and their colleagues (Brakke and Savage-Rumbaugh 1995, 1996). Panzee currently knows more than 120 lexigrams, which are arbitrarily designated visual forms that correspond to particular foods, objects, locations, and actions.

BASIC PROCEDURE

In C. Menzel's (1999) basic procedure, Panzee watches while an experimenter hides a prized object in a randomly selected location in the woods outside Panzee's outdoor enclosure. The item is completely hidden under ground cover, not simply placed behind a bush. The type of object varies across trials. After hiding the item, the experimenter leaves the area. Panzee cannot enter the woods or go get the object by herself. To obtain the item, she must in effect "tell" someone about it and direct them to recover it for her. Panzee reliably recruits the assistance of an otherwise uninformed caregiver through vocalization and gesture, touches the lexigram corresponding to the type of item hidden, and directs the caregiver to the precise location of the hidden item. She directs the person through a variety of actions, including pointing at the location where the item is hidden. By controlled vocalizations, wrist shakes, and other movements, she conveys to the person recruited that the person is on track and, finally, about to find the item. Panzee's caregivers say they are more influenced by the overall organization of her behavior than by a list of specific signals (C. Menzel 1999).

While directing her caregivers, Panzee adjusts to several complications. First, her caregiver does not know the type, location, or sometimes even the existence of Panzee's goal object. The caregiver has other routine duties, and Panzee has to intervene to change that. Second, the caregiver recruited cannot simply move toward the goal in a straight line, due to physical barriers. If Panzee initiates the interaction inside the ape building, for example, then her caregiver must make a detour to go outside. Once outdoors, the person may need to move around a tree, thicket, cage wire, or other barrier. If the person moves farther away from the goal,

Panzee corrects for this by adjusting the direction of her manual pointing and continues to provide directional signals and feedback. She does not break off the interaction and quit simply because the person makes a detour. Third, after Panzee has directed the caregiver to the vicinity of the hidden item, she continues to provide necessary feedback. Thus, she watches the caregiver's manual pointing queries and visual orientation, and she signals that the caregiver is about to succeed when he or she is oriented precisely toward the correct location.

MULTIPLE DELAYED RESPONSE

In a delayed-response variation of a test of "rank ordering" food quantities (E. Menzel and Draper 1965) Panzee tends to optimize the order of recovery of hidden goals. Figure 16.3 shows a sample trial in which Panzee directed her caregiver from her outdoor tower to 17 different locations, each of which contained a bag of grapes. (She had seen 20 bags hidden in the woods more than 20 minutes earlier. The bags were transparent

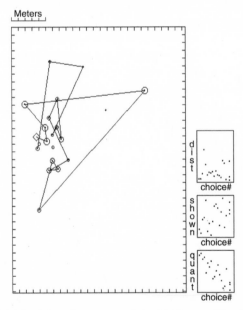

Figure 16.3. Multiple-delayed-response trial with Panzee. The diamond is Panzee's tower, where she sat while directing her caregiver to recover the hidden bags of grapes. Each circle represents a hidden bag. The size of the circle indicates the number of grapes (1 to 20) in the bag. The three scatter plots show the relationship between Panzee's choice order and three variables: the distance of the bag from the tower, the order in which the bag originally was shown to Panzee during the cue-giving phase, and the quantity of grapes in the bag, respectively.

and contained from one to 20 grapes, in 1-grape increments. During the cue giving, the bags were shown to Panzee and then hidden in randomly selected locations, one at a time, in a random order over a 40-minute period.) It may be seen from the figure that Panzee prioritized the order of recovery of the bags according to the quantity of grapes, from memory, and placed no importance on how far the bags were from her tower or on how long it had been since she had seen them being hidden. The Pearson correlation r between food quantity and recovery order for this individual trial was -0.84 (p < .0001), and the average Pearson r for Panzee's first six trials was -0.56 (p < .0001). We have obtained comparable findings on two additional adult chimpanzees, males Sherman and Mercury.

One of our experimental aims is to develop improved methods for querying Panzee about past events and distant locations and to determine the different types of information that she can recall and report about past events in a single trial. As already mentioned, Panzee recalls and reports hidden objects without prompting, and she is highly accurate in reporting the locations and types of objects hidden, after delays of at least several days (C. Menzel 1999). To study the flexibility of her recall in more detail, we are examining how well she can report which type of object is hidden in each of multiple locations, in response to human-initiated queries. The findings will clarify whether Panzee has simultaneous information (analogous to a snapshot overview), at a given moment in time, of what type of object is hidden in each one of multiple locations. The approach assumes that Panzee can understand the human's query behaviors, such as his directional manual pointing, well enough to connect these movements with events that she has seen at an earlier time and in a different place. This requires her to attach significance to the actions and to view them as something more than mere motion, and to adapt her own responses accordingly.

UNDERSTANDING DIRECTIONAL POINTING

Even on the very first times that they were tested, E. Menzel's wild-born juvenile "leaders" accurately interpreted human pointing (manual and other) as a signal that food or some other object of interest was hidden in a given direction. "Followers" in turn readily picked up the same information from the leader's behavior, and often beat him or her to the goal. Distances of 50 meters or more between cue and goal posed no great problem, except for the leader, whose success in outrunning and outguessing several followers dropped toward zero level. We cannot be certain why so many recently tested zoo- and laboratory-raised adult chimps should perform at barely better-than-chance levels even when experimenters point to one of two objects from a distance of a few centimeters (reviewed in Lyn et al.

2010), but our guess is spatially constrained test apparatus and less-than-obvious procedures, at least from the point of view of the chimpanzees, as opposed to the experimenters. Of course, differences in experience, especially early experience, should not be overlooked either.

Panzee's comprehension of her caregivers' behavior was, not surprisingly, substantially more sophisticated than one would expect of any wild-born juveniles. The aim of the experiments described next in more detail is to characterize the flexibility of her recall and her ability to read a human's expressive behavior in relation to objects that are displaced at considerable distances in space and time. The question addressed is whether Panzee can convey which type of object is hidden in any randomly selected location in the woods, in response to queries posed by a human's directional pointing. Panzee's ability to communicate with a human about the environment is undoubtedly due, in part, to her early experience and exposure to humans, but at the same time we must ask why these capabilities would exist in the first place if they were not used in some way by chimpanzees in their natural habitat. Here, we study Panzee's reading of human behavior across spatial contexts. In the experimental series described next, the human queries became increasingly displaced in space and time from the original event. The series provided increasingly challenging tests of generalized spatial knowledge. At the same time, the human's queries became more indirect and potentially more difficult to understand as goal-oriented signals. Their interpretation by Panzee required memory of the broader spatial context and an ability to exchange signals about possible goals. In effect, it required an interactive form of communication that could hold up across spatial contexts. The final two variations of the task required Panzee to connect a person's manual gestures with objects that were not only distant and invisible but also located on the opposite side of the solid walls of a building.

EXPERIMENT: QUERY FROM A DISPLACED SPATIAL CONTEXT

Before the main experiment began, we conducted three preliminary experiments to confirm that Panzee would indicate the type of item hidden in a nearby location in response to a human manual pointing query. The simplest test involved pointing at a location from 10 centimeters away; the most complex involved pointing toward a location that was several meters away on the other side of an opaque curtain. Two to four objects were hidden per trial. From trial 1 of each of these three variations, Panzee accurately indicated the types of item hidden in the designated locations. In the main experiment, while Panzee watched from her outdoor enclosure, the experimenter hid two objects in the woods, one at a time. Each trial used different locations. The test locations differed in direction by at least 25 degrees (average difference 80 degrees), as measured from indoors. After the cue giving, the experimenter confined Panzee to a cage in the Lanson building. From inside the Lanson building, no view of the outdoor enclosures or woods was possible for Panzee or her caregiver. Thus, none of the environmental features that Panzee had seen when the objects were hidden were visible during the query.

After a retention interval of 10–30 minutes, the caregiver stood next to Panzee's indoor cage and pointed with an arm extended toward the location of each object one at a time and said, "What is out there?" The query entailed pointing toward a solid opaque wall. The query also sometimes involved pointing in the direction of the indoor cage where Panzee was located, as well as toward the wall itself. The caregiver knew the locations but not the types of objects. The caregiver's task was to deduce from Panzee's lexigram use and behavior which type of object was in each location. Figures 16.4 and 16.5 show a sample trial. In her first 10 trials, Panzee's lexigram use reliably corresponded to the type of item

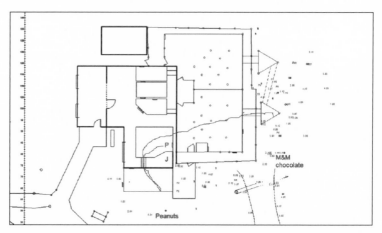

Figure 16.4. Sample trial with overview of indoor and outdoor enclosures at the Lanson building. The dark line is the cinder-block wall of the Lanson building. Light lines are caging. The irregular line shows Panzee's path of travel during the cue giving. During the cue-giving phase of the trial, Panzee watched from her outdoor enclosures while an experimenter hid peanuts and M&M's chocolates in the woods. During the delay phase of the trial, Panzee was confined to an indoor cage. During the reporting phase of the trial, Panzee and John Kelley interacted indoors at locations P and J, respectively (see also Figure 16.5). Tic marks on the scale are 30-centimeter units. Small light numerals outside the outdoor enclosures show some of the locations in the woods where objects were hidden across trials. North is approximately to the left.

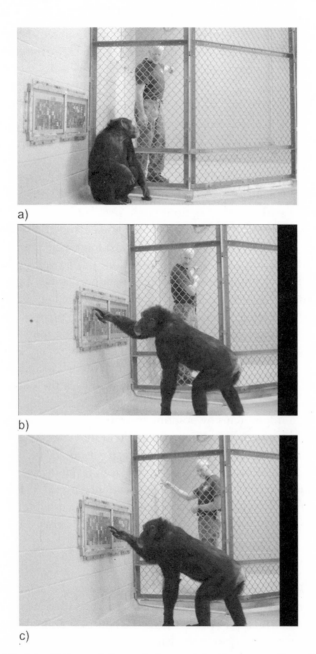

Figure 16.5. John Kelley and Panzee indoors. a) J points manually to the west and says, "What's out there?" Unknown to J, but known to P, peanuts lie beyond the cinder-block wall of the building in the direction in which J is pointing. b) P responds to J's query by touching the PEANUT lexigram on her keyboard. c) J now points to the south. M&M's candies lie beyond the wall in that direction. P responds by touching the M&M's lexigram on her keyboard.

hidden in the direction in which the caregiver pointed. For 19 of 20 locations, including both locations on trial 1, the caregiver correctly deduced the type of item in the location.

EXPERIMENT: QUERY ABOUT MULTIPLE SPATIAL CONTEXTS

The procedure was basically the same as in the preceding experiment, except that three items were hidden per trial, and, more important, a combination of indoor and outdoor locations was used each trial. In principle, the use of both indoor and outdoor locations required Panzee to connect the caregiver's manual pointing and expressive behavior with two different spatial frameworks, either locations "right here" or locations "out there." As usual, the types of items and locations varied across trials. Test objects included nonfood items, such as bubbles, balloon, paper, and a rubber snake, as well as various food types. Either one or two items were hidden outdoors. The remaining items were hidden indoors, about 2 meters from Panzee's indoor cage and adjacent to one of the building walls. The indoor items were hidden behind panels or underneath opaque containers.

During the query, the caregiver stood by Panzee's indoor cage and pointed manually toward the three locations, one at a time. Thus, the direction of his pointing varied between the "indoor" locations and the "outdoor" locations. (The caregiver pointed toward the hidden indoor items with his hand about 10 centimeters from the location. He pointed toward the outdoor items by pointing toward the solid opaque wall with his hand 10 centimeters to 1 meter from the wall; Figure 16.5.)

In the first 10 trials, the caregiver correctly deduced the type of item hidden in 28 of 30 locations, based on Panzee's lexigram use and behavior, including the contents of all three locations on trial 1. Panzee conveyed the types of nonfood items, when she was requested to do so, even though she did not always subsequently request the items.

Discussion of New Data

These findings clearly indicate that Panzee perceived a relationship between the expressive behaviors of her caregiver and objects located in invisible areas. That is, she read her caregiver's manual pointing and verbal queries in relation to objects she had seen in a distant, invisible outdoor area up to 30 minutes earlier, not simply in relation to immediately visible, nearby indoor structures. She responded to the human-generated queries flexibly. She had to detect the meaning of her caregiver's signals against the background of other daily interactions, because her caregiver also used manual pointing gestures and speech sounds during his routine activities with Panzee throughout the day. Thus, she had to realize that the human's pointing gestures were not part of normal daily interactions and that they had nothing to do with moving between cages or being asked to go together with other chimpanzees or being offered food from the refrigerator. In one experiment, Panzee adjusted on trial 1 to shifts in her caregiver's pointing between two spatial frames of reference, that is, between the "here and now" and "out there." The fact that Panzee's caregiver could infer the type of object in each of several locations, based on Panzee's response to changes in the direction of the query, indicates that Panzee was highly engaged socially and that she displayed flexibility in her use of memory. That is, she did not merely recall the item at the top of her priority list, or most recently seen.

The findings, taken together, suggest that Panzee had a form of generalized spatial knowledge analogous to a simultaneous, snapshot overview or mapping of objects and their locations. She used information about the broader spatial context to interpret human queries and to treat the arm movements and speech sounds as connected to other locations. Panzee's responsiveness to moment-to-moment variations in the direction of her caregiver's queries, especially the timing of her lexigram choices and other responses, are compelling evidence that Panzee engaged in an interactive form of communication about the distant environment. The mutual effectiveness of Panzee and her caregiver in this task is due, in part, to their joint ability to read the timing of the other's differential signals and, conversely, to their mutual ability to produce well-timed signals that were on topic with respect to the other's actions.

Retention intervals in the present experiments were long by the standards of many primate memory studies, but modest compared to our other tests with Panzee, in which delays might exceed three days. Based on these and other observations, we speculate that if the retention intervals were days or weeks and if Panzee had by chance forgotten about an object outdoors, a directional gesture by the right person in the right spatial and temporal context could potentially function as a memory retrieval cue, and that Panzee could infer approximately what the person was "talking about" and engage in further interactive communication accordingly. Furthermore, manual pointing per se is probably not any more necessary for Panzee than it is for a wild chimpanzee, for interactive communication about the environment, and we fully expect that a caregiver could interrogate Panzee effectively by a variety of alternative signals, such as directional walking or facial or whole-body pointing.

We have never had, and never expect to have, any difficulty in telling an ape from a human, and Panzee is not an exception to this rule. At the same time, the present results provide strong evidence against the view that humans are the only animals (aside from honeybees) capable of communicating about displaced objects in distant, invisible parts of the environment. The environment as Panzee perceives and conceives it is extended in space and time and is not limited to the "here and now." It goes beyond the region of immediate seeing, hearing, or smelling; it is particulate and highly detailed. We suggest that Panzee perceives that her caregivers are operating in approximately the same extended space; that she knows a fair amount about the barriers they must circumvent; and that she knows which objects are accessible to them but not her. This is a form of knowledge of another being's capacities and potential goals. It surely is based on extensive learning of the goal-oriented quality of others' behavior; it most likely emerged, in part, from her early experience traveling with her "social group" of apes and humans outdoors; and it was probably fostered in particular ways by her condition of dependency on humans. Why would a captive-born chimpanzee be capable of this understanding if it were not used in some manner by chimpanzees in their natural habitat? The fact that Panzee can read directional movements as well as she does suggests that free-living chimpanzees perceive others' behavior and cognitive capabilities in ways that go far beyond what is widely appreciated at the current time.

Conclusions

Primates, like most animals, need to take into account the behavioral and perceptual capabilities of their predators and conspecifics if they are to survive and reproduce. One approach to studying chimpanzees' understanding of others is to present animals with somewhat arbitrarily designed novel puzzles or tasks, in the tradition of Thorndike (1911) or Premack and Woodruff (1978b). An alternative approach is more indirect and inductive. It is most likely to be used by naturalists (e.g., Wallace 1869), ethologists, primatologists, and anthropologists, as opposed to humanists (e.g., Nagel 1974) or more traditional experimental psychologists. It favors directly observing and recording whatever one's chosen subjects already do, ideally without any special apparatus or training. Some questions that arise can be answered purely by systematic observations; others require a manipulation of an animal's home environment. The latter should strive to be conducted within the context of the particular organism in question, with its personal history, evolutionary history, and physical constraints. A philosophy of animal behavior research based on long-term, firsthand familiarity with one's selected organism, especially as the animal fits into its home environment, exemplifies the approach of Karl von Frisch, Martin Lindauer, and other ethologists (Lorenz 1977; Hölldobler and Wilson 1994, 19). The research described in this chapter is an example of this second approach. To paraphrase Wehner (1996), it did not begin with an attempt to test a preconceived idea about how theories of mind, or a subsidiary memory or symbolic system, might work in principle. Instead, it began with a detailed look at what particular chimpanzees do in their particular ecological and social contexts, and especially their travel and foraging interactions with one another and their caregivers. We suggest that in these directions, along the lines we have sketched, psychological research can advance substantially and successfully.

Acknowledgments

For advice and support E. Menzel thanks his family and Henry Nissen, Richard Davenport, Charles Rogers, William Mason, Naosuke Itoigawa, Donald Sade, Hans Kummer, Arthur Riopelle, Helmut Hofer, Leonard Carmichael, Clarence Ray Carpenter, Duane Rumbaugh, Michael Andrews, Everett Waters, George Williams, and Lawrence Slobodkin. C. Menzel thanks John Kelley, Michael Beran, Mary Beran, Stephanie Berger, Christopher Elder, Cameron Hastie, Sarah Hunsberger, John Mustaleski, Isabel Sanchez, Ken Sayers, and Shelly Williams for assistance with experiments on Panzee. Research at the Language Research Center was supported by NIH grants HD-056352, HD-060563, MH-58855, and HD-38051, NSF grant SBR-9729485, the L. S. B. Leakey Foundation, and the Wenner Gren Foundation.

CHAPTER SEVENTEEN

What Is Uniquely Human? A View from Comparative Cognitive Development in Humans and Chimpanzees

Tetsuro Matsuzawa

THERE ARE FOUR genera in the family Hominidae. I have been studying the chimpanzee mind since 1978. The "Ai project," now in its fourth decade (Matsuzawa 2003), focuses on a female chimpanzee named Ai, her son Ayumu, and other members of a community of chimpanzees living at the Primate Research Institute of Kyoto University (KUPRI; Matsuzawa 2001; Matsuzawa et al. 2006). The project is complemented by parallel research efforts in the wild: a long-running study of wild chimpanzees at Bossou, Republic of Guinea, West Africa (Matsuzawa et al. 2011).

It is not well recognized that brain volume triples between birth and adulthood in both humans and chimpanzees (Matsuzawa 2007). Infants of both species do a great deal of learning and undergo extensive cognitive maturation during postnatal development.

Humans and chimpanzees look different at a glance. However, the two species are very close at the genomic level: they exhibit only 1.23 percent difference in their DNA sequences (Chimpanzee Genome Consortium 2005). In other words, humans are 98.77 percent chimpanzee. The opposite is then also true; chimpanzees are 98.77 percent human. Progress in cognitive science, brain science, and evolutionary genomics has revealed that the two species are closer in many respects than previously assumed.

Humans, *Homo sapiens,* belong to the family Hominidae. However, it must be noted that Hominidae includes not only humans, but also chimpanzees and gorillas. We share the common ancestor born in Africa. Humans are not special creatures. International laws such as CITES take this into account: they treat chimpanzees as a member of the family Hominidae.

Humans and chimpanzees last shared a common ancestor about 6 million years ago; we are close evolutionary neighbors.

From this common ancestor, how did we become human? Among the most exciting news in paleoanthropology in recent years was the discovery in the fossil record of *Ardipithecus ramidus* (White 2009), a creature that lived approximately 4.4 million years ago. Nicknamed Ardi, it was the subject of 11 papers in a single issue of the journal *Science* in late 2009.

Ardi's brain size was slightly smaller than that of present-day chimpanzees. It had bipedal upright posture, but also long upper limbs, a short thumb, and feet that could grasp objects just like hands. These morphological features tell us that Ardi walked bipedally on the ground but was also a skilled arboreal climber. Future discoveries of additional hominid specimens will tell us many more interesting stories about the history of human evolution.

What is uniquely human? This is an old question, but one that we persist in trying to answer with fresh evidence and ideas. We continue to look for answers to "What is human nature?" "Where did it come from?" and "How did we get here?" through discerning brain functions, analyzing genomic information, excavating new fossils, and so forth. However, the mind and the brain cannot be excavated—they are not preserved in the fossil record. Therefore, to explore the evolutionary history of the human mind, we have to examine the minds of other living creatures. This paper aims to answer questions related to the evolution of human cognition through comparative and developmental approaches.

Birth and Life History in Chimpanzees

A recent study compiled information on 534 wild chimpanzee births recorded over the past 46 years of field research in Africa (Emery-Thomson et al. 2007). The longevity of females in the wild was estimated at around 50 years. The average interbirth interval was about five years.

Female chimpanzees usually give birth to their first offspring at around 12 years of age. They can continue to produce offspring until the end of their lives. Taking into account interbirth intervals, a female chimpanzee can therefore give birth to a maximum of seven infants. However, infant mortality rates are as high as 30 percent in the first five years of young chimpanzees' lives. As a result, the average life expectancy in females is about 15 years.

The chimpanzee way of rearing infants is somewhat similar to the case of the single mother among humans. The mother is the primary caretaker of the infant, with the biological father taking no direct role in the rearing of offspring. The next sibling arrives when the infant reaches the age of about five years. Unlike with humans, there are no siblings two or three years apart. Mothers take care of their infants one at a time.

The biological father is one of the males in the community. Chimpanzee society is referred to as patrilinial: males are philopatric, while females emigrate from their natal community to join neighboring groups. Males make alliances to patrol the territory of their community, protecting mother-infant pairs and gaining access to reproductively receptive females without infants. Intercommunity interactions are violent.

Biological fathers contribute no specific paternal care. However, all males in the community collaborate to protect mother-infant pairs within their community (Hockings et al. 2006). Viewed from the infants' perspective, they each have a mother and multiple fathers. From the viewpoint of a particular male, an infant in the community may be his own son/daughter, or a young brother/sister fathered by his own father, or a nephew/niece fathered by his brother, or another form of kin. Such is the nature of chimpanzee society.

Collaborative Breeding and the Role of the Grandmother among Humans

The gestation period in humans is about 280 days, while in wild chimpanzees it is about 225 days (Wallis 1997). Human infants are born weighing around three kilograms; chimpanzees at birth are a little less than two kilograms. Humans need a longer period for rearing offspring. Suppose that a human child needs the mother's full-time care for at least eight years or so. Suppose also that a woman starts giving birth at the age of 18. She would then continue to give birth at the ages of 26, 34, 42, and then 50—although by 50 years she may find it difficult to conceive. This means that women could have four or five offspring at most. Now suppose that the infant mortality rate is as high as that among wild chimpanzees. Then the number of offspring who reach maturity would be two or three. This way of rearing children may not be viable.

It is not well recognized that chimpanzees have no brothers or sisters two to three years younger or older than themselves. This is true in gorillas and orangutans too. In contrast, only in humans do we see the two- to three-year age gap (or indeed only a year's gap) between siblings (Blurton-Jones 1986; Howie and McNeilly 1982). The rate of twin births is also higher in humans than in the other three genera of Hominidae. Humans are unique in terms of their capacity to rear multiple children at the same time.

Humans have invented special foods for babies that facilitate weaning (Hawkes et al. 1990). This invention makes interbirth intervals shorter.

Human mothers can have extra offspring within their limited period of fertility, even though the rearing of each offspring takes a long time.

The disadvantage of this human reproductive strategy is clear. Raising multiple children at a time incurs a lot of costs and requires extra energy (Blurton-Jones 1986). A single mother like a chimpanzee cannot raise multiple children concurrently. In human evolution, the solution has involved the emergence of a role for the father. Having a spouse meant that he could join in the rearing of multiple offspring.

Chimpanzee females advertise their period of ovulation. Females in estrus have greatly enlarged, pink sexual skin in their anogenital area, maximally swollen at the peak of sexual receptivity. The swelling overlaps with ovulation but less than perfectly. This signal attracts all males in the community. In general, they copulate with the estrous female one after the other. The females thus mate with many males and rear the offspring by themselves, while males in general make no specific investment in any of the offspring born into their community.

In contrast to chimpanzees, humans have evolved to conceal estrus. It is difficult to tell whether a particular female is ovulating simply from her external appearance. It is speculated that this might be a strategy by human females for maintaining a long-term consortship with a specific male (Hrdy 1999, 2009). Both sexes want to maximize their reproductive success by having as many offspring as possible. However, females are constrained by the long period of gestation, breast feeding, and infant attachment prior to independence. Males may instead develop a strategy of fathering offspring with many different females—just like chimpanzees. The females' counterstrategy was to conceal their estrus and thereby give no clue to males as to their receptivity, nor regarding the likely paternity of any given mate. In humans, males have to keep a constant eye on their partner, otherwise they risk being cuckolded and may in fact unwittingly end up raising the offspring of other males. Such "mate guarding" is also popular, to some extent, among chimpanzees, but only in humans has it become particularly prolonged—in some cases lasting a lifetime (van Schaik and Dunbar 1990). Among primates, humans have the clearest tendency to maintain strong male-female pair bonds and as such are referred to as collaborative breeders. As mentioned above, this likely derives from the necessity of rearing multiple children at one time.

Human innovations in reproductive strategy include not only long-term spouses, but also a role for the grandmother. Females who stop breeding by themselves and invest in taking care of their son's or daughter's offspring are unique to humans among primates (Hawkes et al. 1990; Emery-Thompson et al. 2007). Grandfathers, too, may contribute to raising the second generation. In the case of chimpanzees, the participation of older males to group hunting is known to result in higher rates of success at capturing prey (Boesch 1994).

The collaborative raising of multiple children may have contributed to the prolonged post-reproductive life span in humans. Similar principles apply to close kin who serve as helpers, and may also explain the prevalence of community-based mutual support among humans. The burden of having multiple children at one time may have facilitated group living and mutual support in early hominids. This kind of group living, incorporating mutual bidirectional support of males and females, might be uniquely human (Hrdy 2009).

Simple male-female bonds and collaborative breeding by parents who defend a common territory are features of many species in the animal kingdom, in particular among birds and mammals (E. Wilson 1975). The uniqueness of human families lies in collaborative breeding among individuals of at least three generations: the grandparents and the parents of offspring. Breeding and offspring rearing continue in a transgenerational way. As a result, so-called exogamy—marriage between individuals from different communities—is essential to avoid inbreeding. Relationships between neighboring communities are highly violent in chimpanzees (*Pan troglodytes;* see Goodall 1986) but quite peaceful in bonobos (*Pan paniscus;* see Kano 1992). Two neighboring bonobo communities are able to assume peaceful coexistence in the wild (Idani 1991). This coexistence is often accompanied by feeding together and frequent sexual interaction between members of the two groups. This may provide opportunities for female transfer from one community to another. Immigrant females are known to establish a special bond to a specific resident elder female (SSF—specific senior female). The SSF often accompanies her sons. The sons may in turn develop consort relationships with the newly joined female—this might provide a mechanism for creating a nuclear family within the community. The process by which human families are created is still unknown. However, further study on intercommunity relationships in chimpanzee and bonobo societies may shed light on the emergence of the uniquely human three-generation family.

The process of rearing multiple children in humans is characterized by a unique form of collaboration among individuals of three generations and other extended networks. This means that the human way of rearing children is based on mutual and reciprocal support. In short, humans are collaborative breeders, with family units often consisting of three generations of both sexes, for whom social communication is essential for survival.

Education by Master-Apprenticeship

The absolute brain size of humans is almost three times that of chimpanzees. As mentioned above, brain size triples during development from neonate to adult in both species (3.26 times in humans vs. 3.20 in chimpanzees; Matsuzawa 2007). This illuminates the importance of postnatal development and extensive learning, scaffolded by a strong maternal bond in chimpanzees and parental (and grandparental) care in humans.

There are no schools among chimpanzees, of course, but young chimpanzees do get an education of sorts within the community. Each community has its own cultural traditions of tool use, ways of greeting, and so forth (McGrew 2004; Nakamura and Nishida 2004; Whiten et al. 1999). For example, chimpanzees at Gombe use twigs to "fish" for termites at termite mounds (Goodall 1986). Chimpanzees at Bossou do not practice this form of tool use, although they do eat termites by picking them up directly by hand as the termites emerge from their mounds. On the other hand, Bossou chimpanzees use a pair of stones to crack open the hard shell of oil-palm nuts to obtain the edible kernel (Figure 17.1; Matsuzawa 1994). Chimpanzees at Gombe are not known to use stone tools in this way, although they do consume the outer soft tissue of the oil-palm

Figure 17.1. Bossou chimpanzees use a pair of stones to crack open the hard shell of oil-palm nuts to obtain the edible kernel. Photo by Etsuko Nogami.

nut, and stones for cracking are also readily available in their environment. Japanese people use chopsticks to eat sashimi, but not all human societies use a pair of sticks as a tool to eat raw fish. Just like humans, each chimpanzee community develops its unique set of cultural traditions.

The chimpanzee way of education has been referred to as "education by master-apprenticeship" (Matsuzawa et al. 2001), or "bonding and identification-based observational learning (BIOL)" (de Waal 2001). The most important master for the infant is the mother (Hirata and Celli 2003). The mother and the infant are constantly together, at least for the first four years of life, until the infant in weaned. Chimpanzee mothers do not teach: they do not hand the infant a good stone tool, they do not demonstrate the behavior specifically for the benefit of the infant, nor they do mold the hand of the infant who tries to use a tool.

Nonetheless, chimpanzee mothers make very good models for infants to learn from. An infant will copy its mother following intensive observation of her behavior. The mothers are highly tolerant toward their infants: they allow infants to watch from close range, and even to take freshly cracked nuts from their anvil. Later on, infants start paying attention to other members of the community as well, particularly older individuals. Youngsters carefully watch older group mates, but such observations are rarely reciprocated: in the case of stone tool use, elder chimpanzees pay little attention to young individuals' behavior and are therefore unlikely to learn from them (Biro et al. 2003). Young chimpanzees learn how to use stone tools through long-term observation and their own practice. It takes them three to five years to learn to use stone hammers and anvils correctly (Matsuzawa 1994).

Once you recognize the chimpanzee way of education, you can appreciate the unique features of human education. The latter has at its core humans' collaborative rearing of young by multiple elder individuals in addition to the mother. From the beginning, the teacher is not limited to the mother. The father, the grandparents, elder brothers and sisters, uncles and aunts, and so forth, can all contribute. Through collaborative efforts of the three generation to raise young, supported by a broad social network, each community member may at times take on the role of the teacher.

Active teaching is, of course, a fundamental feature of the human approach to education. This includes unique behaviors such as social praise, scolding, verbal instruction, and molding, among others. Moreover, there are many subtle variants of, for example, social praise, such as nodding or smiling. Human children in turn seem to have a strong desire to be socially praised.

Let us imagine a mother and her child visiting a park, and the child coming across a sandbox for the first time. Before starting to play, the child looks up at his mother. The mother smiles. Suppose that the child successfully scoops some sand into a bucket. I'll bet that the child will look up at his mother again. The mother will smile, nod, or even applaud the child so as to give social praise. In English, the word "educate" originates from the Greek meaning "extract." Education extracts something already within a child. Mothers and other teachers do not educate by "putting something into" children, but by extracting something from them. Subtle social praise, and the child's intrinsic motivation to be praised, have important roles in human education.

The Mother-Infant Relationship Is Also an Evolutionary Product

Let us think about the close mother-infant relationship that gives a firm foundation to education. The human body is a product of evolution, as are the mind, society, and the mother-infant relationship. Throughout my experience of observing chimpanzees both in the wild and in the laboratory, I have always been fascinated by the strong bond between chimpanzee mothers and infants.

Chimpanzee mothers never scold their infants. They never hit them, nor do they ever neglect them. In the first three months of life, the infant always clings to the mother, and the mother embraces the infant. It may be a little-known fact that chimpanzee infants never cry at night—only humans do. Chimpanzee infants have no need to cry to attract attention, because they are constantly held in the mother's embrace.

Bearing in mind the nature of the mother-infant relationship in chimpanzees, a number of aspects unique to humans become evident. The mother-infant relationship has an evolutionary basis just like the body, mind, education, culture, and so forth. Let us think about the evolution of the mother-infant relationship. First of all, we know that many species make no parental investment into the next generation, contrary to our intuition. Ever since the beginning of life on our planet about 3.8 billion years ago, parental care of offspring has been rare. For example, most fishes lay large numbers of eggs but make no effort to raise the young that hatch. Most fishes and amphibians make no parental investment whatsoever.

The mother-infant relationship and parental investment may have their origins about 0.3 billion years ago, in the common ancestor of mammals, birds, and dinosaurs. Parents began to invest in the next generation, taking care of their young by utilizing their own energy and efforts: mammals provided their young with food in the form of milk, and many birds raised their hatchlings in the relative safety of nests by bringing back food until the young fledged.

Clinging and embracing is a unique feature of the mother-infant relationship in primates. Primates can grasp objects because each of their four limbs is adapted to an arboreal lifestyle (Figure 17.2). In cats, dogs, cows, horses, and so forth, the infants do not cling to their mothers, and the mothers do not embrace their infants. Monkeys and apes do—the anatomy of their limbs allows it. Primates (about 350 species) are unique among mammals (5,417 species as of 2005) in having four hands. We do not pay much attention to the simple fact that primates do not have four

Figure 17.2. Clinging and embracing is a unique feature of the mother-infant relationship in primates. Primates, even chimpanzees, can grasp objects because each of their four limbs is adapted to an arboreal lifestyle. We do not pay much attention to the simple fact that primates do not have four legs but four hands: the shape and function of primate feet are very similar to the shape and function of their hands. Photo by Akihiro Hirata.

Figure 17.3. Primate infants can cling to their mothers, while the mothers can embrace their infants. Photo by Tomomi Ochiai.

legs but four hands: the shape and function of primate feet are very similar to the shape and function of the hand. Primate infants can cling to their mothers, while the mothers can embrace their infants (Figure 17.3). Strictly speaking, it is likely that clinging evolved first, later followed by embracing. In many species of prosimians and New World monkeys, the infants cling to the mothers, but the mothers do not embrace the infants.

Eye-to-eye contact is also an interesting and important point. Gazing at someone directly has no positive meaning among Japanese monkeys: it conveys a mild threat. When you stare into the eyes of a Japanese monkey, it will display a "grin face" and assume a submissive posture, or respond aggressively by opening its mouth. Gazing into another's eyes has only one meaning, that of hostility. Chimpanzees look into each other's eyes with affection or friendship; however, they generally still avoid staring at others directly. Only humans routinely gaze into each other's eyes with a great deal of affection (see the evolutionary basis in macaque infants; Ferrari, Paukner, Ionica, and Suomi 2009).

While maintaining such direct gazes, humans smile. Smiling, easily recognized, is the facial expression that involves pulling back the lips horizontally and lifting them up at both ends. There are similarities and differences between the primate silent-bared-teeth face and the human smile—see especially van Hooff's milestone work on the origin of laughter and smile (van Hooff 1967). Chimpanzees are also capable of smiling, and adopt the expression in similar contexts to humans. Human infants perform neonatal smiling: newborns spontaneously smile during their sleep with their eyes closed. Neonatal smiling disappears in the first two to three months of life and is replaced by social smiling (performed with the eyes open). Our work has shown that neonatal smiling is also found in chimpanzee infants (Mizuno et al. 2006). Both human and chimpanzee infants thus show neonatal smiling, followed by a subsequent shift to social smiling. We have, for example, observed a chimpanzee mother lifting up her infant and tickling it on the neck, while mother and infant smiled at each other.

These studies with chimpanzees (Myowa-Yamakoshi et al. 2004) have triggered parallel observations in macaques, including rhesus monkeys (Ferrari, Paukner, Ruggiero, et al. 2009) and Japanese monkeys (Kawakami et al. 2006). They revealed that monkeys may also show this kind of face-to-face communication, at least during a limited period in the early days of infant development. The evolutionary basis of eye-to-eye contact and smiling may thus go as far back as the common ancestor of humans, chimpanzees, and monkeys. However, it is also true that this kind of face-to-face communication is most prevalent in human mother-infant interaction.

In terms of the mother-infant relationship, humans are the only hominoid in whom mother and infant are physically separated from each other for extended periods, from right after birth. There are some prosimians who temporarily leave their infants behind while they are foraging (Ross 2002). However, this does not happen in monkeys and apes. In primate evolution, infants began to cling to the mother, and the mother began embracing the infant in return. Clinging-embracing is a major evolutionary trend in primates, but humans are the exception.

Human mothers frequently place their infants in a supine position, lying face up, already from a very young age right after birth. Primate habitats tend to be forested, with relatively constant temperatures. During the daytime, trees create shade, while at night they prevent the warm air from escaping. The ancestors of humans left the forest behind to expand their niche into the savanna, encountering for the first time considerable temperature differences between day and night. Nights could be very cold indeed; as a result, human infants are equipped with much thicker layers of fat than other primates. Fat constitutes only about 4 percent of body weight in chimpanzee infants; in humans it is around 20 percent (Hamada and Udono 2006). The insulation afforded by this fat allowed human infants to survive the low temperatures of the savanna nights, even when they were physically isolated from their mothers.

Having so much fat is also helpful in terms of energy supply. A large brain needs a lot of energy in comparison to other organs. To keep the brain adequately functioning, it is important to have a stable supply of energy. Human babies are unique in having very rich fat stores. This is probably due to the two reasons outlined above: keeping warm and constantly supplying the brain with energy.

Stable Supine Posture Makes Us Human

It is not bipedal upright posture and bipedal locomotion, but the stable supine posture that makes us human (Matsuzawa 2007; Takeshita et al. 2009). This is a new proposal framework on human cognitive development. Suppose that you put a primate infant on its back (i.e., in the supine posture). Macaque infants show the so-called righting reflex immediately, and turn themselves over in order to assume the prone posture. I have also tested chimpanzee infants and orangutan infants by placing them on their back. They cannot turn themselves over, but instead slowly lift up one arm and the contra-lateral leg (Figure 17.4). Several seconds later, the infants will slowly move the two limbs down and lift up the opposite arm and leg. These movements alternate spontaneously. This means that ape infants cannot assume a stable supine posture, as they are constantly struggling to cling. Chimpanzee infants, up to the age of three months, always cling to their mothers and are virtually never separated. Only human infants can take the stable supine posture.

The stable supine posture is the primary impetus behind many unique features of human social intelligence. Bipedalism was thought to have been an important driving force in human evolution. According to the bipedalism hypothesis, when our four-legged ancestor stood up on its hind limbs, it freed up its forelimbs from having to support the body. The freed forelimbs then allowed the ancestor to manipulate objects. Manipulation by the hands stimulated the brain, and the brain in turn facilitated more-complex object manipulation, leading to tool use and manufacture. However, this story may be far from the truth.

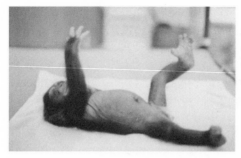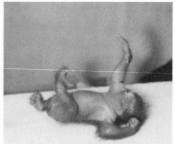

Figure 17.4. Chimpanzee infants and orangutan infants were placed on their back. They cannot turn themselves over but instead slowly lift up one arm and the contra-lateral leg. Several seconds later, the infants will slowly move the two limbs down and lift up the opposite arm and leg. These movements alternate spontaneously. This means that ape infants cannot assume a stable supine posture, as they are constantly struggling to cling. Only human infants can take the stable supine posture. Photos by Hideko Takeshita and Tetsuro Matsuzawa.

The common ancestor of primates was arboreal. Primates are adapted to life in the forest, with four hands able to grasp branches during locomotion in the trees. The trunk of the primate body is already upright, adapted to vertical climbing in the forest. This is a preadaptation to human upright posture. When the hominid ancestor left the forest and expanded its niche to open savanna, bidepal locomotion became necessary. This means that the four-handed creature had to "create" two feet. In other words, it is not the two hands but the two feet that were invented during human evolution, as part of the transition from an arboreal to a terrestrial lifestyle.

Humans also changed their reproductive strategy. They were able to rear multiple children at one time through the collaboration between male-female pairs, grandparents, and others. Human infants are physically separated from the mother and can assume the stable supine posture. Infants separated may need a lot of communication that replaces body contact (Falk 2004). The stable supine posture facilitated face-to-face communication, vocal exchange, and also the manipulation of objects already during the early phases of development.

It is not well recognized that the supine posture made our hands free. Human infants can stand on their feet at around one year of age. However, well before this ability to assume the bipedal upright posture, the stable supine posture makes our hands free to manipulate various objects, such as rattles. The early onset of object manipulation is a precursor of tool technology (Hayashi and Matsuzawa 2003).

A Trade-off between Memory and Language

A recent study has revealed that young chimpanzees outperform human adults in a working memory test (Inoue and Matsuzawa 2007; Matsuzawa 2009ab). The task was to memorize Arabic numerals that were flashed up briefly on a computer monitor. Young chimpanzees were very good at remembering nine numerals after presentations as short as 0.7 second (Figure 17.5). They were able to retain this memory for at least 10 seconds (possibly even longer). Their performance was far superior to that of both chimpanzee and human adults.

This kind of memory capability may have been favored by the demands of life in chimpanzees' natural habitat. Arriving, for example, at the base of a huge fig tree, they must make some rapid decisions regarding which branch to go to: they have to quickly scan the tree and note where the ripe red figs are among the many unripe green ones, as well as locate where the first, second, and third most dominant males are feeding. It might also be important to quickly detect and keep track of enemies hiding in the bush during hostile encounters with neighboring communities.

The chimpanzee may be good at memorizing things at a glance. Suppose you catch a glimpse of a creature passing through the vegetation in front of you. You can see many details: the white spot on the forehead, the short brown hair covering the trunk, the four black legs, and so forth. Memorizing this image quickly may be useful, but being able to sum up the experience in a single word, "duiker," also has its advantages. The human ancestor was a collaborative breeder, creating a division of roles and practicing group hunting. If you are a member of such a community,

Figure 17.5. Young chimpanzees outperform human adults in a working memory test. The task was to memorize Arabic numerals that were flashed up briefly on a computer monitor. Young chimpanzees were very good at remembering nine numerals after presentations as short as 0.7 seconds. Photos by Tetsuro Matsuzawa.

you can bring your experience back and share it with others, by telling them exactly what you saw: a duiker. What is the advantage of being able to transform perception into representation in this way? The invention of symbolic representation has allowed us to share private experiences with other community members. Such information is mobile, so it can easily be carried to others. Early hominids practiced group hunting, just like modern chimpanzees. The information brought back by those witnessing prey out of sight of others may have increased the success rate of group hunts.

If the case of computers, new functions and expansions can be added easily by wiring in new modules. However, in the case of the brain, the total brain volume reached its limit at some point during evolution. In order to acquire a new function, the brain may have to lose an existing one, if it is not so important. Broadly speaking, we may have lost the kind of photographic working memory that we see in chimpanzee infants, but instead acquired a new function: symbolic representation. This is referred to as the trade-off theory of intelligence (Matsuzawa 2009b). Once the perceptual world becomes complicated, you have to integrate various sources of information from different sensory inputs such as vision, audition, olfaction, and touch. You have to compare what you perceive and what you do. It might be advantageous to identify invariant components that remain constant from the past to the present, and to the future. Symbolic representation may have originated in this effort to combine multiple different sources into a single unified output.

A Lesson from Chimpanzee Drawing

A recent study on chimpanzee drawing brought forth an interesting fact. It was a discovery concerning imagination. Let me introduce the summary of the finding (Saitoh 2008; Saitoh et al., in preparation). Previous studies have revealed that chimpanzees are prepared to draw without any food reward: they appear to enjoy the experience and have an intrinsic motivation to draw (Morris 1962). For example, they love to scribble on children's books. Based on these observations, a free-drawing test was given in which chimpanzees were presented with white sheets of paper onto which chimpanzee faces had been sketched in advance. Specifically, the drawings showed the outer contours of a chimpanzee face, but other parts of the face, such as eyes, nose, and mouth, were missing.

Chimpanzee subjects had a strong tendency to draw on what had already been drawn. All seven tested chimpanzees spontaneously drew along the outer contour of the faces. Human children less than three

Figure 17.6. A free-drawing test was given in which chimpanzees were presented with white sheets of paper onto which chimpanzee faces had been sketched in advance (Saitoh 2008). Specifically, the drawings showed the outer contours of a chimpanzee face, but other parts of the face, such as eyes, nose, and mouth, were missing. Chimpanzee subjects had a strong tendency to draw on what had already been drawn. All seven tested chimpanzees spontaneously drew along the outer contour of the faces. Human children less than three years old showed a similar tendency. In contrast, human children above the age of three years instead filled in the missing detail: the eyes, nose, and mouth. Figures provided by Aya Saitoh.

years old showed a similar tendency. In contrast, human children above the age of three years instead filled in the missing detail: the eyes, nose, and mouth (Figure 17.6). They often spontaneously commented on the fact that the facial features were missing. Human children, but not chimpanzees, uniquely draw on blank space to complete a partial image.

Chimpanzees may be good at remembering what is present. This is shown by their extraordinary memory capability when recalling briefly presented numerals. However, human children are good at thinking about what is absent. They can identify what is missing, and conceive of ways of filling the empty space.

Imagination

The results of the drawing test reminded me of a recent experience of taking care of a handicapped chimpanzee for three years. At our institute, an adult chimpanzee named Reo suddenly became paralyzed from the neck down. The symptoms were first noted on September 26, 2006, when he was 24 years old. He could move his head normally, but everything below the neck was completely paralyzed. He was diagnosed with a spinal inflammation. He was able to eat and drink to some extent but was unable to move his body. Day by day his weight continued to drop, and as he was forced to lie motionless he developed extensive bedsores (Miyabe-Nishiwaki et al. 2010).

Young veterinarians, graduate students, and caretakers voluntarily organized a care team and performed round-the-clock treatment for six months. Reo slowly recovered from the paralysis and regained movement in his body (Figure 17.7). One thing very clear to everyone involved in his care was that Reo's attitude to things did not change throughout his period of almost complete paralysis. He often teased young students by spitting water at them—just as he had done before his illness. His outlook on life was the same after he fell ill as it was before; we did not notice any perceptible change even as he lay there thin as a rake and covered in sores. Bluntly put, he did not seem worried about his future. He did not become depressive, even though the situation looked, to us, extremely grave.

Imagination—this may make us human. Of course, chimpanzees too can memorize things and have expectations about the future. For example, there are observations of pretend play in apes (Hirata et al. 2001), and there are several studies on future orientation (planning) in apes that clearly indicate the rudimentary form of imagination (Goodall 1986). However, their range of mental time travel and mental space travel is likely to be rather limited compared to ours. Chimpanzees live in the world of the here and now. In contrast, humans can project millennia into the

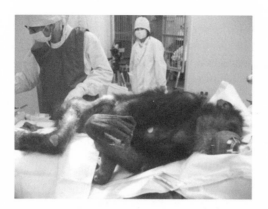

Figure 17.7. An adult chimpanzee named Reo suddenly became paralyzed from the neck down. He could move his head normally, but everything below the neck was completely paralyzed. Day by day his weight continued to drop, and as he was forced to lie motionless he developed extensive bedsores. Thanks to intensive care, Reo slowly recovered from the paralysis and regained movement in his body. One thing very clear to everyone involved in his care was that Reo's attitude to things did not change throughout his period of almost complete paralysis. Photos by Tetsuro Matsuzawa.

past and the future and imagine places on the other side of the planet, or even of the universe.

Once you recognize this difference, the extraordinary memory capability of young chimpanzees is not so surprising. They are good at memorizing things in front of them, but they do not, for example, contemplate the lives of their ancestors. Nor do they grow anxious about the future, not even about tomorrow.

For over a decade, the number of suicides in Japan has continuously exceeded 30,000 per year. Among people in their twenties, suicides account for 49 percent of all deaths. These young people become depressive and desperate. They become desperate because they cannot easily forget past troubles, and because they worry about the future. Such thinking is based on our powerful imagination. Chimpanzees do not become desperate. Humans, sometimes, do. However, because of the power of their imagination, humans can have hope too.

Acknowledgments

This study was financially supported by the grants to the author from the Japanese government (MEXT-16002001, 20002001, COE-A06-D07, JSPS-ITP-HOPE). Thanks are due to my colleagues both in the laboratory and in the field, particularly Drs. Masaki Tomonaga, Misato Hayashi, Ikuma Adachi, Imura Tomoko, Shinya Yamamoto, Satoshi Hirata, and Ms. Michiko Sakai. I thank Dr. Dora Biro for her help in polishing up the English text. Finally, I would like to thank Drs. Pier Ferrari and Frans de Waal, who gave me the opportunity of writing this chapter and also the important suggestion for the improvement.

References

Abramovich, D. R., and P. Rowe. 1973. Foetal plasma testosterone levels at mid-pregnancy and at term: Relationship to foetal sex. *Journal of Endocrinology* 56 (3):621–622.

Adams-Curtis, L. E., and D. M. Fragaszy. 1994. Development of manipulation in capuchin monkeys during the first 6 months. *Developmental Psychobiology* 27:123–136.

———. 1995. Influence of a skilled model on the behavior of conspecific observers in capuchin monkeys. *American Journal of Primatology* 37:65–71.

Adamson, L. B., and R. Bakeman 1991. The development of shared attention during infancy. In R. Vasta, ed. *Annals of child development,* vol. 8, 1–41. London, Jessica Kingsley.

Addessi, E., and E. Visalberghi. 2001. Social facilitation of eating novel food in tufted capuchin monkeys *(Cebus apella):* Input provided by group members and responses affected in the observer. *Animal Cognition* 4:297–303.

Adolphs, R. 1999. Social cognition and the human brain. *Trends in Cognitive Sciences* 3:469–479.

———. 2003. Cognitive neuroscience of human social behaviour. *Nature Reviews Neuroscience* 4:165–178.

Adolphs, R., D. Tranel, H. Damasio, and A. R. Damasio. 1994. Impaired recognition of emotion in facial expressions following bilateral damage to the human amygdala. *Nature* 372:669–672.

Ainsworth, M. D. S., M. C. Blehar, E. Waters, and S. Wall. 1978. *Patterns of attachment.* Hillsdale, NJ: Lawrence Erlbaum.

Aitken, P. G., and D. A. Wilson. 1979. Discriminative vocal conditioning in rhesus monkeys: Evidence for volitional control? *Brain and Language* 8, 227–240.

Allen, M. H., A. J. Lincoln, and A. S. Kaufman. 1991. Sequential and simultaneous processing abilities of high-functioning autistic and language-impaired children. *Journal of Autism and Developmental Disorders* 21 (4):483–502.

Allison, T., A. Puce, and G. McCarthy. 2000. Social perception from visual cues: Role of the STS region. *Trends in Cognitive Sciences* 4:267–278.

Alterman, A. I., P. A. McDermott, J. S. Cacciola, and M. J. Rutherford. 2003. Latent structure of the Davis Interpersonal Reactivity Index in methadone maintenance patients. *Journal of Psychopathology and Behavioral Assessment*, 25:257–265.

Amsterdam, B. 1972. Mirror self-image reactions before age two. *Developmental Psychobiology* 5:297–305.

Amunts, K., A. Schleicher, U. Bürgel, H. Mohlberg, H. B. Uylings, and K.Zilles. 1999. Broca's region revisited: Cytoarchitecture and intersubject variability. *Journal of Comparative Neurology* 412:319–341.

Anderson, J. R. 2006. Tool use: A discussion of diversity. In K. Fujita and S. Itakura, eds. *Diversity of cognition: Evolution, development, domestication and pathology*, 367–400. Kyoto: Kyoto University Press.

Anderson, J. R., M. Montant, and D. Schmitt. 1996. Rhesus monkeys fail to use gaze direction as an experimenter-given cue in an object-choice task. *Behavioural Processes* 37:47–55.

Anderson, J. R., M. Myowa-Yamakoshi, and T. Matsuzawa. 2004. Contagious yawning in chimpanzees. *Proceedings of the Royal Society B: Biological Sciences* 271:S468–S470.

Anderson, J. R., P. Sallaberry, and H. Barbier. 1995. Use of experimenter-given cues during object-choice tasks by capuchin monkeys. *Animal Behaviour* 49:201–208.

Andreoni, J. 1989. Giving with impure altruism: Applications to charity and Ricardian equivalence. *Journal of Political Economy* 97:1447–1458.

APA. 1994. *DSM-IV Diagnostic and Statistical Manual of Mental Disorders*, 4th ed. Washington, DC: American Psychiatric Association.

Apperly, I. A., D. Samson, C. Chiavarino, W. L. Bickerton, and G. W. Humphreys. 2007. Testing the domain-specificity of a theory of mind deficit in brain-injured patients: Evidence for consistent performance on non-verbal, "reality-unknown" false belief and false photograph task. *Cognition* 103:300–321.

Apperly, I. A., D. Samson, C. Chiavarino, and G. W. Humphreys. 2004. Frontal and temporo-parietal lobe contributions to theory of mind: Neuropsychological evidence from a false-belief task with reduced language and executive demands. *Journal of Cognitive Neuroscience* 16:1773–1784.

Arbib, M. A., K. Liebal, and S. Pika. 2008. Primate vocalization, gesture and the evolution of human language. *Current Anthropology* 49:1053–1076.

Arnold, K., O. N. Fraser, and F. Aureli. 2010. Postconflict reconciliation. In C. J. Campbell, A. Fuentes, K. C. MacKinnon, S. K. Bearder, and R. M. Stumpf, eds. *Primates in perspective*, 2nd ed., 608–625. Oxford: Oxford University Press.

Augustine, J. R. 1985. The insular lobe in primates including humans. *Neurological Research* 7:2–10.

———. 1996. Circuitry and functional aspects of the insular lobe in primates including humans. *Brain Research Reviews* 22:229–244.

Aureli, F. 1997. Post-conflict anxiety in nonhuman primates: The mediating role of emotion in conflict resolution. *Aggressive Behavior* 23:315–328.

Aureli, F., M. Cords, and C. P. van Schaik. 2002. Conflict resolution following aggression in gregarious animals: A predictive framework. *Animal Behaviour* 64:325–343.

Aureli, F., R. Cozzolino, C. Cordischi, and S. Scucchi. 1992. Kin-oriented redirection among Japanese macaques: An expression of a revenge system. *Animal Behaviour* 44:283–291.

Aureli, F., S. D. Preston, and F. B. M. de Waal. 1999. Heart rate responses to social interactions in free-moving rhesus macaques *(Macaca mulatta):* A pilot study. *Journal of Comparative Psychology* 113:59–65.

Aureli, F., and C. Schaffner. 2002. Empathy as a special case of emotional mediation of social behavior. *Behavioral and Brain Sciences* 25:23–24.

———. 2007. Aggression and conflict management at fusion in spider monkeys. *Biology Letters* 3:147–149.

Aureli, F., and D. A. Smucny. 2000. The role of emotion in conflict and conflict resolution. In F. Aureli and F. B. M. de Waal, eds. *Natural conflict resolution,* 199–224. Berkeley: University of California Press.

Aureli, F., and C. P. van Schaik. 1991a. Post-conflict behaviour in long-tailed macaques *(Macaca fascicularis):* I. The social events. *Ethology* 89:89–100.

———. 1991b. Post-conflict behaviour in long-tailed macaques *(Macaca fascicularis):* II. Coping with uncertainty. *Ethology* 89:101–114.

Aureli, F., C. P. van Schaik, and J. A. R. A. M. van Hooff. 1989. Functional aspects of reconciliation among captive long-tailed macaques *(Macaca fascicularis). American Journal of Primatology* 19:39–51.

Aureli, F., and A. Whiten. 2003. Emotions and behavioural flexibility. In D. Maestripieri, ed. *Primate psychology,* 289–323. Cambridge, MA: Harvard University Press.

Aureli, F., and K. Yates. 2009. Distress prevention by grooming others in crested black macaques. *Biology Letters* 6:27–29.

Auyeung, B., S. Baron-Cohen, E. Chapman, R. C. Knickmeyer, K. Taylor, and G. Hackett. 2006. Foetal testosterone and the child systemizing quotient. *European Journal of Endocrinology* 155 (suppl. no. 1):S123–S130.

———. 2009. Fetal testosterone and autistic traits. *British Journal of Psychology* 100:1–22.

Auyeung, B., S. Baron-Cohen, S. Wheelwright, and C. Allison. 2008. The autism spectrum quotient: Children's version (AQ-child). *Journal of Autism and Developmental Disorders* 38:1230–1240.

Auyeung, B., S. Baron-Cohen, S. Wheelwright, N. Samarawickrema, and M. Atkinson. 2009. The children's empathy quotient (EQ-C) and systemizing quotient (SQ-C): Sex differences in typical development and of autism spectrum conditions. *Journal of Autism and Developmental Disorders* 39:1509–1521.

Auyeung, B., K. Taylor, G. Hackett, and S. Baron-Cohen. 2010. Fetal testosterone and autistic traits in 18- to 24-month-old children. *Molecular Autism* 1 (11).

Avenanti, A., D. Bueti, G. Galati, and S. M. Aglioti. 2005. Transcranial magnetic stimulation highlights the sensorimotor side of empathy for pain. *Nature Neuroscience* 8:955–960.

Avenanti, A., A. Sirigu, and S. M. Aglioti. 2010. Racial bias reduces empathic sensorimotor resonance with other-race pain. *Current Biology* 20:1018–1022

Avian Brain Nomenclature Consortium. 2005. Avian brains and a new understanding of vertebrate brain evolution. *Nature Reviews Neuroscience* 6:151–159.

Babb, S. J., and J. D. Crystal. 2006. Episodic-like memory in the rat. *Current Biology* 16:1317–1321.

Baird, G., E. Simonoff, A. Pickles, S. Chandler, T. Loucas, D. Meldrum, and T. Charman. 2006. Prevalence of disorders of the autism spectrum in a population cohort of children in South Thames: The Special Needs and Autism Project (SNAP). *Lancet* 368 (9531):210–215.

Bakeman, R., L. B. Adamson, M. Konner, and R. Barr. 1990. !Kung infancy: the social context of object exploration. *Child Development* 61:794–809.

Baldwin, D. A., and J. A. Baird. 2001. Discerning intentions in dynamic human action. *Trends in Cognitive Sciences* 5:171–178.

Bard, K. A. 1992. Intentional behavior and intentional communication in young free-ranging orangutans. *Child Development* 63:1186–1197.

———. 1994. Evolutionary roots of intuitive parenting: Maternal competence in chimpanzees. *Early Development and Parenting* 3:19–28.

———. 1996. *Responsive care: A behavioral intervention program for nursery-reared chimpanzees.* Tucson, AZ: Jane Goodall Institute.

———. 2000. Crying in infant primates: Insights into the development of crying in chimpanzees. In R. G. Barr, B. Hopkins, and J. A. Green, eds. *Crying as a sign, a symptom, and a signal: Clinical emotional and developmental aspects of infant and toddler crying,* 157–175. New York: Cambridge University Press.

———. 2002. Primate parenting. In M. H. Bornstein, ed. *Handbook of parenting,* vol. 2, *Biology and ecology of parenting,* 2nd ed., 99–140. Mahwah, NJ: Lawrence Erlbaum.

———. 2003. Development of emotional expression in chimpanzees *(Pan troglodytes). Annals of the New York Academy of Sciences* 1000:88–90.

———. 2005. Emotions in chimpanzee infants: The value of a comparative developmental approach to understand the evolutionary bases of emotion. In J. Nadel and D. Muir, eds. *Emotional development: Recent research advances,* 31–60. New York: Oxford University Press.

———. 2007. Neonatal imitation in chimpanzees *(Pan troglodytes)* tested with two paradigms. *Animal Cognition* 10:233–242.

———. 2009. Social cognition: Evolutionary history of emotional engagements with infants. *Current Biology* 19:R941–R943.

Bard, K. A., L. Brent, B. Lester, J. Worobey, and S. J. Suomi. 2011. Neurobehavioral integrity of chimpanzee newborns: Comparisons across groups and across species reveal gene-environment interaction effects. *Infant and Child Development* 20: 47–93.

Bard, K. A., and K. H. Gardner. 1996. Influences on development in infant chimpanzees: Enculturation, temperament, and cognition. In A. E. Russon, K. A.

Bauman, M. L., and T. L. Kemper. 2005. Neuroanatomic observations of the brain in autism: A review and future directions. *International Journal of Developmental Neuroscience* 23:183–187.

Bayley, N. 1969. *Bayley scales of infant development.* New York: Psychological Corp.

Beaton, A. A. 1997. The relation of planum temporale asymmetry and morphology of the corpus callosum to handedness, gender and dyslexia: A review of the evidence. *Brain and Language* 60:255–322.

Beer, J. S., J. P. Mitchell, and K. N. Ochsner. 2006. Special issue: Multiple perspectives on the psychological and neural bases of social cognition. *Brain Research* 1079:1–3.

Behne, T., M. Carpenter, J. Call, and M. Tomasello. 2005. Unwilling versus unable: Infants' understanding of intentional action. *Developmental Psychology* 41:328–337.

Bekkering, H., A. Wohlschläger, and M. Gattis. 2000. Imitation of gestures in children is goal-directed. *Quarterly Journal of Experimental Psychology A* 53:153–164.

Bellagamba, F. M., and M. Tomasello. 1999. Re-enacting intended acts: Comparing 12- and 18-month-olds. *Infant Behavior and Development* 22:277–282.

Bengtsson, S. L., J. D. Haynes, K. Sakai, M. J. Buckley, and R. E. Passingham. 2009. The representation of abstract task rules in the human prefrontal cortex. *Cerebral Cortex* 19:1929–1936.

Berenbaum, S. A. 1999. Effects of early androgens on sex-typed activities and interests in adolescents with congenital adrenal hyperplasia. *Hormones and Behavior* 35 (1):102–110.

Berenbaum, S. A., and M. Hines. 1992. Early androgens are related to childhood sex-typed toy preferences. *Psychological Science* 3:203–206.

Berenbaum, S. A., and E. Snyder. 1995. Early hormonal influences on childhood sex-typed activity and playmate preferences: Implications for the development of sexual orientation. *Developmental Psychology* 31:31–42.

Bermejo, M., and G. Illera. 1999. Tool-set for termite fishing and honey extraction by wild chimpanzees in the Lossi Forest, Congo. *Primates* 40:619–627.

Binder, J. R., S. J. Swanson, T. A. Hammeke, G. L. Morris, W. M. Mueller, M. Fischer, S. Benbadis, J. A. Frost, S. M. Rao, and V. M. Haughton, 1996. Determination of language dominance using functional MRI: A comparison with the Wada test. *Neurology* 46:978–984.

Birbaumer, N., R. Veit, M. Lotze, M. Erb, C. Hermann, W. Grodd, and H. Flor. 2005. Deficient fear conditioning in psychopathy: A functional magnetic resonance imaging study. *Archives of General Psychiatry* 62:799–805.

Bird, G., G. Silani, R. Brindley, S. White, U. Frith, and T. Singer. 2010. Empathic brain responses in insula are modulated by levels of alexithymia but not autism. *Brain* 133:1515–1525.

Biro, D., N. Inoue-Nakamura, R. Tonooka, G. Yamakoshi, C. Sousa, and T. Matsuzawa. 2003. Cultural innovation and transmission of tool use in wild chimpanzees: Evidence from field experiments. *Animal Cognition* 6:213–223.

Bischof-Köhler, D. 1988. Über den Zusammenhang von Empathie und der Fähigkeit sich im Spiegel zu erkennen [On the connection between empathy and the ability to recognize oneself in the mirror]. *Schweizer Zeitschrift für Psychologie* 47:147–159.

———. 1991. The development of empathy in infants. In M. E. Lamb and M. Keller, eds. *Infant development: Perspectives from German-speaking countries*, 245–273. Hillsdale, NJ: Lawrence Erlbaum.

Bjorklund, D. F. 2006. Mother knows best: Epigenetic inheritance, maternal effects, and the evolution of human intelligence. *Developmental Review* 26:213–242.

Bjorklund, D. F., J. M. Bering, and P. Ragan. 2000. A two-year longitudinal study of deferred imitation of object manipulation in a juvenile chimpanzee *(Pan troglodytes)* and orangutan *(Pongo pygmaeus)*. *Developmental Psychobiology* 37:229–237.

Blair, J., D. R. Mitchell, and K. Blair. 2005. *The psychopath: Emotion and the brain.* Oxford: Blackwell.

Blair, R. J., E. Colledge, L. Murray, and D. G. Mitchell. 2001. A selective impairment in the processing of sad and fearful expressions in children with psychopathic tendencies. *Journal of Abnormal Child Psychology* 29:491–498.

Blurton-Jones, N. 1986. Bushman birth spacing: A test for optimal interbirth intervals. *Ethology and Sociobiology* 7:91–105.

Boccia, M. L., M. L. Laudenslager, and M. L. Reite. 1995. Individual differences in macaques' responses to stressors based on social and physiological factors: Implications for primate welfare and research outcomes. *Laboratory Animals* 29:250–257.

Boccia, M. L., M. Reite, and M. L. Laudenslager. 1989. On the physiology of grooming in a pigtail macaque. *Physiology and Behavior* 45:667–670.

Boehm, C. 1999. *Hierarchy in the forest: The evolution of egalitarian behavior.* Cambridge, MA: Harvard University Press.

Boesch, C. 1994. Cooperative hunting in wild chimpanzees. *Animal Behavior* 48:653–667.

———. 2007. What makes us human *(Homo sapiens)?* The challenge of cognitive cross-species comparison. *Journal of Comparative Psychology* 121:227–240.

———. 2008. Taking development and ecology seriously when comparing cognition: Reply to Tomasello and Call (2008). *Journal of Comparative Psychology* 122:453–455.

Boesch, C., and H. Boesch. 1989. Hunting behavior of wild chimpanzees in the Taï National Park. *American Journal of Physical Anthropology* 78:547–573.

Boesch, C., and H. Boesch-Achermann. 2000. *The chimpanzees of the Taï forest: Behavioural ecology and evolution.* Oxford: Oxford University Press.

Bonini, L., S. Rozzi, F. Ugolotti Serventi, L. Simone, P. F. Ferrari, and L. Fogassi. 2009. Ventral premotor and inferior parietal cortices make distinct contribution to action organization and intention understanding. *Cerebral Cortex,* bhp200v1–bhp200.

Bonner, J. T. 1980. *The evolution of culture in animals*. Princeton, NJ: Princeton University Press.

Bonnie, K. E., and F. B. M. de Waal. 2004. Primate social reciprocity and the origin of gratitude. In R. A. Emmons and M. E. McCullough, eds. *The psychology of gratitude*, 213–229. Oxford: Oxford University Press.

———. 2007. Copying without rewards: Socially influenced foraging decisions among brown capuchin monkeys. *Animal Cognition* 10 (3): 283–292.

Bonnie, K. E., V. Horner, A. Whiten, and F. B. M. de Waal. 2007. Spread of arbitrary conventions among chimpanzees: A controlled experiment. *Proceedings of the Royal Society B: Biological Sciences* 274:367–372.

Boyd, R., and P. J. Richerson. 1985. *Culture and the evolutionary process*. London: University of Chicago Press.

———. 1996. Why culture is common, but cultural evolution is rare. *Proceedings of the British Academy* 88:77–93.

Brakke, K. E., and E. S. Savage-Rumbaugh. 1995. The development of language skills in bonobo and chimpanzee—I. Comprehension. *Language and Communication* 15:121–148.

———. 1996. The development of language skills in *Pan*—II. Production. *Language and Communication* 16:361–380.

Brass, M., and C. Heyes. 2005. Imitation: Is cognitive neuroscience solving the correspondence problem? *Trends in Cognitive Sciences* 9:489–495.

Bräuer, J., J. Call, and M. Tomasello. 2005. All great ape species follow gaze to distant locations and around barriers. *Journal of Comparative Psychology* 119:145–154.

Brazelton, T. B., and J. K. Nugent. 1995. *Neonatal behavioral assessment scale*, 3rd ed. London: Mac Keith.

Breedlove, S. M. 1992. Sexual dimorphism in the vertebrate nervous system. *Journal of Neuroscience* 12:4133–4142.

Brewer, S. M., and W. C. McGrew. 1990. Chimpanzee use of a tool-set to get honey. *Folia Primatologica* 54:100–104.

Brosnan, S. F., and M. J. Beran. 2009. Trading behavior between conspecifics in chimpanzees, *Pan troglodytes*. *Journal of Comparative Psychology* 123:181–194.

Brosnan, S. F., and F. B. M. de Waal. 2002. A proximate perspective on reciprocal altruism. *Human Nature* 13:129–152.

Brosnan, S. F., J. B. Silk, J. Henrich, M. C. Mareno, S. P. Lambeth, and S. J. Schapiro. 2009. Chimpanzees *(Pan troglodytes)* do not develop contingent reciprocity in an experimental task. *Animal Cognition* 12:587–589.

Brown, G. R., C. M. Nevison, H. M. Fraser, and A. F. Dixson. 1999. Manipulation of postnatal testosterone levels affect phallic and clitoral development in infant rhesus monkeys. *International Journal of Andrology* 22:119–128.

Brown, W. M., M. Hines, B. Fane, and S. M. Breedlove. 2002. Masculinized finger length patterns in human males and females with congenital adrenal hyperplasia. *Hormones and Behavior* 42:380–386.

Brownell, C. A., M. Svetlova, and S. Nichols. 2009. To share or not to share: When do toddlers respond to another's needs? *Infancy* 14:117–130.

Brugger, A., L. A. Lariviere, D. L. Mumme, and E. W. Bushnell. 2007. Doing the right thing: Infants' selection of actions to imitate from observed event sequences. *Child Development* 78:806–824.

Brune, M., U. Brune-Cohrs, W. C. McGrew, and S. Preuschoft. 2006. Psychopathology in great apes: Concepts, treatment options and possible homologies to human psychiatric disorders. *Neuroscience and Biobehavioral Reviews* 30:1246–1259.

Bruner, J. 1975. From communication to language—a psychological perspective. *Cognition* 3:255–287.

———. 1983. Play, thought, and language. *Peabody Journal of Education* 60: 60–69.

Brunet-Gouet, E., and J. Decety. 2006. Social brain dysfunctions in schizophrenia: a review of neuroimaging studies. *Psychiatry Research* 148:75–92.

Buccino, G., S. Vogt, A. Ritzl, G. R. Fink, K. Zilles, H. J. Freund, and G. Rizzolatti. 2004. Neural circuits underlying imitation learning of hand actions: An event-related fMRI study. *Neuron* 42:323–334.

Buckley, M. J., F. A. Mansouri, H. Hoda, M. Mahboubi, P. G. Browning, S. C. Kwok, A. Phillips, and K. Tanaka. 2009. Dissociable components of rule-guided behavior depend on distinct medial and prefrontal regions. *Science* 325:52–58.

Bufalari, I., T. Aprile, A. Avenanti, F. Di Russo, and S. M. Aglioti. 2007. Empathy for pain and touch in the human somatosensory cortex. *Cerebral Cortex* 17:2553–2561.

Bugnyar, T., and L. Huber. 1997. Push or pull: An experimental study on imitation in marmosets. *Animal Behaviour* 54:817–831.

Bugnyar, T., and K. Kotrschal. 2002. Observational learning and the raiding of food caches in ravens, *Corvus corax:* Is it "tactical deception"? *Animal Behaviour* 64:185–195.

Bulloch, M. J., S. T. Boysen, and E. E. Furlong. 2008. Visual attention and its relation to knowledge states in chimpanzees, *Pan troglodytes. Animal Behaviour* 76:1147–1155.

Bullowa, M. 1979. *Before speech: The beginnings of interpersonal communication.* New York: Cambridge University Press.

Bunge, S. A., J. D. Wallis, A. Parker, M. Brass, E. A. Crone, E. Hoshi, and K. Sakai. 2005. Neural circuitry underlying rule use in humans and nonhuman primates. *Journal of Neuroscience* 25:10347–10350.

Burghardt, G. 2009. Darwin's legacy to comparative psychology and ethology. *American Psychologist* 64:102–110.

Burkart, J. M., E. Fehr, C. Efferson, and C. P. van Schaik. 2007. Other-regarding preferences in a non-human primate: Common marmosets provision food altruistically. *Proceedings of the National Academy of Sciences USA* 104:19762–19766.

Burkart, J. M., and A. Heschl. 2007. Understanding visual access in common marmosets, *Callithrix jacchus:* Perspective taking or behaviour reading? *Animal Behaviour* 73:457–469.

Burnham, T. C., and B. Hare. 2007. Engineering human cooperation: Does involuntary neural activation increase public goods contributions? *Human Nature* 18:88–108.

Butovskaya, M. L. 2008. Reconciliation, dominance and cortisol levels in children and adolescents (7–15-year-old boys). *Behaviour* 145:1557–1576.

Buttelmann, D., J. Call, and M. Tomasello. 2008. Behavioral cues that great apes use to forage for hidden food. *Animal Cognition* 11:117–128.

Buttelmann, D., M. Carpenter, J. Call, and M. Tomasello. 2007. Enculturated chimpanzees imitate rationally. *Developmental Science* 10:31–38.

———. 2008. Rational tool use and tool choice in human infants and great apes. *Child Development* 79:609–626.

Byrne, R. W. 1994. The evolution of intelligence. In J. B. Slater and T. R. Halliday, eds. *Behaviour and evolution*, 223–265. Cambridge: Cambridge University Press.

———. 1998a. Comment on Boesch and Tomasello's "chimpanzee and human cultures." *Current Anthropology* 39:604–605.

———. 1998b. Imitation: The contributions of priming and program-level copying. In S. Bråten, ed. *Intersubjective communication and emotion in early ontogeny*, 228–244. Cambridge: Cambridge University Press.

———. 2002. Imitation of novel complex actions: What does the evidence from animals mean? *Advances in the Study of Behavior* 31:77–105.

Byrne, R. W., and A. E. Russon. 1998. Learning by imitation: A hierarchical approach. *Behavioral and Brain Sciences* 21:667–684.

Byrne, R. W., and J. E. Tanner. 2006. Gestural imitation by a gorilla: Evidence and nature of the capacity. *International Journal of Psychology and Psychological Therapy* 6:215–231.

Byrne, R. W., and A. Whiten. 1988. *Machiavellian intelligence*. Oxford: Oxford University Press.

———. 1992. Cognitive evolution in primates: Evidence from tactical deception. *Man* 27:609–627.

Byrnit, J. T. 2004. Nonenculturated orangutans' *(Pongo pygmaeus)* use of experimenter-given manual and facial cues in an object-choice task. *Journal of Comparative Psychology* 118:309–315.

Caggiano, V., L. Fogassi, G. Rizzolatti, P. Their, and A. Casile. 2009. Mirror neurons differentially encode the peripersonal and extrapersonal space of monkeys. *Science* 324:403–406.

Caldwell, C. A., and A. Whiten. 2004. Testing for social learning and imitation in common marmosets, *Callithrix jacchus,* using an artificial fruit. *Animal Cognition* 7:77–85.

———. 2010. Social learning in monkeys and apes: Cultural animals? In C. J. Campbell, A. Fuentes, K. C. MacKinnon, S. K. Bearder, and R. M. Stumpf, eds. *Primates in perspective*, 2nd ed., 652–662. Oxford: Oxford University Press.

Call, J. 2001. Body imitation in an enculturated orangutan *(Pongo pygmaeus)*. *Cybernetics and Systems* 32:97–119.

Call, J., B. Agnetta, and M. Tomasello. 2000. Cues that chimpanzees do and do not use to find hidden objects. *Animal Cognition* 3:23–34.

Call, J., and M. Carpenter. 2002. Three sources of information in social learning. In K. Dautenhahn and C. L. Nehaniv, eds. *Imitation in animals and artifacts,* 211–228. Cambridge, MA: MIT Press.

———. 2003. On imitation in apes and children. *Infancia y Aprendizaje* 26: 325–349.

Call, J., M. Carpenter, and M. Tomasello. 2005. Copying results and copying actions in the process of social learning: Chimpanzees *(Pan troglodytes)* and human children *(Homo sapiens). Animal Cognition* 8:151–163.

Call, J., B. Hare, and M. Tomasello. 1998. Chimpanzee gaze following in an object-choice task. *Animal Cognition* 1:89–99.

Call, J., and L. R. Santos. In press. In J. C. Mitani, J. Call, P. Kappeler, R. Palombit, and J. B. Silk, eds. *The evolution of primate societies.* Chicago: University of Chicago Press.

Call, J., and M. Tomasello. 1995. Use of social information in the problem solving of orangutans *(Pongo pygmaeus)* and human children *(Homo sapiens). Journal of Comparative Psychology* 109:308–320.

———. 1996. The effect of humans on cognitive development of apes. In A. E. Russon, K. A. Bard, and S. T. Parker, eds. *Reaching into thought: The minds of the great apes,* 371–403. New York: Cambridge University Press.

———. 1999. A nonverbal false belief task: The performance of children and apes. *Child Development* 70:381–395.

———. 2008. Does the chimpanzee have a theory of mind? 30 years later. *Trends in Cognitive Sciences* 12:187–192.

Calvo-Merino, B., D. E. Glaser, J. Grezes, R. E. Passingham, and P. Haggard. 2005. Action observation and acquired motor skills: An fMRI study with expert dancers. *Cerebral Cortex* 15:1243–1249.

Calvo-Merino, B., J. Grezes, D. E. Glaser, R. E. Passingham, and P. Haggard. 2006. Seeing or doing? Influence of visual and motor familiarity in action observation. *Current Biology* 16:1905–1910.

Campbell, D. T. 1974. Evolutionary epistemology. In P. A. Schilpp, ed. *The philosophy of Karl Popper,* 413–463. LaSalle, IL: Open Court Press.

Campbell, M. W., J. D. Carter, D. Proctor, M. L. Eisenberg, and F. B. M. de Waal. 2009. Computer animations stimulate contagious yawning in chimpanzees. *Proceedings of the Royal Society B: Biological Sciences* 276:4255–4259.

Canale, G. R., C. E. Guidorizzi, M. C. M. Kierulff, and C. A. F. R. Gatto. 2009. First record of tool use by wild populations of the yellow-breasted capuchin monkey *(Cebus xanthosternos)* and new records for the bearded capuchin *(Cebus libidinosus). American Journal of Primatology* 71:1–7.

Cantalupo, C., and W. D. Hopkins. 2001. Asymmetric Broca's area in great apes. *Nature* 414:505–507.

Cappa, S. F., C. Guariglia, C. Papagno, L. Pizzamiglio, G. Vallar, P. Zoccolotti, B. Ambrosi, and V. Santiemma. 1988. Patterns of lateralization and performance levels for verbal and spatial tasks in congenital androgen deficiency. *Behavioural Brain Research* 31:177–183.

Carlstead, K., and D. Shepherdson. 2000. Alleviating stress in zoo animals with environmental enrichment. In G. P. Moberg and J. A. Mench, eds. *The biol-*

ogy of animal stress: Basic principles and implications for animal welfare, 337–354. Wallingford, CT: CABI International.

Carpenter, C. R. 1964. *Naturalistic behavior of nonhuman primates.* University Park: Pennsylvania State University Press.

Carpenter, M., N. Akhtar, and M. Tomasello. 1998. Fourteen- through 18-month-old infants differentially imitate intentional and accidental actions. *Infant Behavior and Development* 21:315–330.

Carpenter, M., and M. Nielsen. 2008. Tools, TV, and trust: Introduction to the special issue on imitation in typically-developing children. *Journal of Experimental Child Psychology* 101:335–227.

Carr, L., M. Iacoboni, M. C. Dubeau, J. C. Mazziotta, and G. L. Lenzi. 2003. Neural mechanisms of empathy in humans: A relay from neural systems for imitation to limbic areas. *Proceedings of the National Academy of Sciences USA* 100:5497–5502.

Cartmill, E., and R. W. Byrne. 2007. Orangutans modify their gestural signaling according to their audience's comprehension. *Current Biology* 17:1–14.

Caruana, F., A. Jezzini, M. Rochat, B. Sbriscia Fioretti, I. Stoianov, M. A. Umiltà, G. Rizzolatti, and V. Gallese. 2008. Intracortical microstimulation mapping of the inner perisylvian regions: Effects on behavior and ECG outcome. *Society for Neuroscience Abstracts* F.03.

Castelli, F., C. Frith, F. Happé, and U. Frith. 2002. Autism, Asperger syndrome and brain mechanisms for the attribution of mental states to animated shapes. *Brain* 125:1839–1849.

Castles, D. L., and A. Whiten. 1998. Post-conflict behaviour of wild olive baboons. II. Stress and self-directed behaviour. *Ethology,* 104:148–160.

Castles, D. L., A. Whiten, and F. Aureli. 1999. Social anxiety, relationships and self-directed behaviour among wild female olive baboons. *Animal Behaviour* 58:1207–1215.

Catchpole, C. K., and P. J. B. Slater. 1995. *Bird song: Biological themes and variations.* Cambridge: Cambridge University Press.

Chakrabarti, S., and E. Fombonne. 2005. Pervasive developmental disorders in preschool children: Confirmation of high prevalence. *American Journal of Psychiatry* 162 (6):1133–1141.

Chalmeau, R. 1994. Do chimpanzees cooperate in a learning task? *Primates* 35:385–392.

Chapman, E., S. Baron-Cohen, B. Auyeung, R. Knickmeyer, K. Taylor, and G. Hackett. 2006. Fetal testosterone and empathy: Evidence from the Empathy Quotient (EQ) and the "Reading the Mind in the Eyes" test. *Social Neuroscience* 1:135–148.

Charman, T., J. Swettenham, S. Baron-Cohen, A. Cox, G. Baird, and A. Drew. 1997. Infants with autism: An investigation of empathy, pretend play, joint attention, and imitation. *Developmental Psychology* 33:781–789.

Chartrand, T. L., and J. A. Bargh. 1999. The chameleon effect: The perception-behavior link and social interaction. *Journal of Personality and Social Psychology* 76:893–910.

Chemero, A. 2003. An outline of a theory of affordances. *Ecological Psychology* 15:181–195.

Chemero, A., C. Klein, and W. Cordeiro. 2003. Events as changes in the layout of affordances. *Ecological Psychology* 15:19–28.

Cheney, D. L., and R. M. Seyfarth. 1989. Redirected aggression and reconciliation among vervet monkeys, *Cercopithecus aethiops. Behaviour* 110:258–275.

———. 2009. Stress and coping mechanisms in female primates. *Advances in the Study of Behavior* 39:1–44.

Cheney, D. L., R. M. Seyfarth, and J. B. Silk. 1995. The role of grunts in reconciling opponents and facilitating interactions among adult female baboons. *Animal Behaviour* 50:249–257.

Cheng, Y., C. P. Lin, H. L. Liu, Y. Y. Hsu, K. E. Lim, D. Hung, and J. Decety. 2007. Expertise modulates the perception of pain in others. *Current Biology* 17:1708–1713.

Chimpanzee Genome Consortium. 2005. Initial sequence of the chimpanzee genome and comparison with the human genome. *Nature* 437:69–87.

Chomsky, N. 2006. *Language and mind,* 3rd ed. New York: Cambridge University Press.

Church, R. M. 1959. Emotional reactions of rats to the pain of others. *Journal of Comparative and Physiological Psychology* 52:132–134.

Cisek, P., and J. F. Kalaska. 2005. Neural correlates of reaching decisions in dorsal premotor cortex: Specification of multiple direction choices and final selection of action. *Neuron* 45:801–814.

Clayton, N. S., T. J. Bussey, and A. Dickinson. 2003. Can animals recall the past and plan for the future? *Nature Reviews Neuroscience* 4:685–691.

Cohen-Bendahan, C. C., C. van de Beek, and S. A. Berenbaum. 2005. Prenatal sex hormone effects on child and adult sex-typed behavior: Methods and findings. *Neuroscience and Biobehavioral Reviews* 29 (2):353–384.

Collaer, M. L., and M. Hines. 1995. Human behavioural sex differences: A role for gonadal hormones during early development? *Psychological Bulletin* 118:55–107.

Colman, A. D., K. E. Liebold, and J. J. Boren. 1969. A method for studying altruism in monkeys. *Psychological Record* 19:401–405.

Colmenares, F. 1990. Greeting behaviour in male baboons, I: Communication, reciprocity and symmetry. *Behaviour* 113:81–116.

Connellan, J., S. Baron-Cohen, S. Wheelwright, A. Batki, and J. Ahluwalia. 2000. Sex differences in human neonatal social perception. *Infant Behavior and Development* 23 (1):113–118.

Constantino, J. N., and R. D. Todd. 2003. Autistic traits in the general population. *Archives of General Psychiatry* 60:524–530.

Cools, A. K. A., A. J. M. van Hout, and M. H. J. Nelissen. 2008. Canine reconciliation and third-party-initiated postconflict affiliation: Do peacemaking social mechanisms in dogs rival those of higher primates? *Ethology* 114: 53–63.

Cooper, M. A., F. Aureli, and M. Singh. 2007. Sex differences in reconciliation and post-conflict anxiety in bonnet macaques. *Ethology* 113:26–38.

Cordoni, G., and E. Palagi. 2008. Reconciliation in wolves *(Canis lupus)*: New evidence for a comparative perspective. *Ethology* 114:298–308.

Cordoni, G., E. Palagi, and S. B. Tarli. 2004. Reconciliation and consolation in captive western gorillas. *International Journal of Primatology* 27:1365–1382.

Couchet, H., and J. Vauclair. 2010. Features of spontaneous pointing gestures in toddlers. *Gesture* 10:86–107.

———. In press. Pointing gestures produced by toddlers from 15 to 30 months: Different functions, hand shapes and laterality patterns. *Infant Behavior and Development.*

Coudé, G., P. F. Ferrari, F. Roda, M. Maranesi, S. Rozzi, and L. Fogassi. 2009. Ventral premotor cortex of the macaque monkey controls conditioned vocalization. *Society for Neuroscience Abstracts* 82.9.

Coussi-Korbel, S., and D. M. Fragaszy. 1995. On the relation between social dynamics and social learning. *Animal Behaviour* 50:1441–1453.

Crast, J., J. M., Hardy, and D. Fragaszy. 2010. Inducing traditions in captive capuchin monkeys. *Animal Behaviour* 80: 955–964.

Crawford, M. P. 1937. The cooperative solving of problems by young chimpanzees. *Comparative Psychology Monographs* 14:1–88.

———. 1941. The cooperative solving by chimpanzees of problems requiring serial responses to color cues. *Journal of Social Psychology* 13:259–280.

Creel, S. 2001. Social dominance and stress hormones. *Trends in Ecology and Evolution* 19:491–497.

Crockford, C., I. Herbinger, L. Vigilant, and C. Boesch. 2004. Wild chimpanzees produce group-specific calls: A case for vocal learning? *Ethology* 110:221–243.

Crockford, C., R. M. Wittig, P. L. Whitten, R. M. Seyfarth, and D. L. Cheney. 2008. Social stressors and coping mechanisms in wild female baboons *(Papio hamadryas ursinus)*. *Hormones and Behavior* 53:254–265.

Cronin, K. A., and C. T. Snowdon. 2007. The effects of unequal reward distributions on cooperative problem solving by cottontop tamarins, *Saguinus oedipus*. *Animal Behaviour* 75:245–257.

Cross, E. S., A. F. Hamilton, and S. T. Grafton. 2006. Building a motor simulation de novo: Observation of dance by dancers. *Neuroimage* 31:1257–1267.

Csibra, G., and G. Gergely. 2006. Social learning and social cognition: The case for pedagogy. *Processes of Change in Brain and Cognitive Development: Attention and Performance* 21:249–274.

Custance, D. M., A. Whiten, and K. A. Bard. 1995. Can young chimpanzees *(Pan troglodytes)* imitate arbitrary actions? Hayes and Hayes (1952) revisited. *Behaviour* 132:837–859.

Custance, D. M., A. Whiten, and T. Fredman. 1999. Social learning of an artificial fruit task in capuchin monkeys *(Cebus apella)*. *Journal of Comparative Psychology* 113:13–25.

Damasio, A. R. 1994. *Descartes' error: Emotion, reason, and the human brain.* New York: Putnam.

———. 1995. On some functions of the human prefrontal cortex. *Annals of the New York Academy of Science* 769:241–251.

Damasio, A. R., and N. Geschwind. 1984. The neural basis of language. *Annual Review of Neuroscience* 7:127–147.

Damasio, H., T. Grabowski, R. Frank, A. M. Galaburda, and A. R. Damasio. 1994. The return of Phineas Gage: Clues about the brain from the skull of a famous patient. *Science* 264:1102–1105.

Dapretto, M., M. S. Davies, J. H. Pfeifer, A. A. Scott, M. Sigman, S. Y. Bookheimer, and M. Iacoboni. 2006. Understanding emotions in others: Mirror neuron dysfunction in children with autism spectrum disorders. *Nature Neuroscience* 9:28–30.

Darwin, C. [1871] 1982. *The descent of man, and selection in relation to sex.* Princeton, NJ: Princeton University Press.

———. 1872. *The expression of the emotions in man and animals.* London: John Murray.

Darwin, C., P. Ekman, and P. Prodger. 2002. *The expression of the emotions in man and animals.* Oxford: Oxford University Press.

Das, M. 2000. Conflict management via third parties: Post-conflict affiliation of the aggressor. In F. Aureli and F. B. M. de Waal, eds. *Natural conflict resolution,* 263–280. Berkeley: University of California Press.

Das, M., Z. Penke, and J. A. R. A. M. van Hooff. 1998. Postconflict affiliation and stress-related behavior of long-tailed macaque aggressors. *International Journal of Primatology* 19:53–71.

Dasser, V., I. Ulbaek, and D. Premack. 1989. The perception of intention. *Science* 243:365–367.

Davis, M. H. 1983. Measuring individual differences in empathy: Evidence for a multidimensional approach. *Journal of Personality and Social Psychology* 44:113–126.

Dawkins, R. 1976. *The Selfish Gene.* Oxford: Oxford University Press.

Decety, J., and T. Chaminade. 2003. When the self represents the other: A new cognitive neuroscience view on psychological identification. *Consciousness and Cognition* 12:577–596.

Decety, J., and C. Lamm. 2007. The role of the right temporoparietal junction in social interaction: How low-level computational processes contribute to meta-cognition. *The Neuroscientist* 13:580–593.

de Kort, S. R., N. J. Emery, and N. S. Clayton. 2006. Food sharing in jackdaws, *Corvus monedula:* What, why and with whom? *Animal Behaviour* 72:297–304.

De Marco, A., R. Cozzolino, F. Dess-Fulgheri, and B. Thierry. 2010. Conflicts induce affiliative interactions among bystanders in a tolerant species of macaque *(Macaca tonkeana). Animal Behaviour* 80:197–203.

Dennett, D. C. 1978. Beliefs about beliefs. *Behavioral and Brain Sciences* 1:568–570.

DeSilva, J., and J. Lesnik. 2006. Chimpanzee neonatal brain size: Implications for brain growth in *Homo erectus. Journal of Human Evolution* 51:207–212.

Desrochers, S., P. Morrisette, and M. Ricard. 1995. Two perspectives on pointing in infancy. In C. Moore and P. J. Dunham, eds. *Joint attention: Its origins and role in development,* 85–103. Hillsdale, NJ: Lawrence Erlbaum.

de Vignemont, F., and T. Singer. 2006. The empathic brain: How, when and why? *Trends in Cognitive Sciences* 10:435–441.

de Waal, F. B. M. 1986. Deception in the natural communication of chimpanzees. In R. Mitchell and N. Thompson, eds. *Deception: Perspectives on human and nonhuman deceit*, 221–244. New York: SUNY Press.

———. 1989. Food sharing and reciprocal obligations among chimpanzees. *Journal of Human Evolution* 18:433–459.

———. 1992. Appeasement, celebration, and food sharing in the two *Pan* species. In T. Nishida, W. C. McGrew, P. R. Marler, M. Piekford, and F. B. M. de Waal, eds. *Topics in primatology*, 37–50. Tokyo: University of Tokyo Press.

———. 1996. *Good natured: The origins of right and wrong in humans and other animals*. Cambridge, MA: Harvard University Press.

———. 1997a. The chimpanzee's service economy: Food for grooming. *Evolution and Human Behavior* 18:375–386.

———. 1997b. Food-transfers through mesh in brown capuchins. *Journal of Comparative Psychology* 111:370–378.

———. 1998. No imitation without identification. *Behavioral and Brain Sciences* 21:689.

———. 1999. Anthropomorphism and anthropodenial: Consistency in our thinking about humans and other animals. *Philosophical Topics* 27:255–280.

———. 2001. *The ape and the sushi master: Cultural reflections of a primatologist*. New York: Basic Books.

———. 2005. *Our inner ape*. New York: Riverhead Books.

———. 2007 (orig. 1982). *Chimpanzee politics: Power and sex among apes*. New York: Harper & Row.

———. 2008. Putting the altruism back into altruism: The evolution of empathy. *Annual Review of Psychology* 59: 279–300.

———. 2009a. *The age of empathy*. New York: Harmony.

———. 2009b. Darwin's last laugh. *Nature* 460:175.

de Waal, F. B. M., and F. Aureli. 1996. Consolation, reconciliation, and a possible cognitive difference between macaque and chimpanzee. In A. E. Russon, K. A. Bard, and S. T. Parker, eds. *Reaching into thought: The minds of the great apes*, 80–110. Cambridge: Cambridge University Press.

de Waal, F. B. M., and M. Berger. 2000. Payment for labour in monkeys. *Nature* 404:583.

de Waal, F. B. M., C. Boesch, V. Horner, and A. Whiten. 2008. Comparing social skills of children and apes. *Science* 319:569.

de Waal, F. B. M., and J. M. Davis. 2003. Capuchin cognitive ecology: Cooperation based on projected returns. *Neuropsychologia* 41:221–228.

de Waal, F. B. M., M. Dindo, C. A. Freeman, and M. Hall. 2005. The monkey in the mirror: Hardly a stranger. *Proceedings of the National Academy of Sciences USA* 102:11140–11147.

de Waal, F. B. M., and P. F. Ferrari. 2010. Toward a bottom-up perspective on animal and human cognition. *Trends in Cognitive Sciences* 14:201–207.

de Waal, F. B. M., K. Leimgruber, and A. R. Greenberg. 2008. Giving is self-rewarding for monkeys. *Proceedings of the National Academy of Sciences USA* 105:13685–13689.

de Waal, F. B. M., L. M. Luttrell, and M. E. Canfield. 1993. Preliminary data on voluntary food sharing in brown capuchin monkeys. *American Journal of Primatology* 29:73–78.

de Waal, F. B. M., and P. L. Tyack, eds. 2003. *Animal social complexity: Intelligence, culture and individualized societies.* Cambridge, MA: Harvard University Press.

de Waal, F. B. M., and A. van Roosmalen. 1979. Reconciliation and consolation among chimpanzees. *Behavioral Ecology and Sociobiology* 5:55–66.

Dewsbury, D. 2006. *Monkey farm: A history of the Yerkes Laboratories of Primate Biology, Orange Park, Florida, 1930–1965.* Lewisburg, PA: Bucknell University Press.

Diamond, J., and A. B. Bond. 1999. *Kea, bird of paradox: The evolution and behaviour of a New Zealand parrot.* Berkeley: University of California Press.

Dias, P. A. D., E. Rodriguez-Luna, and D. Canales-Espinosa. 2008. The functions of the "greeting ceremony" among male mantled howlers *(Alouatta palliata)* on Agaltepec Island, Mexico. *American Journal of Primatology* 70:621–628.

Dijksterhuis, A. 2005. Why we are social animals: The high road to imitation as social glue. In S. Hurley and N. Chater, eds. *Perspective on imitation: From neuroscience to social science,* 207–220. Cambridge, MA: MIT Press.

Dimberg, U., M. Thunberg, and K. Elmehed. 2000. Unconscious facial reactions to emotional facial expressions. *Psychological Science* 11:86–89.

Dindo, M., and F. B. M. de Waal. 2006. Partner effects on food consumption in brown capuchin monkeys. *American Journal of Primatology* 69:1–9.

Dindo, M., A. Whiten, and F. B. M. de Waal. 2009. In-group conformity sustains different foraging traditions in capuchin monkeys *(Cebus apella). PLoS One* 4:e7858.

Di Pellegrino, G., L. Fadiga, L. Fogassi, V. Gallese, and G. Rizzolatti. 1992. Understanding motor events: A neurophysiological study. *Experimental Brain Research* 91:176–180.

Donohue, S. E., C. Wendelken, E. A. Crone, and S. A. Bunge. 2005. Retrieving rules for behavior from long-term memory. *Neuroimage* 26:1140–1149.

Drapier, M., E. Addessi, and E. Visalberghi. 2003. Response of *Cebus apella* to foods flavored with familiar and novel odor. *International Journal of Primatology* 24:295–315.

Drewe, E. 1974. The effect of type and area of brain lesion on Wisconsin Card Sorting Test performance. *Cortex* 10:159–170.

Dufour, V., M. Pelé, M. Neumann, B. Thierry, and J. Call. 2008. Calculated reciprocity after all: Computation behind token transfers in orang-utans. *Biology Letters* 5:172–175.

Dunbar, R. I. M. 1998. The social brain hypothesis. *Evolutionary Anthropology* 6:178–190.

Ehrhardt, A. A., and S. W. Baker. 1974. Fetal androgens, human central nervous system differentiation, and behavior sex differences. In *Sex differences in behavior,* edited by R. C. Friedman, R. R. Richart, and R. L. Vande Wiele. New York: Wiley.

Eisenberg, N. 2000. Emotion, regulation, and moral development. *Annual Review of Psychology* 51:665–697.

Eisenberg, N., R. A. Fabes, P. A. Miller, J. Fultz, R. Shell, R. M. Mathy, and R. R. Reno. 1989. Relation of sympathy and personal distress to prosocial behavior: A multimethod study. *Journal of Personality and Social Psychology* 57:55–66.

Eisenberg, N., and P. Miller. 1987. The relation of empathy to prosocial and related behaviors. *Psychological Bulletin* 101:91–119.

Eisenberg, N., and J. Strayer. 1987. *Empathy and its development.* Cambridge: Cambridge University Press.

Ekman, L. 1992. Facial expressions of emotion: An old controversy and new findings. *Philosophical Transactions of the Royal Society B: Biological Sciences* 335:63–69.

Emery, N. J. 2000. The eyes have it: The neuroethology, function and evolution of social gaze. *Neuroscience and Biobehavioral Reviews* 24:581–604.

Emery, N. J., and N. S. Clayton. 2004. The mentality of crows: Convergent evolution of intelligence in corvids and apes. *Science* 306:1903–1907.

Emery, N. J., N. S. Clayton, and C. D. Frith. 2007. Introduction. Social intelligence: From brain to culture. *Philosophical Transactions of the Royal Society B: Biological Sciences* 362:485–488.

Emery-Thompson, M., J. H. Jones, A. E. Pusey, S. Brewer-Marsden, J. Goodall, D. Marsden, T. Matsuzawa, T. Nishida, V. Reynolds, Y. Sugiyama, and R. W. Wrangham. 2007. Aging and fertility in wild chimpanzees provide insights into the evolution of menopause. *Current Biology* 17:2150–2156.

Engh, A. L., J. C. Beehner, T. J. Bergman, P. L. Whitten, R. R. Hoffmeier, R. M. Seyfarth, and D. L. Cheney. 2006a. Behavioural and hormonal responses to predation in female chacma baboons *(Papio hamadryas ursinus). Proceedings of the Royal Society B: Biological Sciences* 273:707–712.

———. 2006b. Female hierarchy instability, male immigration and infanticide increase glucocorticoid levels in female chacma baboons. *Animal Behaviour* 71:1227–1237.

Eriksson, K., M. Enquist, and S. Ghirlanda. 2007. Critical points in current theory of conformist social learning. *Journal of Evolutionary Psychology* 5:67–87.

Fadiga, L., L. Craighero, G. Buccino, and G. Rizzolatti. 2002. Speech listening specifically modulates the excitability of tongue muscles: A TMS study. *European Journal of Neuroscience* 15:399–402.

Fadiga, L., L. Fogassi, G. Pavesi, and G. Rizzolatti. 1995. Motor facilitation during action observation: A magnetic stimulation study. *Journal of Neurophysiology* 73:2608–2611.

Falk, D. 2004. Prelinguistic evolution in early hominins: Whence motherese? *Behavioral and Brain Sciences* 27:491–503.

Fausto-Sterling, A. 1992. *Myths of gender*. New York: Basic Books.

Feenders, G., M. Liedvogel, M. Rivas, M. Zapka, H. Horita, E. Hara, K. Wada, H. Mouritsen, and E. D. Jarvis. 2008. Molecular mapping of movement-associated areas in the avian brain: A motor theory for vocal learning origin. *PLoS One* 3:e1768.

Fehr, E., and U. Fischbacher. 2003. The nature of human altruism. *Nature* 425: 785–791.

Fehr, E., and S. Gächter. 2002. Altruistic punishment in humans. *Nature* 415: 137–140.

Fein, D., L. Waterhouse, D. Lucci, B. Pennington, and M. Humes. 1985. Handedness and cognitve functions in pervasive developmental disorders. *Journal of Autism and Developmental Disorders* 15:323–333.

Feistner, A. T. C., and W. C. McGrew. 1989. Food-sharing in primates: A critical review. In P. K. Seth and S. Seth, eds. *Perspectives in primate biology,* 3:21–36. New Delhi: Today and Tomorrow's.

Ferrari, P. F., L. Bonini, and L. Fogassi. 2009. From monkey mirror neurons to mirror-related behaviours: Possible "direct" and "indirect" pathways. *Philosophical Transactions of the Royal Society B: Biological Sciences* 364:2311–2323.

Ferrari, P. F., V. Gallese, G. Rizzolatti, and L. Fogassi. 2003. Mirror neurons responding to the observation of ingestive and communicative mouth actions in the monkey ventral premotor cortex. *European Journal of Neuroscience* 17:1703–1714.

Ferrari, P. F., E. Kohler, L. Fogassi, and V. Gallese. 2000. The ability to follow eye gaze and its emergence during development in macaque monkeys. *Proceedings of the National Academy of Sciences USA* 97:13997–14002.

Ferrari, P. F., C. Maiolini, E. Addessi, L. Fogassi, and E. Visalberghi. 2005. The observation and hearing of eating actions activates motor programs related to eating in macaque monkeys. *Behavioural Brain Research* 161:95–101.

Ferrari, P. F., A. Paukner, C. Ionica, and S. J. Suomi. 2009. Reciprocal face-to-face communication between rhesus macaque mothers and their newborn infants. *Current Biology* 19:1768–1772.

Ferrari, P. F., A. Paukner, A. Ruggiero, L. Darcey, S. Unbehagen, and S. J. Suomi. 2009. Interindividual differences in neonatal imitation and the development of action chains in rhesus macaques. *Child Development* 80:1057–1068.

Ferrari, P. F., S. Rozzi, and L. Fogassi. 2005. Mirror neurons responding to observation of actions made with tools in monkey ventral premotor cortex. *Journal of Cognitive Neuroscience* 17:212–226.

Ferrari, P. F., R. Vanderwert, K. Herman, A. Paukner, N. A. Fox, and S. J. Suomi. 2008. EEG activity in response to facial gestures in 1–7 days old infant rhesus macaques. *Society for Neuroscience Abstracts* 297.13.

Ferrari, P. F, E. Visalberghi, A. Paukner, L. Fogassi, A. Ruggiero, and S. J. Suomi. 2006. Neonatal imitation in rhesus macaques. *PLoS Biology* 4: e302.

Finegan, J. K., G. A. Niccols, and G. Sitarenios. 1992. Relations between prenatal testosterone levels and cognitive abilities at 4 years. *Developmental Psychology* 28 (6):1075–1089.

Flombaum, J. I., and L. R. Santos. 2005. Rhesus monkeys attribute perceptions to others. *Current Biology* 15:447–452.

Flor, H., N. Birbaumer, C. Hermann, S. Ziegler, and C. J. Patrick. 2002. Aversive Pavlovian conditioning in psychopaths: Peripheral and central correlates. *Psychophysiology* 39:505–518.

Fogassi, L., and P. F. Ferrari. 2007. Mirror neurons and the evolution of embodied language. *Current Directions in Psychological Science* 16:136–141.

———. 2010. Mirror systems. *WIREs Cognitive Science*, DOI: 10.1002/wcs.89.

Fogassi, L, P. F. Ferrari, B. Gesierich, S. Rozzi, F. Chersi, and G. Rizzolatti. 2005. Parietal lobe: From action organization to intention understanding. *Science* 308:662–667.

Fogassi, L., and V. Gallese. 2002. The neural correlates of action understanding in non-human primates. In M. I. Stamenov and V. Gallese, eds. *Mirror neurons and the evolution of brain and language,* 13–35. Philadelphia: John Benjamins.

Forest, M. G., P. C. Sizonenko, A. M. Cathiard, and J. Bertrand. 1974. Hypophyso-gonadal function in humans during the first year of life: I. Evidence for testicular activity in early infancy. *Journal of Clinical Investigation* 53:819–828.

Foundas, A., C. Leonard, and K. Heilman. 1995. Morphological cerebral asymmetries and handedness: The pars triangularis and planum temporale. *Archives of Neurology* 52:501–508.

Foundas, A. L., K. F. Eure, L. F. Luevano, and D. R. Weinberger. 1998. MRI asymmetries of Broca's area: The pars triangularis and pars opercularis. *Brain and Language* 64:282.

Fouts, R., and T. Mills. 1997. *Next of kin.* New York: Morrow.

Fragaszy, D. M. 1986. Time budgets and foraging behavior in wedge-capped capuchins *(Cebus olivaceus):* Age and sex differences. In D. M. Taub and F. A. King, eds. *Current perspectives in primate social dynamics,* 159–174. New York: Van Nostrand Reinhold.

Fragaszy, D. M., and K. A. Bard. 1997. Comparisons of development and life history in *Pan* and *Cebus. International Journal of Primatology* 18:683–701.

Fragaszy, D. M., and S. E. Cummins-Sebree. 2005. Relational spatial reasoning by a nonhuman: The example of capuchin monkeys. *Behavioral and Cognitive Neuroscience Reviews* 4:282–306.

Fragaszy, D. M., J. M. Feuerstein, and D. Mitra. 1997. Transfers of food from adults to infants in tufted capuchins *(Cebus apella). Journal of Comparative Psychology* 111:194–200.

Fragaszy, D. M., R. Greenberg, E. Visalberghi, E. B. Ottoni, P. Izar, and Q. Liu. 2010. How wild bearded capuchin monkeys select stones and nuts to minimize the number of strikes per nut cracked. *Animal Behaviour* 80 (2):205–214.

Fragaszy, D. M., and S. Perry, eds. 2003. *The biology of traditions: Models and evidence.* Cambridge: Cambridge University Press.

Fragaszy, D. M., T. Pickering, Q. Liu, P. Izar, E. B. Ottoni, and E. Visalberghi. 2010. Bearded capuchin monkeys' and a human's efficiency at cracking palm nuts with stone tools: Field experiments. *Animal Behaviour* 79:321–332.

Fragaszy, D. M., and E. Visalberghi. 2001. Recognizing a swan: Socially-biased learning. *Psychologia* 44:82–98.

Fragaszy, D. M., E. Visalberghi, and L. M. Fedigan. 2004. *The complete capuchin: The biology of the genus* Cebus. Cambridge: Cambridge University Press.

Fraser, O. N., and F. Aureli. 2008. Reconciliation, consolation and postconflict behavioral specificity in chimpanzees. *American Journal of Primatology* 70: 1114–1123.

Fraser, O. N., and T. Bugnyar. 2010. Do ravens show consolation? Responses to distressed others. *PLoS One* 5:e10605.

Fraser, O. N., S. E. Koski, R. M. Wittig, and F. Aureli. 2009. Why are bystanders friendly to recipients of aggression? *Communicative and Integrative Biology* 2:285–291.

Fraser, O. N., D. Stahl, and F. Aureli. 2008. Stress reduction through consolation in chimpanzees. *Proceedings of the National Academy of Sciences USA* 105:8557–8562.

———. 2010. Function and determinants of reconciliation in *Pantroglodytes*. *International Journal of Primatology* 31:39–57.

Fredman, T., and A. Whiten. 2008. Observational learning from tool using models by human-reared and mother-reared capuchin monkeys *(Cebus apella)*. *Animal Cognition* 11 (2): 295–309.

Frith, C. D., and T. Singer. 2008. The role of social cognition in decision making. *Philosophical Transactions of the Royal Society B: Biological Sciences* 363:3875–3886.

Frith, U. 2001. Mind blindness and the brain in autism. *Neuron* 32:969–979.

Frith, U., and C. D. Frith. 2003. Development and neurophysiology of mentalizing. *Philosophical Transactions of the Royal Society B: Biological Sciences* 358:459–473.

Fritz, J., and K. Kotrschal. 1999. Social learning in common ravens, *Corvus corax*. *Animal Behaviour* 57:785–793.

Frost, J. A., J. R. Binder, J. A. Springer, T. A. Hammeke, P. S. F. Bellgowan,. S. M. Rao, and R. W. Cox. 1999. Language processing is strongly left lateralized in both sexes: Evidence from functional MRI. *Brain* 122:199–208.

Fruth, B., and G. Hohmann. 2002. How bonobos handle hunts and harvests: Why share food? In C. Boesch, G. Hohmann, and L. Marchant, eds. *Behavioural diversity in chimpanzees and bonobos,* 231–243. Cambridge: Cambridge University Press.

Fuchs, F., and A. Klopper. 1983. *Endocrinology of pregnancy*. Philadelphia: Harper & Row.

Fujii, N., S. Hihara, and A. Iriki. 2007. Dynamic social adaptation of motion-related neurons in primate parietal cortex. *PLoS One* 2:e397.

———. 2008. Social cognition in premotor and parietal cortex. *Social Neuroscience* 3:250–260.

Fujii, N., S. Hihara, Y. Nagasaka, and A. Iriki. 2009. Social state representation in prefrontal cortex. *Social Neuroscience* 4:73–84.

Fujisawa, K. K., N. Kutskake, and T. Hasegawa. 2005. Reconciliation pattern after aggression among Japanese preschool children. *Aggressive Behavior* 31:138–152.

Funahashi, S. 2008. Neural mechanisms of decision making. *Brain Nerve* 60: 1017–10127.

Funahashi, S., M. V. Chafee, and P. S. Goldman-Rakic. 1993. Prefrontal neuronal activity in rhesus monkeys performing a delayed anti-saccade task. *Nature* 365:753–756.

Funk, J., C. Fox, M. Chan, and K. Curtiss. 2008. The development of the children's empathic attitudes questionnaire using classical and Rasch analyses. *Journal of Applied Developmental Psychology* 29:187–196.

Fuster, J. M. 2000. Executive frontal functions. *Experimental Brain Research* 133:66–70.

Galef, B. B. 2005. Breathing new life into the study of imitation by animals: What and when do chimpanzees imitate? In S. Hurley and N. Chater, eds. *Perspectives on imitation: From neuroscience to social science*, 1:295–297. Cambridge, MA: MIT Press.

Galef, B. G. 1993. Function of social learning about food: A causal analysis of effects of diet novelty on preference transmission. *Animal Behaviour* 46:257–265.

Galef, B. G., and K. N. Laland. 2005. Social learning in animals: Empirical studies and theoretical models. *BioScience* 55:489–499.

Galef, B. G., and E. E. Whiskin. 2008. "Conformity" in Norway rats? *Animal Behaviour* 75:2035–2039.

Galis, F., C. M. Ten Broek, S. Van Dongen, and L. C. Wijnaendts. 2010. Sexual dimorphism in the prenatal digit ratio (2D:4D). *Archives of Sexual Behavior* 39 (1):57–62.

Gallagher, H. L., and C. D. Frith. 2003. Functional imaging of "theory of mind." *Trends in Cognitive Sciences* 7:77–83.

———. 2004. Dissociable neural pathways for the perception and recognition of expressive and instrumental gestures. *Neuropsychologia* 42:1725–1736.

Gallese, V. 2001. The "shared manifold" hypothesis: From mirror neurons to empathy. *Journal of Consciousness Studies* 8:33–50.

Gallese, V., L. Fadiga, L. Fogassi, and G. Rizzolatti. 1996. Action recognition in the premotor cortex. *Brain* 119:593–609.

———. 2002. Action representation and the inferior parietal lobule. In W. Prinz and B. Hommel, eds. *Common mechanisms in perception and action: Attention and performance*, 19:334–355. Oxford: Oxford University Press.

Gallese, V., and A. Goldman. 1998. Mirror neurons and the simulation theory of mind-reading. *Trends in Cognitive Sciences* 12:493–501.

Gallup, G. G., Jr. 1970. Chimpanzees: Self-recognition. *Science* 167: 86–87.

———. 1977. Self-recognition in primates: A comparative approach to the bidirectional properties of consciousness. *American Psychologist* 32:329–338.

———. 1983. Toward a comparative psychology of mind. In R. L. Mellgren, ed. *Animal cognition and behavior*, 473–510. Amsterdam: North-Holland.

Gallup, G. G., Jr., D. J. Povinelli, S. D. Suarez, J. R. Anderson, J. Lethmate, and E. W. Menzel. 1995. Further reflections on self-recognition in primates. *Animal Behavior* 50:1525–1532.

Gardner, B. T., and R. A. Gardner. 1971. Two-way communication with an infant chimpanzee. In A. M. Schrier and F. Stollnitz, eds. *Behavior of nonhuman primates: Modern research trends,* 117–184. New York: Academic Press.

Gardner, R. A., and B. T. Gardner. 1969. Teaching sign language to a chimpanzee. *Science* 165:664–672.

———. 1989. A cross-fostering laboratory. In R. A. Gardner, B. T. Gardner, and T. E. Van Cantfort, eds. *Teaching sign language to chimpanzees,* 1–28. Albany, NY: SUNY Press.

Gattis, M. 2002. Imitation is mediated by many goals, not just one. *Developmental Science* 5:27–29.

Gazzola, V., L. Aziz-Zadeh, and C. Keysers. 2006. Empathy and the somatotopic auditory mirror system in humans. *Current Biology* 16:1824–1829.

George, F. W., and J. D. Wilson. 1992. Embryology of the genital tract. In P. C. Walsh, A. B. Retik, and T. A. Stamey, eds. *Campbell's urology,* 6th ed. Philadelphia: W. B. Saunders.

Gergely, G., H. Bekkering, and I. Király. 2002. Rational imitation in preverbal infants. *Nature* 415:755.

Gergely, G., and G. Csibra. 2003. Teleological reasoning in infancy: The naive theory of rational action. *Trends in Cognitive Sciences* 7:287–292.

———. 2005. The social construction of the cultural mind: Imitative learning as a mechanism of human pedagogy. *Interaction Studies* 6:463–481.

———. 2006. Sylvia's recipe: The role of imitation and pedagogy in the transmission of cultural knowledge. In N. J. Enfield and S. C. Levenson, eds. *The roots of human sociality: Culture, cognition, and human interaction,* 229–255. Oxford: Berg.

Gerloff, U., B. Hartung, B. Fruth, G. Hohmann, and D. Tautz. 1999. Intracommunity relationships, dispersal pattern and paternity success in a wild living community of bonobos *(Pan paniscus)* determined from DNA analysis of faecal samples. *Proceedings of the Royal Society B: Biological Sciences* 266:1189–1195.

Ghazanfar, A. A., and L. R. Santos. 2004. Primate brains in the wild: The sensory bases for social interactions. *Nature Reviews Neuroscience* 5: 603–616.

Gibson, J. J. 1979. *The ecological approach to visual perception.* Boston: Houghton Mifflin.

Gilby, I. C. 2006. Meat sharing among the Gombe chimpanzees: Harassment and reciprocal exchange. *Animal Behaviour* 71:953–963.

Gillberg, C. 1983. Autistic children's hand preferences: Results from an epidemiological study of infantile autism. *Psychiatry Research* 10 (1):21–30.

Gillberg, C., M. Cederlund, K. Lamberg, and L. Zeijlon. 2006. Brief report: "The autism epidemic"; The registered prevalence of autism in a Swedish urban area. *Journal of Autism and Developmental Disorders* 36 (3):429–435.

Gogtay, N., J. N. Giedd, L. Lusk, K. M. Hayashi, D. Greenstein, A. C. Vaituzis, T. F. Nugent, D. H., Herman, L. S. Clasen, A. W. Toga, J. L. Rapoport, and P. M. Thompson. 2004. Dynamic mapping of human cortical development during childhood through early adulthood. *Proceedings of the National Academy of Sciences USA* 101:8174–8179.

Golan, O., S. Baron-Cohen, and J. Hill. 2006. The Cambridge Mindreading (CAM) Face-Voice Battery: Testing complex emotion recognition in adults with and without Asperger syndrome. *Journal of Autism and Developmental Disorders* 36 (2):169–183.

Goldenberg, G., and H. O. Karnath. 2006. The neural basis of imitation is body part specific. *Journal of Neuroscience* 26:6282–6287.

Goldman, A., and F. de Vignemont. 2009. Is social cognition embodied? *Trends in Cognitive Sciences* 13:154–159.

Goldman, A. I. 2006. *Simulating minds: The philosophy, psychology, and neuroscience of mindreading.* Oxford: Oxford University Press.

Gómez, J. C. 1991. Visual behaviour as a window for reading the mind of others in primates. In A. Whiten, ed. *Natural theories of mind: Evolution, development and simulation of everyday mindreading,* 195–207. Oxford: Basil Blackwell.

———. 1996. Non-human primate theories of (non-human primate) minds: Some issues concerning the origins of mind reading. In P. Carruthers and P. K. Smith, eds. *Theories of theories of mind,* 330–343. Cambridge: Cambridge University Press.

———. 2005. Requesting gestures in captive monkeys and apes. *Gesture* 5:89–103.

Good, C. D., I. Johnstrude, J. Ashburner, R. N. A. Henson, K. J. Friston, and R. S. J. Frackowiak. 2001. Cerebral asymmetry and the effects of sex and handedness on brain structure: A voxel-based morphometric analysis of 465 normal human brains. *NeuroImage* 14:685–700.

Goodall, J. 1968. The behaviour of free-living chimpanzees in the Gombe stream area. *Animal Behaviour Monographs* 1:161–311.

———. 1986. *The chimpanzees of Gombe: Patterns of behavior.* Cambridge, MA: Belknap Press of Harvard University Press.

Goosen, C. 1987. Social grooming in primates. In G. Mitchell and J. Erwin, eds. *Comparative primate biology.* Vol. 2B, *Behavior, cognition and motivation,* 107–131. New York: Alan R. Liss.

Gothard, K. M., F. P. Battaglia, C. A. Erickson, K. M. Spitler, and D. G. Amaral. 2007. Neural responses to facial expression and face identity in the monkey amygdala. *Journal of Neurophysiology* 97:1671–1683.

Goy, R. W., F. B. Bercovitch, and M. C. McBrair. 1988. Behavioral masculinization is independent of genital masculinization in prenatally androgenized female rhesus macaques. *Hormones and Behavior* 22:552–571.

Goy, R. W., and B. S. McEwen. 1980. *Sexual differentiation of the brain.* Cambridge, MA: MIT Press.

Grafton, S. T. 2009. Embodied cognition and the simulation of action to understand others. *The Year in Cognitive Neuroscience 2009: Annals of the New York Academy of Sciences* 1156:97–117.

Grandin, T., and C. Johnson. 2005. *Animals in translation: Using the mysteries of autism to decode animal behavior.* New York: Scribner.

Greenberg, J. R., K. Hamann, F. Warneken, and M. Tomasello. 2010. Chimpanzee helping in collaborative and noncollaborative contexts. *Animal Behaviour* 82 (5):873–880.

Grezes, J., and J. Decety. 2001. Functional anatomy of execution, mental simulation, observation and verb generation actions: A meta-analysis. *Human Brain Mapping* 12:1–19.

Griffin, D. R. 1976. *The question of animal awareness: Evolutionary continuity of mental experience.* New York: Rockefeller University Press.

Grimshaw, G. M., M. P. Bryden, and J. K. Finegan. 1995. Relations between prenatal testosterone and cerebral lateralization in children. *Neuropsychology* 9 (1):68–79.

Grimshaw, G. M., G. Sitarenios, and J. K. Finegan. 1995. Mental rotation at 7 years: Relations with prenatal testosterone levels and spatial play experiences. *Brain and Cognition* 29 (1):85–100.

Grosbras, M. H., A. R. Laird, and T. Paus. 2005. Cortical regions involved in eye movements, shifts of attention, and gaze perception. *Human Brain Mapping* 25:140–154.

Grühn, D., K. Rebucal, M. Diehl, M. Lumley, and G. Labouvie-Vief. 2008. Empathy across the adult lifespan: Longitudinal and experience-sampling findings. *Emotion* 8:753–765.

Gu, X., and S. Han. 2007. Attention and reality constraints on the neural processes of empathy for pain. *Neuroimage* 36:256–267.

Gunst, N., S. Boinski, and D. M. Fragaszy. 2008. Acquisition of foraging competence in wild brown capuchins *(Cebus apella)*, with special reference to conspecifics' foraging artefacts as an indirect social influence. *Behaviour* 145: 195–229.

———. 2010. Development of skilled detection and extraction of embedded prey by wild brown capuchin monkeys *(Cebus apella apella). Journal of Comparative Psychology* 124 (2): 194–204.

Guthrie, E. R., and G. P. Horton. 1946. *Cats in a puzzle box.* New York: Rinehart.

Haldane, J. B. S. 1963. The truth about death. *Journal of Genetics* 58:464.

Hallock, M. B., J. Worobey, and P. Self. 1989. Behavioral development in chimpanzees *(Pan troglodytes)* and human newborns across the first month of life. *International Journal of Behavioral Development* 12:527–540.

Hamada, Y., and T. Udono. 2006. Understanding the growth pattern of chimpanzees: Does it conserve the pattern of the common ancestor of humans and chimpanzees? In T. Matsuzawa, M. Tomonaga, and M. Tanaka, eds. *Cognitive development in chimpanzees,* 96–112. Tokyo: Springer-Verlag.

Hampson, E., J. F. Rovet, and D. Altmann. 1998. Spatial reasoning in children with congenital adrenal hyperplasia due to 21-hydroxylase deficiency. *Developmental Neuropsychology* 14 (2):299–320.

Hanus, D., and J. Call. 2007. Discrete quantity judgments in the great apes *(Pan paniscus, Pan troglodytes, Gorilla gorilla, Pongo pygmaeus)*: The effect of presenting whole sets versus item-by-item. *Journal of Comparative Psychology* 121:241–249.

Happé, F., and U. Frith. 1996. The neuropsychology of autism. *Brain* 119: 1377–1400.

Harbaugh, W. T., U. Mayr, and D. R. Burghart. 2007. Neural responses to taxation and voluntary giving reveal motives for charitable donations. *Science* 316:1622–1625.

Harcourt, A. H., and F. B. M. de Waal. 1992. *Coalitions and alliances in humans and other animals*. Oxford: Oxford University Press.

Hare, B. 2001. Can competitive paradigms increase the validity of experiments on primate social cognition? *Animal Cognition* 4:269–280.

———. 2007. From nonhuman to human mind. What changed and why? *Current Directions in Psychological Science* 16:60–64.

Hare, B., M. Brown, C. Williamson, and M. Tomasello. 2002. The domestication of social cognition in dogs. *Science* 298:1634–1636.

Hare, B., J. Call, B. Agnetta, and M. Tomasello. 2000. Chimpanzees know what conspecifics do and do not see. *Animal Behaviour* 59:771–785.

Hare, B., J. Call, and M. Tomasello. 2001. Do chimpanzees know what conspecifics know? *Animal Behaviour* 61:139–151.

———. 2006. Chimpanzees deceive a human competitor by hiding. *Cognition* 101:495–514.

Hare, B., and S. Kwetuenda. 2010. Bonobos voluntarily share their own food with others. *Current Biology* 20:R230–R231.

Hare, B., A. P. Melis, V. Woods, S. Hastings, and R. Wrangham. 2007. Tolerance allows bonobos to outperform chimpanzees on a cooperative task. *Current Biology* 17:619–623.

Hare, B., and M. Tomasello. 2004. Chimpanzees are more skilful in competitive than in cooperative cognitive tasks. *Animal Behaviour* 68:571–581.

———. 2005. Human-like social skills in dogs? *Trends in Cognitive Sciences* 9:439–444.

Hare, R. D. 1991. *The Hare psychopathy checklist—revised*. Toronto: Multi-Health Systems.

Hari, R., N. Forss, S. Avikainen, E. Kirveskari, S. Salenius, and G. Rizzolatti. 1998. Activation of human primary motor cortex during action observation: A neuroimaging study. *Proceedings of the National Academy of Sciences USA* 95:15061–15065.

Hattori, Y., F. Kano, and M. Tomonaga. 2010. Differential sensitivity to conspecific and allospecific cues in chimpanzees and humans: A comparative eye-tracking study. *Biology Letters,* DOI: 10.1098/rsbl.2010.0120.

Hattori, Y., H. Kuroshima, and K. Fujita. 2007. I know you are not looking at me: Capuchin monkeys' *(Cebus apella)* sensitivity to human attentional states. *Animal Cognition* 10:141–148.

Haun, D. B., and J. Call. 2008. Imitation recognition in great apes. *Current Biology* 18:R288–290.

Hauser, M. D. 2009. The possibility of impossible cultures. *Nature* 460:190–196.

Hauser, M. D., T. Williams, J. D. Kralik, and D. Moskovitz. 2001. What guides a search for food that has disappeared? Experiments on cotton-top tamarins *(Saguinus oedipus). Journal of Comparative Psychology* 115:140–151.

Hawkes, K., J. O'Connell, N. Blurton-Jones, H. Alvarez, and E. Charnov. 1990. Grandmothering, menopause, and the evolution of human life histories. *Proceedings of the National Academy of Sciences USA* 95:1336–1339.

Hayashi, M., and T. Matsuzawa. 2003. Cognitive development in object manipulation by infant chimpanzees. *Animal Cognition* 6:225–233.

Hayes, K., and C. Hayes. 1952. Imitation in a home-raised chimpanzee. *Journal of Comparative and Physiological Psychology* 45:450–459.

Hayes, K. J., and C. H. Nissen. 1971. Higher mental functions of a home-raised chimpanzee. In A. M. Schrier and F. Stollnitz, eds. *Behavior of nonhuman primates: Modern research trends,* 59–115. New York: Academic Press.

Hein, G., G. Silani, K. Preuschoff, C. D. Batson, and T. Singer. 2009. *Ingroup bias in neural empathy predicts ingroup favoritism in costly helping.* Manuscript submitted for publication.

Heinrich, B. 1989. *Ravens in winter.* New York: Summit Books of Simon & Schuster.

Heiser, M., M. Iacoboni, F. Maeda, J. Marcus, and J. C. Mazziotta. 2003. The essential role of Broca's area in imitation. *European Journal of Neuroscience* 17:1123–1128

Henrich, J., S. J. Heine, and N. Norenzayan. 2010. The weirdest people in the world? *Behavioral and Brain Sciences* 33:61–135.

Herculano-Houzel, S. 2009. The human brain in numbers: A linearly scaled-up primate brain. *Frontiers in Human Neuroscience* 3:1–11.

Herrmann, E., J. Call, M. V. Hernandez-Lloreda, B. Hare, and M. Tomasello. 2007. Humans have evolved specialized skills of social cognition: The cultural intelligence hypothesis. *Science* 317:1360–1366.

Herrmann, E., M. V. Hernandez-Lloreda, J. Call, B. Hare, and M. Tomasello. 2010. The structure of individual differences in the cognitive abilities of children and chimpanzees. *Psychological Science* 21:102–110.

Herrmann, E., A. P. Melis, and M. Tomasello. 2006. Apes' use of iconic cues in the object-choice task. *Animal Cognition* 9:118–130.

Heyes, C. M. 1994. Reflections on self-recognition in primates. *Animal Behaviour* 47:909–919.

———. 1998. Theory of mind in non-human primates. *Behavioral and Brain Sciences* 21:101–148.

———. 2001. Causes and consequences of imitation. *Trends in Cognitive Sciences* 5:253–261.

Hier, D. B., and W. F. Crowley. 1982. Spatial ability in androgen-deficient men. *New England Journal of Medicine* 306:1202–1205.

Higham, J. P., A. M. Maclarnon, M. Heistermann, C. Ross, and S. Semple. 2009. Rates of self-directed behaviour and faecal glucocorticoid levels are not correlated in female wild olive baboons *(Papio hamadryas anubis). Stress* 12:526–532.

Hill, E., S. Berthoz, and U. Frith. 2004. Brief report: Cognitive processing of own emotions in individuals with autistic spectrum disorder and in their relatives. *Journal of Autism and Developmental Disorders* 34:229–235.

Hill, K. 2002. Altruistic cooperation during foraging by the ache, and the evolved human predisposition to cooperate. *Human Nature* 13:105–128.

Hines, M. 2004. *Brain gender.* New York: Oxford University Press.

Hines, M., S. F. Ahmed, and I. A. Hughes. 2003. Psychological outcomes and gender-related development in complete androgen insensitivity syndrome. *Archives of Sexual Behavior* 32 (2):93–101.

Hines, M., C. Brook, and G. S. Conway. 2004. Androgen and psychosexual development: Core gender identity, sexual orientation and recalled childhood gender role behavior in women and men with congenital adrenal hyperplasia (CAH). *Journal of Sex Research* 41 (1):75–81.

Hines, M., B. A. Fane, V. L. Pasterski, G. A. Matthews, G. S. Conway, and C. Brook. 2003. Spatial abilities following prenatal androgen abnormality: Targeting and mental rotations performance in individuals with congenital adrenal hyperplasia. *Psychoneuroendocrinology* 28:1010–1026.

Hines, M., and C. Shipley. 1984. Prenatal exposure to diethylstilbestrol (DES) and the development of sexually dimorphic cognitive abilities and cerebral lateralization. *Developmental Psychology* 20 (1):81–94.

Hirata, S. 2003. Cooperation in chimpanzees [in Japanese]. *Hattatsu* 95: 103–111.

———. 2006. Tactical deception and understanding of others in chimpanzees. In T. Matsuzawa, M. Tomanaga, and M. Tanaka, eds. *Cognitive development in chimpanzees,* 265–276. Tokyo: Springer-Verlag.

Hirata, S., and M. Celli. 2003. Role of mothers in the acquisition of tool-use behaviours by captive infant chimpanzees. *Animal Cognition* 6:235–244.

Hirata, S., and K. Fuwa. 2007. Chimpanzees *(Pan troglodytes)* learn to act with other individuals in a cooperative task. *Primates* 48:13–21.

Hirata, S., G. Yamakoshi, S. Fujita, G. Ohashi, and T. Matsuzawa. 2001. Capturing and toying with hyraxes *(Dendrohyrax dorsalis)* by wild chimpanzees *(Pan troglodytes)* at Bossou, Guinea. *American Journal of Primatology* 53:93–97.

Hockings, K. J., J. R. Anderson, and T. Matsuzawa. 2006. Road crossing in chimpanzees: A risky business. *Current Biology* 16:668–670.

Hoffman, K. L., K. M. Gothard, M. C. Schmid, and N. K. Logothetis. 2007. Facial-expression and gaze-selective responses in the monkey amygdala. *Current Biology* 17:766–772.

Hoffman, M. L. 1975. Developmental synthesis of affect and cognition and its implications for altruistic motivation. *Developmental Psychology* 11:607–622.

———. 1981a. The development of empathy. In J. P. Rushton and R. M. Sorrentino, eds. *Altruism and helping behavior: Social, personality, and developmental perspectives,* 41–63. Hillsdale, NJ: Lawrence Erlbaum.

———. 1981b. Is altruism part of human nature? *Journal of Personality and Social Psychology* 40:121–137.

————. 1982. Development of prosocial motivation: Empathy and guilt. In N. Eisenberg, ed. *The development of prosocial behavior*, 281–38. New York: Academic Press.

————. 2000. *Empathy and moral development: Implications for caring and justice*. Cambridge: Cambridge University Press.

Hohmann, G., R. Mundry, and T. Deschner. 2008. The relationship between socio-sexual behavior and salivary cortisol in bonobos: Tests of the tension regulation hypothesis. *American Journal of Primatology* 70:1–10.

Hölldobler, B., and E. O. Wilson. 1994. *Journey to the ants*. Cambridge, MA: Belknap Press of Harvard University Press.

Honess, P. E., and C. M. Marin. 2006. Behavioural and physiological aspects of stress and aggression in nonhuman primates. *Neuroscience and Biobehavioral Reviews* 30:390–412.

Hopkins, W. D., and M. Cantero. 2003. From hand to mouth in the evolution of language: The influence of vocal behavior on lateralized hand use in manual gestures by chimpanzees *(Pan troglodytes)*. *Developmental Science* 6:55–61.

Hopkins, W. D., L. Marino, J. K. Rilling, and L. A. Macgregor. 1998. Planum temporale asymmetries in great apes as revealed by magnetic resonance imaging (MRI). *NeuroReport* 9:2913–2918.

Hopkins, W. D., J. Russell, H. Freeman, N. Buehler, E. Reynolds, and S. Schapiro. 2005. The distribution and development of handedness for manual gestures in captive chimpanzees *(Pan troglodytes)*. *Psychological Science* 16:487.

Hopkins, W. D., J. P. Taglialatela, and D. A. Leavens. 2007. Chimpanzees differentially produce novel vocalizations to capture the attention of a human. *Animal Behaviour* 73:281–286.

Hopkins, W. D., J. P. Taglialatela, A. Meguerditchian, T. Nir, N. M. Schenker, and C. C. Sherwood. 2008. Gray matter asymmetries in chimpanzees as revealed by voxel-based morphometry. *NeuroImage* 42:491–497.

Hopper, L. M. 2010. "Ghost" experiments and the dissection of social learning in humans and animals. *Biological Reviews* 85:685–701.

Hopper, L. M., E. G. Flynn, L. A. N. Wood, and A. Whiten. 2010. Observational learning of tool use in children: Investigating cultural spread through diffusion chains and learning mechanisms through ghost displays. *Journal of Experimental Child Psychology* 106:82–97.

Hopper, L. M., S. P. Lambeth, S. J. Schapiro, and A. Whiten. 2008. Observational learning in chimpanzees and children studied through "ghost" conditions. *Proceedings of the Royal Society B: Biological Sciences* 275:835–840.

Hopper, L. M., S. J. Schapiro, S. P. Lambeth, and S. F. Brosnan. 2011. Chimpanzees' socially maintained food preferences indicate both conservatism and conformity. *Animal Behaviour* 81:1195–1202.

Hopper, L. M., A. Spiteri, S. P. Lambeth, S. J. Schapiro, V. Horner, and A. Whiten. 2007. Experimental studies of traditions and underlying transmission processes in chimpanzees. *Animal Behaviour* 73:1021–1032.

Hopper, L. M., and A. Whiten. In press. The evolutionary and comparative psychology of social learning and culture. In J. Vonk and T. Shackelford, eds. *The*

Oxford handbook of comparative evolutionary psychology. Oxford: Oxford University Press.

Hoppitt, W. J. E., G. R. Brown, R. Kendal, L. Rendell, A. Thornton, M. M. Webster, and K. N. Laland. 2008. Lessons from animal teaching. *Trends in Ecology and Evolution* 23:486–493.

Horner, V., and de Waal, F. B. M. 2009. Controlled studies of chimpanzee cultural transmission. *Progress in Brain Research* 178: 3–15.

Horner, V., D. Proctor, K. E. Bonnie, A. Whiten, and F. B. M. de Waal. 2010. Prestige affects cultural learning in chimpanzees. *PLoS One* 5:e10625.

Horner, V., and A. Whiten. 2005. Causal knowledge and imitation/emulation switching in chimpanzees *(Pan troglodytes)* and children *(Homo sapiens)*. *Animal Cognition* 8:164–181.

Horner, V., A. Whiten, E. Flynn, and F. B. M. de Waal. 2006. Faithful replication of foraging techniques along cultural transmission chains by chimpanzees and children *Proceedings of the National Academy of Sciences USA* 103:13878–13883.

Horowitz, A. C. 2003. Do humans ape? Or do apes human? Imitation and intention in humans *(Homo sapiens)* and other animals. *Journal of Comparative Psychology* 117:325–336.

Horton, K. E., and C. A. Caldwell. 2006. Visual co-orientation and expectations about attentional orientation in pileated gibbons *(Hylobates pileatus)*. *Behavioural Processes* 72:65–73.

Hostetter, A. B., M. Cantero, and W. D. Hopkins. 2001. Differential use of vocal and gestural communication by chimpanzees *(Pan troglodytes)* in response to the attentional status of a human *(Homo sapiens)*. *Journal of Comparative Psychology* 115, 337–343.

Hostetter, A. B., J. L. Russell, H. Freeman, and W. D. Hopkins. 2007. Now you see me, now you don't: Evidence that chimpanzees understand the role of the eyes in attention. *Animal Cognition* 10:55–62.

Howie, P., and A. McNeilly. 1982. Effect of breast-feeding patterns on human birth intervals. *Journal of Reproduction and Fertility* 65:545–557.

Hrdy, S. 1999. *Mother nature: A history of mothers, infants and natural selection.* New York: Pantheon.

———. 2009. *Mothers and others: The evolutionary origin of mutual understanding.* Cambridge, MA: Harvard University Press.

Hrubesch, C., S. Preuschoft, and C. P. van Schaik. 2009. Skill mastery inhibits adoption of observed alternative solutions among chimpanzees *(Pan troglodytes)*. *Animal Cognition* 12:209–216.

Huang, C.-T., C. M. Heyes, and T. Charman. 2002. Infants' behavioral reenactment of "failed attempts": Exploring the roles of emulation learning, stimulus enhancement, and understanding of intentions. *Developmental Science* 38:840–855.

———. 2006. Preschoolers' behavioural reenactment of "failed attempts": The roles of intention-reading, emulation and mimicry. *Cognitive Development* 21:36–45.

Huber, L. 1998. Movement imitation as faithful copying in the absence of insight. *Behavioral and Brain Sciences* 22:694.

Huber, L., and G. Gajdon. 2007. Technical intelligence in animals: The kea model. *Animal Cognition* 9:295–305.

Huber, L., G. Gajdon, I. Federspiel, and D. Werdenich. 2008. Cooperation in keas: Social and cognitive factors. In S. Itakura and K. Fujita, eds. *Origins of the social mind: Evolutionary and developmental views,* 99–119. Tokyo: Springer-Verlag.

Huber, L., F. Range, B. Voelkl, A. Szucsich, Z. Viranyi, and Á. Míklósi. 2009. The evolution of imitation: What do the capacities of non-human animals tell us about the mechanisms of imitation? *Philosophical Transactions of the Royal Society B: Biological Sciences* 364:2299–2309.

Huber, L., S. Rechberger, and M. Taborsky. 2001. Social learning affects object exploration and manipulation in keas, *Nestor notabilis. Animal Behaviour* 62:945–954.

Huber, L., and B. Voelkl. 2009. Social and physical cognition in marmosets and tamarins. In S. M. Ford, L. C. Davis, and L. Porter, eds. *The smallest anthropoids: The callimico/marmoset radiation,* 183–201. New York: Springer-Verlag.

Hull, C. L. 1952. *A behavior system.* New Haven, CT: Yale University Press.

Humphrey, N. K. 1976. The social function of intellect. In P. P. G. Bateson and R. A. Hinde, eds. *Growing points in ethology,* 303–317. Cambridge: Cambridge University Press.

Iacoboni, M. 2008. The role of premotor cortex in speech perception: Evidence from fMRI and rTMS. *Journal of Physiology* 102:313–4.

———. 2009. Imitation, empathy, and mirror neurons. *Annual Review of Psychology* 60:653–670.

Iacoboni, M., and M. Dapretto. 2006. The mirror neuron system and the consequences of its dysfunction. *Nature Review Neuroscience* 7:942–951.

Iacoboni, M., L. M. Koski, M. Brass, H. Bekkering, R. P. Woods, M. C. Dubeau, J. C. Mazziotta, and G. Rizzolatti. 2001. Reafferent copies of imitated actions in the right superior temporal cortex. *Proceedings of the National Academy of Sciences USA* 98:13995–13999.

Iacoboni, M., I. Molnar-Szakacs, V. Gallese, G. Buccino, J. C. Mazziotta, and G. Rizzolatti. 2005. Grasping the intentions of others with one's own mirror neuron system. *PLoS Biology* 3:e79.

Iacoboni, M., R. P. Woods, M. Brass, H. Bekkering, J. C. Mazziotta, and G. Rizzolatti. 1999. Cortical mechanisms of human imitation. *Science* 286:2526–2528.

Idani, G. 1991. Social relationships between immigrant and resident bonobos *(Pan paniscus)* females at Wamba. *Folia Primatologica* 57:83–95.

Imanishi, K. 1957. Identification: A process of enculturation in the subhuman society of *Macaca fuscata. Primates* 1:1–29.

Ingersoll, B. 2010. Brief report: Pilot randomized controlled trial of Reciprocal Imitation Training for teaching elicited and spontaneous imitation to children with autism. *Journal of Autism and Developmental Disorders* 40: 1154–1160

Ingersoll, B., and S. Gergans. 2007. The effect of a parent-implemented imitation intervention on spontaneous imitation skills in young children with autism. *Research on Developmental Disabilities* 28:163–175

Ingersoll, B., E. Lewis, and E. Kroman. 2007. Teaching the imitation and spontaneous use of descriptive gestures in young children with autism using a naturalistic behavioral intervention. *Journal of Autism and Developmental Disorders* 37:1446–1456

Ingersoll, B., and L. Schreibman. 2006. Teaching reciprocal imitation skills to young children with autism using a naturalistic behavioral approach: Effects on language, pretend play, and joint attention. *Journal of Autism and Developmental Disorders* 36:487–505.

Ingudomnukul, E., S. Baron-Cohen, S. Wheelwright, and R. Knickmeyer. 2007. Elevated rates of testosterone-related disorders in women with autism spectrum conditions. *Hormones and Behavior* 51 (5):597–604.

Inoue, S., and T. Matsuzawa. 2007. Working memory of numerals in chimpanzees. *Current Biology* 17:R1004–R1005.

Inoue-Nakamura, N. 2001. Mirror self-recognition in primates: An ontogenetic and a phylogenetic approach. In T. Matsuzawa, ed., *Primate origins of human cognition and behavior,* 297–312. New York: Springer-Verlag.

Inoue-Nakamura, N., and T. Matsuzawa. 1997. Development of stone tool use by wild chimpanzees *(Pan troglodytes). Journal of Comparative Psychology* 111:159–173.

Iriki, A. 2006. The neural origins and implications of imitation, mirror neurons and tool use. *Current Opinion in Neurobiology* 16:660–667.

Iriki, A., and O. Sakura. 2008. The neuroscience of primate intellectual evolution: Natural selection and passive and intentional niche construction. *Philosophical Transactions of the Royal Society B: Biological Sciences* 363:2229–2241.

Iriki, A., M. Tanaka, and Y. Iwamura. 1996. Coding of modified body schema during tool use by macaque postcentral neurones. *NeuroReport* 7:2325–2330.

Iriki, A., M. Tanaka, S. Obayashi, and Y. Iwamura. 2001. Self-images in the video monitor coded by monkey intraparietal neurons. *Neuroscience Research* 40:163–173.

Ishida, H., K. Nakajima, M. Inase, and A. Murata. 2009. Shared mapping of own and others' bodies in visuotactile bimodal area of monkey parietal cortex. *Journal of Cognitive Neuroscience* 22:83–96.

Itakura, S. 1996. An exploratory study of gaze-monitoring in nonhuman primates. *Japanese Psychological Research* 38:174–180.

Iverson, J. M., and E. Thelen. 1999. Hand, mouth and brain. *Journal of Consciousness Studies* 6:19–40.

Iverson, J. M., and R. H. Wozniak. 2007. Variation in vocal-motor development in infant siblings of children with autism. *Journal of Autism and Developmental Disorders* 37:158–170.

Iwamura, Y. 1998. Hierarchical somatosensory processing. *Current Opinion in Neurobiology* 8:522–528.

Izawa, K. 1980. Social behavior of the wild black-capped capuchin *(Cebus apella). Primates* 21:443–467.

Jabbi, M., and C. Keysers. 2008. Inferior frontal gyrus activity triggers anterior insula response to emotional facial expressions. *Emotion* 8:775–780.

Jabbi, M., M. Swart, and C. Keysers. 2007. Empathy for positive and negative emotions in the gustatory cortex. *Neuroimage* 34:1744–1753.

Jablonka, E., and M. Lamb. 2007. Bridging the gap: The developmental aspects of evolution. *Behavioral and Brain Sciences* 30:353–365.

Jacklin, C. N., E. E. Maccoby, and C. H. Doering. 1983. Neonatal sex-steroid hormones and timidity in 6–18-month-old boys and girls. *Developmental Psychobiology* 16 (3):163–168.

Jacklin, C. N., K. T. Wilcox, and E. E. Maccoby. 1988. Neonatal sex-steroid hormones and cognitive abilities at six years. *Developmental Psychobiology* 21 (6):567–574.

Jackson, P. L., E. Brunet, A. N. Meltzoff, and J. Decety. 2006. Empathy examined through the neural mechanisms involved in imagining how I feel versus how you feel pain. *Neuropsychologia* 44:752–761.

Jackson, P. L., A. N. Meltzoff, and J. Decety. 2005. How do we perceive the pain of others? A window into the neural processes involved in empathy. *Neuroimage* 24:771–779.

Jaeggi, A. V., J. M. G. Stevens, and C. P. van Schaik. 2010. Tolerant food sharing and reciprocity is precluded by despotism among bonobos but not chimpanzees. *American Journal of Physical Anthropology* 143:41–51.

James, W. 1890. *The principles of psychology*, vol. 1. New York: Holt.

Janik, V. M., and P. J. B. Slater. 1997. Vocal learning in mammals. *Advances in the Study of Behavior* 26:59–99.

Janson, C. H. 1996. Toward an experimental socioecology of primates: Examples from Argentine brown capuchin monkeys *(Cebus apella nigritus)*. In M. A. Norconk, A. L. Rosenberger, and P. A. Garber, eds. *Adaptive radiations of neotropical primates,* 309–325. New York: Plenum.

Janson, C. H., and C. P. van Schaik. 1993. Ecological risk aversion in juvenile primates: Slow and steady wins the race. In M. E. Pereira and L. A. Fairbanks, eds. *Juvenile primates: Life history, development, and behavior,* 57–74. New York: Oxford University Press.

Jarvis, E. D., and Avian Brain Nomenclature Consortium. 2005. Avian brains and a new understanding of vertebrate brain evolution. *Nature Reviews Neuroscience* 6:151–159.

Jeannerod, M. 1994. The representing brain: Neural correlates of motor intention and imagery. *Behavioral Brain Science* 17:187–245.

Jellema, T., C. I. Baker, M. W. Oram, and D. I. Perrett. 2002. Cell populations in the superior temporal sulcus of the macaque and imitation. In A. N. Meltzoff and W. Prinz, eds. *The imitative mind: Development, evolution, and brain bases,* 267–290. Cambridge: Cambridge University Press.

Jellema, T., C. I. Baker, B. Wicker, and D. I. Perrett. 2000. Neural representation for the perception of the intentionality of actions. *Brain and Cognition* 44:280–302.

Jensen, K., B. Hare, J. Call, and M. Tomasello. 2006. What's in it for me? Self-regard precludes altruism and spite in chimpanzees. *Proceedings of the Royal Society B: Biological Sciences* 273:1013–1021.

Jeon, D., S. Kim, M. Chetana, D. Jo, H. E. Ruley, S. Y. Lin, D. Rabah, J. P. Kinet and H. S. Shin. 2010. Observational fear learning involves affective pain system and Cav1.2 Ca2+ channels in ACC. *Nature Neuroscience* 13:482–488.

Johansson, G. 1973. Visual perception of biological motion and a model for its analysis. *Perception and Psychophysics* 14:201–211.

Johnson-Pynn, J., D. M. Fragaszy, and S. Cummins-Sebree. 2003. Common territories in comparative and developmental psychology: Quest for shared means and meaning in behavioral investigations. *International Journal of Comparative Psychology* 16:1–27.

Jolliffe, T., and S. Baron-Cohen. 1997. Are people with autism and Asperger syndrome faster than normal on the Embedded Figures Test? *Journal of Child Psychology and Psychiatry* 38 (5):527–534.

Josse, G., and N. Tzouio-Mazoyer. 2004. Hemispheric specialization for language. *Brain Research Reviews* 44:1–12.

Jost, A. 1961. The role of foetal hormones in prenatal development. *Harvey Lectures* 55:201–226.

———. 1970. Hormonal factors in the sex differentiation of the mammalian foetus. *Philosophical Transactions of the Royal Society B: Biological Sciences* 259 (828):119–130.

———. 1972. A new look at the mechanism controlling sexual differentiation in mammals. *Johns Hopkins Medical Journal* 130:38–53.

Judd, H. L., J. D. Robinson, P. E. Young, and O. W. Jones. 1976. Amniotic fluid testosterone levels in midpregnancy. *Obstetrics and Gynecology* 48 (6):690–692.

Judge, P. G. 1991. Dyadic and triadic reconciliation in pigtail macaques *(Macaca nemestrina)*. *American Journal of Primatology* 23:225–237.

———. 2000. Coping with crowded conditions. In F. Aureli and F. B. M. de Waal, eds. *Natural conflict resolution,* 129–154. Berkeley: University of California Press.

Judge, P. G., and S. H. Mullen. 2005. Quadratic post-conflict affiliation among bystanders in a hamadryas baboon group. *Animal Behaviour* 69:1345–1355.

Kable, J. W., and P. W. Glimcher. 2009. The neurobiology of decision: Consensus and controversy. *Neuron* 63:733–745.

Kagan, J. 2000. Human morality is distinctive. *Journal of Consciousness Studies* 7:46–48.

Kalcher, E., C. Franz, K. Crailsheim, and S. Preuschoft. 2008. Differential onset of infantile deprivation produces distinctive long-term effects in adult ex-laboratory chimpanzees *(Pan troglodytes)*. *Developmental Psychobiology* 50:777–788.

Kaminski, J., J. Call, and M. Tomasello. 2004. Body orientation and face orientation: Two factors controlling apes' begging behavior from humans. *Animal Cognition* 7:216–223.

———. 2008. Chimpanzees know what others know but not what they believe. *Cognition* 109:224–234.

Kano, T. 1992. *The last ape: Pygmy chimpanzee behavior and ecology.* Stanford, CA: Stanford University Press.

Kaplan, J. T., and M. Iacoboni. 2006. Getting a grip on other minds: Mirror neurons, intention understanding, and cognitive empathy. *Social Neuroscience* 1:175–183.

Kawakami, K., K. Takai-Kawakami, M. Tomonaga, J. Suzuki, T. Kusaka, and T. Okai. 2006. Origins of smile and laughter: A preliminary study. *Early Human Development* 82:61–66.

Kazem, A. J. N., and F. Aureli. 2005. Redirection of aggression: Multiparty signalling within a network? In P. K. McGregor, ed. *Animal communication networks,* 191–218. Cambridge: Cambridge University Press.

Keller, G. B., and R. H. Hahnloser. 2009. Neural processing of auditory feedback during vocal practice in a songbird. *Nature* 457:187–190.

Keller, H. 2003. Socialization for competence: Cultural models of infancy. *Human Development* 46:288–311.

———. 2007. Cultures of infancy. Mahwah, NJ: Lawrence Erlbaum.

Keller, H., R. Yovsi, J. Borke, J. Kartner, H. Jensen, and Z. Papaligoura. 2004. Developmental consequences of early parenting experiences: Self-recognition and self-regulation in three cultural communities. *Child Development* 75: 1745–1760.

Keller, S. S., T. J. Crow, A. L. Foundas, K. Amunts, and N. Roberts. 2009. Broca's area: Nomenclature, anatomy, typology and asymmetry. *Brain and Language* 109:29–48.

Keller, S. S., J. R. Highley, M. Garcia-Finana, V. Sluming, R. Rezaie, and N. Roberts. 2007. Sulcal variability, stereological measurement and asymmetry of Broca's area on MR images. *Journal of Anatomy* 211:534–555.

Keller, S. S., N. Roberts, and W. D. Hopkins. 2009. A comparative magnetic resonance imaging study of the anatomy, variability and asymmetry of Broca's area in the human and chimpanzee brain. *Journal of Neuroscience* 29:14607–14616.

Kellogg, W. N., and L. A. Kellogg. 1933. *The ape and the child: A study of environmental influence upon early behavior,* 341. New York: Hafner.

Kendal, R. L., I. Coolen, Y. van Bergen, and K. N. Laland. 2005. Trade-offs in the adaptive use of social and asocial learning. *Advances in the Study of Behavior* 35:333–379.

Keverne, E. B., N. D. Martensz, and B. Tuite. 1989. Beta-endorphin concentrations in cerebrospinal-fluid of monkeys are influenced by grooming relationships. *Psychoneuroendocrinology* 14:155–161.

Keysers, C., B. Wicker, V. Gazzola, J. L. Anton, L. Fogassi, and V. Gallese. 2004. A touching sight: SII/PV activation during the observation and experience of touch. *Neuron* 42:335–346.

Kikusui, T., J. T. Winslow, and Y. Mori. 2006. Social buffering: Relief from stress and anxiety. *Philosophical Transactions of the Royal Society B: Biological Sciences* 361:2215–2228.

Kimura, D. 1999. *Sex and cognition.* Cambridge, MA: MIT Press.

Klein, E. D., and T. R. Zentall. 2003. Imitation and affordance learning by pigeons *(Columba livia). Journal of Comparative Psychology* 117:414–419.

Knaus, T. A., A. M. Bollich, D. M. Corey, L. C. Lemen, and A. L. Foundas. 2004. Sex-linked differences in the anatomy of the persi-sylvian language cortex: A volumetric MRI study of gray matter volumes. *Neuropsychology* 18:738–747.

———. 2006. Variability in perisylvian brain anatomy in healthy adults. *Brain and Language* 97:219–232.

Knickmeyer, R. C., and S. Baron-Cohen. 2006a. Fetal testosterone and sex differences. *Early Human Development* 82:755–760.

———. 2006b. Fetal testosterone and sex differences in typical social development and in autism. *Journal of Child Neurology* 21 (10):825–845.

Knickmeyer, R. C., S. Baron-Cohen, B. A. Fane, S. Wheelwright, G. A. Mathews, G. S. Conway, C. G. Brook, and M. Hines. 2006. Androgens and autistic traits: A study of individuals with congenital adrenal hyperplasia. *Hormones and Behavior* 50 (1):148–153.

Knickmeyer, R. C., S. Baron-Cohen, P. Raggatt, and K. Taylor. 2005. Foetal testosterone, social relationships, and restricted interests in children. *Journal of Child Psychology and Psychiatry* 46:198–210.

Knickmeyer, R. C., S. Baron-Cohen, P. Raggatt, K. Taylor, and G. Hackett. 2006. Fetal testosterone and empathy. *Hormones and Behavior* 49:282–292.

Knickmeyer, R. C., S. Wheelwright, R. Hoekstra, and S. Baron-Cohen. 2006. Age of menarche in females with autism spectrum conditions. *Developmental Medicine and Child Neurology* 48 (12):1007–1008.

Knörnschild, M., M. Nagy, M. Metz, F. Mayer, and O. von Helversen. 2009. Complex vocal imitation during ontogeny in a bat. *Biology Letters* 6:156–159.

Kohler, E., C. Keysers, M. A. Umiltà, L. Fogassi, V. Gallese, and G. Rizzolatti. 2002. Hearing sounds, understanding actions: Action representation in mirror neurons. *Science* 297:846–848.

Köhler, W. 1925. *The mentality of apes.* London: Routledge and Kegan Paul.

Koolhaas, J. M., S. F. de Boer, B. Buwalda, and K. van Reenen. 2007. Individual variation in coping with stress: A multidimensional approach of ultimate and proximate mechanisms. *Brain, Behavior and Evolution* 70:218–226.

Koski, L., M. Iacoboni, M. C. Dubeau, R. P. Woods, and J. C. Mazziotta. 2003. Modulation of cortical activity during different imitative behaviors. *Journal of Neurophysiology* 89:460–471.

Koski, L., A. Wohlschläger, H. Bekkering, R. P. Woods, M. C. Dubeau, J. C. Mazziotta, and M. Iacoboni. 2002. Modulation of motor and premotor activity during imitation of target-directed actions. *Cerebral Cortex* 12:847–855.

Koski, S. E., K. Koops, and E. H. M. Sterck. 2007. Reconciliation, relationship quality, and postconflict anxiety: Testing the integrated hypothesis in captive chimpanzees. *American Journal of Primatology* 69:158–172.

Koski, S. E., and E. H. M. Sterck. 2007. Triadic postconflict affiliation in captive chimpanzees: Does consolation console? *Animal Behaviour* 73:133–142.

Koyama, N. F. 2001. The long-term effects of reconciliation in Japanese macaques *Macaca fuscata*. *Ethology* 107:975–987.

Kraskov, A., N. Dancause, M. M. Quallo, S. Shepherd, and R. N. Lemon. 2009. Corticospinal neurons in macaque ventral premotor cortex with mirror properties: A potential mechanism for action suppression? *Neuron* 64:922–930.

Kummer, H. 1968. *Social organization of hamadryas baboons: A field study.* Basel: Karger.

Kutsukake, N. 2003. Assessing relationship quality and social anxiety among wild chimpanzees using self-directed behaviour. *Behaviour* 140:1153–1171.

Kutsukake, N., and D. L. Castles. 2001. Reconciliation and variation in post-conflict stress in Japanese macaques *(Macaca fuscata fuscata):* Testing the integrated hypothesis. *Animal Cognition* 4:259–268.

Kutsukake, N., and T. H. Clutton-Brock. 2006. Social functions of allogrooming in cooperatively breeding meerkats. *Animal Behaviour* 72:1059–1068.

Kutsukake, N., N. Suetsugu, and T. Hasegawa. 2006. Pattern, distribution, and function of greeting behavior among black-and-white colobus. *International Journal of Primatology* 27:1271–1291.

Ladygina-Kohts, N. N. 2001 (orig. 1935). F. B. M de Waal, ed. *Infant chimpanzee and human child: A classic 1935 comparative study of ape emotions and intelligence.* New York: Oxford University Press.

Lakshminarayanan, V. R., and L. R. Santos. 2008. Capuchin monkeys are sensitive to others' welfare. *Current Biology* 18:R999–R1000.

Laland, K. N. 2004. Social learning strategies. *Learning and Behavior* 32:4–14.

Laland, K. N., and B. G. Galef. 2009. *The question of animal culture.* Cambridge, MA: Harvard University Press.

Laland, K. N., J. R. Kendal, and R. L. Kendal. 2009. Animal culture: Problems and solutions. In K. N. Laland and B. G. Galef, eds. *The question of animal culture,* 174–197. Cambridge, MA: Harvard University Press.

Laland, K. N., J. Odling-Smee, and M. W. Feldman. 2000. Niche construction, biological evolution, and cultural change. *Behavioral Brain Sciences* 23:131–146.

Laland, K. N., and K. Williams. 1998. Social transmission of maladaptive information in the guppy. *Behavioral Ecology* 9:493–499.

Lamm, C., C. D. Batson, and J. Decety. 2007. The neural substrate of human empathy: Effects of perspective-taking and cognitive appraisal. *Journal of Cognitive Neuroscience* 19:42–58.

Lamm, C., J. Decety, and T. Singer. 2009. *An image-based meta-analysis of neural circuits involved in the direct experience of pain and empathy for pain.* Manuscript submitted for publication.

Lamm, C., A. N. Meltzoff, and J. Decety. 2010. How do we empathize with someone who is not like us? A functional magnetic resonance imaging study. *Journal of Cognitive Neuroscience* 22:362–376.

Lamm, C., H. C. Nusbaum, A. N. Meltzoff, and J. Decety. 2007. What are you feeling? Using functional magnetic resonance imaging to assess the modulation of sensory and affective responses during empathy for pain. *PLoS One* 2:e1292.

Lamm, C., and T. Singer. 2010. The role of anterior insular cortex in social emotions. *Brain Structure and Function* 214:579–951.

Langergraber, K. E., J. C. Mitani, and L. Vigilant. 2007. The limited impact of kinship in cooperation in wild chimpanzees. *Proceedings of the National Academy of Sciences USA* 104:7786–7790.

Langford, D. J., S. E. Crager, Z. Shehzad, S. B. Smith, S. G. Sotocinal, J. S. Levenstadt, M. L. Chanda, D. J. Levitin, and J. S. Mogil. 2006. Social modulation of pain as evidence for empathy in mice. *Science* 312:1967–1970.

Larsen, P. R., H. M. Kronenberg, S. Melmed, and K. S. Polonsky, eds. 2002. *Williams textbook of endocrinology.* 10th ed. Philadelphia: Saunders.

Lauterbach, O., and D. Hosser. 2007. Assessing empathy in prisoners: A shortened version of the Interpersonal Reactivity Index. *Swiss Journal of Psychology* 66:91–101.

Lavelli, M., and A. Fogel. 2002. Developmental changes in mother-infant face-to-face communication. *Developmental Psychology* 38:288–305.

Lawson, J., S. Baron-Cohen, and S. Wheelwright. 2004. Empathising and systemising in adults with and without Asperger syndrome. *Journal of Autism and Developmental Disorders* 34 (3):301–310.

Leavens, D. A., F. Aureli, and W. D. Hopkins. 2004. Behavioral evidence for the cutaneous expression of emotion in a chimpanzee *(Pan troglodytes). Behaviour* 141:979–997.

Leavens, D. A., and K. A. Bard. 2011. Environmental influences on joint attention in great apes: Implications for human cognition. In H. Keller, ed. *Journal of Cognitive Education and Psychology* 10:9–31.

Leavens, D. A., K. A. Bard, and W. D. Hopkins. 2010. BIZARRE chimpanzees do not represent "the chimpanzee." Commentary on target article by Henrich, Heine, and Norenzayan, "The weirdest people in the world?" *Behavioral and Brain Sciences* 33:100–101.

Leavens, D. A., and W. D. Hopkins. 1998. Intentional communication by chimpanzee *(Pan troglodytes):* A cross-sectional study of the use of referential gestures. *Developmental Psychology* 34:813–822.

Leavens, D. A., W. D. Hopkins, and K. A. Bard. 1996. Indexical and referential pointing in chimpanzees *(Pan troglodytes). Journal of Comparative Psychology* 110:346–353.

———. 2005. Understanding the point of chimpanzee pointing: Epigenesis and ecological validity. *Current Directions in Psychological Science* 14:185–189.

———. 2008. The heterochronic origins of explicit reference. In J. Zlatev, T. P. Racine, C. Sinha and E. Itkonen, eds. *The shared mind: Perspectives on intersubjectivity,* 185–214. Amsterdam: John Benjamins.

Leavens, D. A., W. D. Hopkins, and R. Thomas. 2004. Referential communication by chimpanzees *(Pan troglodytes). Journal of Comparative Psychology* 118:48–57.

Leavens, D. A., A. B. Hostetter, M. J. Wesley, and W. D. Hopkins. 2004. Tactical use of unimodal and bimodal communication by chimpanzees, *Pan troglodytes. Animal Behaviour* 67:467–476.

Leavens, D. A., T. P. Racine, and W. D. Hopkins. 2009. The ontogeny and phylogeny of non-verbal dexis. In C. Botha and C. Knight, eds. *The prehistory of language.* Oxford: Oxford University Press.

Leavens, D. A., J. L. Russell, and W. D. Hopkins. 2005. Intentionality as measured in the persistence and elaboration of communication by chimpanzees *(Pan troglodytes)*. *Child Development* 76:291–306.

Lepage, J. F., and H. Théoret. 2007. The mirror neuron system: Grasping others' actions from birth? *Developmental Science* 10:513–523.

Lester, B. M., L. T. Anderson, C. F. Z. Boukydis, C. T. Garcia Coll, B. Vohr, and M. Peucker. 1989. Early detection of infants at risk for later handicap through acoustic cry analysis. In N. Paul, ed. *Research in infant assessment*. New York: March of Dimes Birth Defects Foundation.

Levenston, G. K., C. J. Patrick, M. M. Bradley, and P. J. Lang. 2000. The psychopath as observer: Emotion and attention in picture processing. *Journal of Abnormal Psychology* 109:373–385.

Levy, R., and P. S. Goldman-Rakic. 2000. Segregation of working memory functions within the dorsolateral prefrontal cortex. *Experimental Brain Research* 133:23–32.

Lewis, M. 2002. Empathy requires the development of the self. *Behavioral and Brain Sciences* 25:42.

Lewis, M., and J. Brooks-Gunn. 1979. *Social cognition and the acquisition of self*. New York: Plenum.

Lhermitte, F., B. Pillon, and M. Serdaru. 1986. Human autonomy and the frontal lobes. Part I: Imitation and utilization behavior: A neuropsychological study of 75 patients. *Annals in Neurology* 19:326–334.

Liberman, A. M., F. S. Cooper, D. P. Shankweiler, and M. Studdert-Kennedy. 1967. Perception of the speech code. *Psychological Review* 74:431–461.

Liberman, A. M., and I. G. Mattingly. 1985. The motor theory of speech perception revised. *Cognition* 21:1–36.

Liberman, A. M., and D. H. Whalen. 2000. On the relation of speech to language. *Trends in Cognitive Science* 4:187–196.

Liebal, K., J. Call, and M. Tomasello. 2004. Use of gesture sequences in chimpanzees. *American Journal of Primatology* 64:377–396.

Lincoln, A. J., E. Courchesne, B. A. Kilman, R. Elmasian, and M. Allen. 1988. A study of intellectual abilities in high-functioning people with autism. *Journal of Autism and Developmental Disorders* 18 (4):505–524.

Lipps, T. 1903. Einfühlung, innere Nachahmung und Organempfindung. *Archiv für die gesammte Psychologie* 1:185–204.

Liu, Q., K. Simpson, P. Izar, E. Ottoni, E. Visalberghi, and D. Fragaszy. 2009. Kinematics and energetics of nut-cracking in wild capuchin monkeys *(Cebus libidinosus)* in Piauí, Brazil. *American Journal of Physical Anthropology* 138:210–220.

Lock, A. 2001. Preverbal communication. In J. G.Bremner and A. Fogel, eds. *The Blackwell handbook of infant development*. Blackwood Handbooks of Developmental Psychology. Malden, MA: Blackwell.

Lonsdorf, E. V., L. E. Eberly, and A. E. Pusey. 2004. Sex differences in learning in chimpanzees. *Nature* 428:715–716.

Lorenz, K. 1935. Der Kumpan in der Umwelt des Vogels. *Journal für Ornithologie* 83:137–413.

————. 1977. *Behind the mirror*. London: Methuen.

Lorincz, E. N., C. I. Baker, and D. I. Perrett. 1999. Visual cues for attention following in rhesus monkeys. *Cahiers de psychologie cognitive (Current Psychology of Cognition)* 18:973–1003.

Lutchmaya, S., S. Baron-Cohen, and P. Raggatt. 2002a. Foetal testosterone and eye contact in 12 month old infants. *Infant Behavior and Development* 25:327–335.

————. 2002b. Foetal testosterone and vocabulary size in 18- and 24-month-old infants. *Infant Behavior and Development* 24 (4):418–424.

Lutchmaya, S., S. Baron-Cohen, P. Raggatt, R. Knickmeyer, and J. T. Manning. 2004. 2nd to 4th digit ratios, fetal testosterone and estradiol. *Early Human Development* 77:23–28.

Lyn, H., J. L. Russell, and W. D. Hopkins. 2010. The impact of the environment on the comprehension of declarative communication in apes. *Psychological Science* 21:360–365.

Lynn, R., A. Raine, P. H. Venables, S. A. Mednick, and P. Irwing. 2005. Sex differences on the WISC-R in Mauritius. *Intelligence* 33:527–533.

Lyons, D. E., D. H. Damrosch, J. K. Lin, D. M. Macris, and F. C. Keil. 2011. The scope and limits of overimitation in the transmission of artefact culture. *Philosophical Transactions of the Royal Society B*, 366: 1158–1167.

Lyons, D. E., W. Phillips, and L. R. Santos. 2005. Motivation is not enough. *Behavioural and Brain Sciences* 28:708.

Lyons, D. E., L. R. Santos, and F. C. Keil. 2006. Reflections of other minds: How primate social cognition can inform the function of mirror neurons. *Current Opinion in Neurobiology* 16:230–234.

Lyons, D. E., A. G. Young, and F. C. Keil. 2007. The hidden structure of overimitation. *Proceedings of the National Academy of Sciences USA* 104: 19751–19756.

Maestripieri, D. 1993. Maternal anxiety in rhesus macaques *(Macaca mulatta)*. II: Emotional bases of individual differences in mothering style. *Ethology* 95:32–42.

Maestripieri, D., G. Schino, F. Aureli, and A. Troisi. 1992. A modest proposal: Displacement activities as an indicator of emotions in primates. *Animal Behaviour* 44:967–979.

Mah, L., M. C. Arnold, and J. Grafman. 2004. Impairment of social perception associated with lesions of the prefrontal cortex. *American Journal of Psychiatry* 161:1247–1255.

Makin, T. R., N. P. Holmes, and H. H. Ehrsson. 2008. On the other hand: Dummy hands and peripersonal space. *Behavioral Brain Research* 191:1–10.

Malas, M. A., S. Dogan, E. H. Evcil, and K. Desdicioglu. 2006. Fetal development of the hand, digits and digit ratio (2D:4D). *Early Human Development* 82 (7):469–475.

Mallavarapu, S., T. S. Stoinski, M. A. Bloomsmith, and T. L. Maple. 2006. Postconflict behavior in captive western lowland gorillas *(Gorilla gorilla gorilla)*. *American Journal of Primatology* 68:789–801.

Manning, J. T., S. Baron-Cohen, S. Wheelwright, and G. Sanders. 2001. The 2nd to 4th digit ratio and autism. *Developmental Medicine and Child Neurology* 43 (3):160–164.

Manning, J. T., P. E. Bundred, and B. F. Flanagan. 2002. The ratio of 2nd to 4th digit length: A proxy for transactivation activity of the androgen receptor gene? *Medical Hypotheses* 59 (3):334–336.

Mansouri, F. A., K. Matsumoto, and K. Tanaka. 2006. Prefrontal cell activities related to monkeys' success and failure in adapting to rule changes in a Wisconsin Card Sorting Test analog. *Journal of Neuroscience* 26:2745–2756.

Maravita, A., and A. Iriki. 2004. Tools for the body (schema). *Trends in Cognitive Science* 8:79–86.

Marcus, J., E. E. Maccoby, C. N. Jacklin, and C. H. Doering. 1985. Individual differences in mood in early childhood: Their relation to gender and neonatal sex steroids. *Developmental Psychobiology* 18 (4):327–340.

Margoliash, D., and H. C. Nusbaum. 2009. Language: The perspective from organismal biology. *Trends in Cognitive Sciences* 13:505–510.

Marshall-Pescini, S., and A. Whiten. 2008a. Chimpanzees *(Pan troglodytes)* and the question of cumulative culture: An experimental approach. *Animal Cognition* 11:449–456.

———. 2008b. Social learning of nut-cracking behavior in East African sanctuary-living chimpanzees *(Pan troglodytes schweinfurthii)*. *Journal of Comparative Psychology* 122:186–194.

Marticorena, D. W., A. M. Ruiz, C. Mukerji, A. Goddu, and L. R. Santos. In press. Monkeys represent others' knowledge but not their beliefs. *Developmental Science*.

Mason, W. A., R. K. Davenport, and E. W. Menzel Jr. 1968. Early experience and the social development of rhesus monkeys and chimpanzees. In G. Newton and S. Levine, eds. *Early experience and behavior: The psychobiology of development,* 1–41. Springfield: Charles C. Thomas.

Masserman, J. H., S. Wechkin, and W. Terris. 1964. "Altruistic" behaviors in rhesus monkeys. *American Journal of Psychiatry* 121:584–585.

Matsumoto, K., and K. Tanaka. 2004. The role of the medial prefrontal cortex in achieving goals. *Current Opinion in Neurobiology* 14:178–185.

Matsuzawa, T. 1994. Field experiments on use of stone tools by chimpanzees in the wild. In R. W. Wrangham, W. C. McGrew, F. B. M. de Waal, and P. Heltne, eds. *Chimpanzee cultures,* 351–370. Cambridge, MA: Harvard University Press.

———. 1996. Chimpanzee intelligence in nature and in captivity: Isomorphism of symbol use and tool use. In W. C. McGrew, L. F. Marchant, and T. Nishida, eds. *Great ape societies,* 196–209. Cambridge: Cambridge University Press.

———, ed. 2001. *Primate origins of human cognition and behavior.* Tokyo: Springer-Verlag.

———. 2003. The Ai project: Historical and ecological contexts. *Animal Cognition* 6:199–211.

———. 2007. Comparative cognitive development. *Developmental Science* 10:97–103.

————. 2009a. Symbolic representation of number in chimpanzees. *Current Opinion in Neurobiology* 19:92–98.

————. 2009b. A trade-off theory of intelligence. In D. Mareschal, P. C. Quinn, and S. E. G. Lea, eds. *The making of human concepts*. Oxford: Oxford University Press.

Matsuzawa, T., D. Biro, T. Humle, N. Inoue-Nakamura, R. Tonooka, and G. Yamakoshi. 2001. Emergence of culture in chimpanzees: Education by master-apprenticeship. In T. Matsuzawa, ed. *Primate origins of human cognition and behavior*, 557–574. Tokyo: Springer-Verlag.

Matsuzawa, T., T. Humle, and Y. Sugiyama. 2011. *The chimpanzees of Bossou and Nimba*. Springer: Tokyo

Matsuzawa, T., M. Tanaka, and M. Tomonaga, eds. 2006. *Cognitive development in chimpanzees*. Tokyo: Springer-Verlag.

Mayr, E. 1982. *The Growth of Biological Thought*. Cambridge, MA: Belknap Press of Harvard University Press.

McGrew, W. C. 1977. Socialization and object-manipulation of wild chimpanzees. In S. Chevalier-Skolnikoff and F. E. Poirier, eds. *Primate bio-social development: Biological, social, and ecological determinants*, 261–288. New York: Garland.

————. 2004. *The cultured chimpanzee: Reflections on cultural primatology*. Cambridge: Cambridge University Press.

McGuigan, N., A. Whiten, E. Flynn, and V. Horner. 2007. Imitation of causally opaque versus causally transparent tool use by 3- and 5-year-old children. *Cognitive Development* 22:353–364.

McManus, I. C., B. Murray, K. Doyle, and S. Baron-Cohen. 1992. Handedness in childhood autism shows a dissociation of skill and preference. *Cortex* 28 (3):373–381.

Meguerditchian, A., and J. Vauclair. 2006. Baboons communicate with their right hand. *Behavioural Brain Research* 171:170–174.

Meguerditchian, A., J. Vauclair, and W. D. Hopkins. 2010. Captive chimpanzees use their right hand to communicate with each other: Implications for the origins of hemispheric specialization for language. *Cortex* 46:40–48.

Mehrabian, A., and N. Epstein. 1972. A measure of emotional empathy. *Journal of Personality* 40:525–543.

Meister, I. G., S. M. Wilson, C. Deblieck, A. D. Wu, and M. Iacoboni. 2007. The essential role of premotor cortex in speech perception. *Current Biology* 17 (19): 1692–1696.

Melis, A. P., J. Call, and M. Tomasello. 2006. Chimpanzees *(Pan troglodytes)* conceal visual and auditory information from others. *Journal of Comparative Psychology* 120:154–162.

Melis, A. P., B. Hare, and M. Tomasello. 2006a. Engineering cooperation in chimpanzees: Tolerance constraints on cooperation. *Animal Behaviour* 72:275–286.

————. 2006b. Chimpanzees recruit the best collaborators. *Science* 311: 1297–1300.

———. 2009. Chimpanzees coordinate in a negotiating game. *Evolution and Human Behaviour* 30:381–392.

Melis, A. P., B. Hare, and M. Tomasello. 2008. Do chimpanzees reciprocate received favours? *Animal Behaviour* 76 (3):951–962.

Melis, A. P., F. Warneken, K. Jensen, A.-C. Schneider, J. Call, and M. Tomasello. 2010. Chimpanzees help conspecifics obtain food and non-food items. *Proceedings of the Royal Society B: Biological Sciences*, 1–10.

Meltzoff, A. N. 1985. Immediate and deferred imitation in fourteen- and twenty-four-month-old infants. *Child Development* 56:62–72.

Meltzoff, A. N., and M. K. Moore. 1977. Imitation of facial and manual gestures by human neonates. *Science* 198:75–78.

———. 1983. Newborn infants imitate adult facial gestures. *Child Development* 54:702–709.

Mendres, K. A., and F. B. M. de Waal. 2000. Capuchins do cooperate: The advantage of an intuitive task. *Animal Behaviour* 60:523–529.

Menzel, C. R. 1980. Head-cocking and visual perception in primates. *Animal Behaviour* 28:151–159.

———. 1991. Cognitive aspects of foraging in Japanese monkeys. *Animal Behaviour* 41:397–402.

———. 1999. Unprompted recall and reporting of hidden objects by a chimpanzee *(Pan troglodytes)* after extended delays. *Journal of Comparative Psychology* 113:426–434.

———. 2005. Progress in the study of chimpanzee recall and episodic memory. In H. S. Terrace and J. Metcalfe, eds. *The missing link in cognition: Origins of self-reflective consciousness*, 188–224. Oxford: Oxford University Press.

Menzel, C. R., E. S. Savage-Rumbaugh, and E. W. Menzel Jr. 2002. Bonobo *(Pan paniscus)* spatial memory and communication in a 20-hectare forest. *International Journal of Primatology* 23:601–619.

Menzel, E. W., Jr. 1963. The effects of cumulative experience on responses to novel objects in young isolation-reared chimpanzees. *Behaviour* 21:1–12.

———. 1964a. Patterns of responsiveness in chimpanzees reared through infancy under conditions of environmental restriction. *Psychologische Forschung* 27:337–365.

———. 1964b. Responsiveness to object-movement in young chimpanzees. *Behaviour* 24:147–160.

———. 1966. Responsiveness to objects in free-ranging Japanese monkeys. *Behaviour* 26:130–150.

———. 1969. Naturalistic and experimental approaches to primate behavior. In E. P. Willems and H. L. Raush, eds. *Naturalistic viewpoints in psychological research*, 78–121. New York: Holt, Rinehart, and Winston.

———. 1973. Leadership and communication in young chimpanzees. In E. W. Menzel Jr., ed. *Symposia of the fourth International Congress of Primatology*, vol. 1: *Precultural primate behavior*, 192–225. Basel: Karger.

———. 1974. A group of young chimpanzees in a one-acre field. In A. M. Schrier and F. Stollnitz, eds. *Behavior of non-human primates*, 5:83–153. New York: Academic Press.

————. 1978. Cognitive mapping in chimpanzees. In S. H. Hulse, H. Fowler, and W. K. Honig, eds. *Cognitive processes in animal behavior,* 375–422. Hillsdale, NJ: Lawrence Erlbaum.

————. 1983. Parlez-vous baboon, Bwana Sherlock? *Behavioral and Brain Sciences* 6:371–372.

————. 1988. Mindless behaviorism, bodiless cognitivism or primatology? *Behavioral and Brain Sciences* 11:258–259.

————. 2010. To be buried in thought, lost in space, or lost in action: Is that the question? In F. L. Dolins and R. W. Mitchell, eds. *Spatial cognition, spatial perception: Mapping the self and space,* 75–96. New York: Cambridge University Press.

Menzel, E. W., Jr., R. K. Davenport Jr., and C. M. Rogers. 1963a. Effects of environmental restriction upon the chimpanzee's responsiveness in novel situations. *Journal of Comparative and Physiological Psychology* 56:329–334.

————. 1963b. The effects of environmental restriction upon the chimpanzee's responsiveness to objects. *Journal of Comparative and Physiological Psychology* 56:78–85.

————. 1970. The development of tool use in wild-born and restriction-reared chimpanzees. *Folia Primatologica* 12:273–283.

————. 1972. Proto-cultural aspects of chimpanzees' responsiveness to novel objects. *Folia Primatologica* 17:161–170.

Menzel, E. W., Jr., and W. A. Draper. 1965. Primate selection of food by size: Visible versus invisible rewards. *Journal of Comparative and Physiological Psychology* 59:231–239.

Menzel, E. W., Jr., and S. Halperin. 1975. Purposive behavior as a basis for objective communication between chimpanzees. *Science* 189:652–654.

Menzel, E. W., Jr., and M. K. Johnson. 1976. Communication and cognitive organization in humans and other animals. *Annals of the New York Academy of Science* 280:131–142.

————. 1978. Should mentalistic concepts be defended or assumed? *Behavioral and Brain Sciences* 4:586–587.

Menzel, E. W., Jr., and C. Juno. 1985. Social foraging in marmoset monkeys and the question of intelligence. *Philosophical Transactions of the Royal Society B: Biological Sciences* 308:145–158.

Menzel, E. W., Jr., and C. R. Menzel. 1979. Cognitive, developmental and social aspects of responsiveness to novel objects in a family group of marmosets *(Saguinus fuscicollis). Behaviour* 70:251–279.

————. 2007. Do primates plan routes? Simple detour problems reconsidered. In D. A. Washburn, ed. *Primate perspectives on behavior and cognition,* 175–206. Washington, D. C.: American Psychological Association.

Messinger, D. 2002. Positive and negative: Infant facial expressions and emotions. *Current Directions in Psychological Science* 11:1–6.

Míklósi, Á. 2007. *Dog behaviour, evolution, and cognition.* Oxford: Oxford University Press.

Míklósi, Á., and K. Soproni 2006. A comparative analysis of animals' understanding of the human pointing gesture. *Animal Cognition* 9.

Miles, H. L., R. W. Mitchell, and S. E. Harper. 1996. Simon says: The development of imitation in an enculturated orangutan. In A. E. Russon, K. A. Bard, and S. T. Parker, eds. *Reaching into thought: The minds of the great apes,* 278–299. Cambridge: Cambridge University Press.

Miller, R. E. 1967. Experimental approaches to the physiological and behavioral concomitants of affective communication in rhesus monkeus. In S. A. Altmann, ed. *Social communication among primates,* 125–134. Chicago: University of Chicago Press.

Milne, E., S. White, R. Campbell, J. Swettenham, P. Hansen, and F. Ramus. 2006. Motion and form coherence detection in autistic spectrum disorder: Relationship to motor control and 2:4 digit ratio. *Journal of Autism and Developmental Disorders* 36 (2):225–237.

Milner, B. 1963. Effect of Different Brain Lesions on Card Sorting. *Archives of Neurology* 9:90–100.

Minio-Paluello, I., S. Baron-Cohen, A. Avenanti, V. Walsh, and S. M. Aglioti. 2009. Absence of embodied empathy during pain observation in Asperger syndrome. *Biological Psychiatry* 65:55–62.

Mitani, J. C. 2006. Reciprocal exchange in chimpanzees and other primates. In P. M. Kappeler and C. P. van Schaik, eds. *Cooperation in primates: Mechanisms and evolution.* Berlin: Springer-Verlag.

Mitani, J. C., and D. P. Watts. 2001. Why do chimpanzees hunt and share meat? *Animal Behaviour* 61:915–924.

Mitchell, J. P., M. R. Banaji, and C. N. Macrae. 2005. The link between social cognition and self-referential thought in the medial prefrontal cortex. *Journal of Cognitive Neuroscience* 17:1306–1315.

Mitchell, R. W. 1987. A comparative developmental approach to understanding imitation. In P. P. G. Bateson and P. H. Klopfer, eds. *Perspectives in ethology* 7:183–215. New York: Plenum.

Mito, S. 1979. *Bosuzaru heno michi* (The way of becoming alpha male) [in Japanese]. Tokyo: Popura-sha.

Miyabe-Nishiwaki, T., A. Kaneko, K. Nishiwaki, A. Watanabe, S. Watanabe, N. Maeda, K. Kumazaki, M. Morimoto, R. Hirokawa, J. Suzuki, Y. Ito, M. Hayashi, M. Tanaka, M. Tomonaga, and T. Matsuzawa. 2010. Tetraparesis resembling acute transverse myelitis in a captive chimpanzee *(Pan troglodytes):* Long-term care and recovery. *Journal of Medical Primatology* 39:336–346.

Miyake, K., S.-J. Chen, and J. Campos. 1985. Infant temperament, mother's mode of interaction, and attachment in Japan: An interim report. *Monographs of the Society for Research in Child Development* 50:276–297.

Mizuno, Y., H. Takeshita, and T. Matsuzawa. 2006. Behavior of infant chimpanzees during the night in the first 4 months of life: Smiling and suckling in relation to behavioral state. *Infancy* 9:221–240.

Moberg, G. P. 2000. The biology of animal stress: Implications for animal welfare. In G. P. Moberg and J. A. Mench, eds. *The biology of animal stress: Basic principles and implications for animal welfare,* 1–21. Wallingford, CT: CABI International.

Moberg, G. P., and J. A. Mench. 2000. *The biology of animal stress: Basic principles and implications for animal welfare.* Wallingford, CT: CABI International.

Moll, H., and M. Tomasello. 2007. Cooperation and human cognition: The Vygotskian intelligence hypothesis. *Philosophical Transactions of the Royal Society B: Biological Sciences* 362:639–648.

Money, J., M. Schwartz, and V. Lewis. 1984. Adult erotosexual status and fetal hormonal masculinization and demasculinization: 46 XX congenital adrenal hyperplasia and 46 XY androgen insensitivity syndrome compared. *Psychoneuroendocrinology* 9:405–414.

Montaigne, M. [1575] 1993. *Essays.* Trans. J. M. Cohen. Harmondsworth, UK: Penguin.

Mori, A., K. Watanabe, and N. Yamaguchi. 1989. Longitudinal changes of dominance rank among the females of the Koshima group of Japanese monkeys. *Primates* 30:147–173.

Morris, D. 1962. *The biology of art.* London: Methuen.

Morrison, I., and P. E. Downing. 2007. Organization of felt and seen pain responses in anterior cingulate cortex. *Neuroimage* 37:642–651.

Mukamel, R., A. D. Ekstrom, J. Kaplan, M. Iacoboni, and I. Fried. 2010. Single-neuron responses in humans during execution and observation of actions. *Current Biology* 27:750–756.

Mulcahy, N. J., and J. Call. 2006. Apes save tools for future use. *Science* 312:1038–1040.

Muller, M. N., and J. C. Mitani. 2005. Conflict and cooperation in wild chimpanzees. *Advances in the Study of Behavior* 35:275–331.

Muroyama, Y. 1991. Mutual reciprocity of grooming in female Japanese macaques. *Behavior* 119:161–170.

Myowa, M. 1996. Imitation of facial gestures by an infant chimpanzee. *Primates* 37:207–213.

Myowa-Yamakoshi, M. 2006. How and when do chimpanzees acquire the ability to imitate? In T. Matsuzawa, M. Tomonaga, and M. Tanaka, eds. *Cognitive development in chimpanzees,* 214–232. Tokyo: Springer-Verlag.

Myowa-Yamakoshi, M., and T. Matsuzawa. 1999. Factors influencing imitation of manipulatory actions in chimpanzees *(Pan troglodytes). Journal of Comparative Psychology* 113:128–136.

Myowa-Yamakoshi, M., M. Tomonaga, M. Tanaka, and T. Matsuzawa. 2004. Imitation in neonatal chimpanzees *(Pan troglodytes). Developmental Science* 7:437–442.

Nadel, J. 2002. Imitation and imitation recognition: Functional use in preverbal infants and nonverbal children with autism. In A. N. Meltzoff and W. Prinz, eds. *The imitative mind: Development, evolution, and brain bases,* 42–62. Cambridge: Cambridge University Press.

Nagel, T. 1974. What is it like to be a bat? *Philosophical Review* 83:435–450.

Nagell, K., R. S. Olguin, and M. Tomasello. 1993. Processes of social learning in the tool use of chimpanzees *(Pan troglodytes)* and human children *(Homo sapiens). Journal of Comparative Psychology* 107:174–186.

Nakamura, M., and T. Nishida. 2004. Subtle behavioral variation in wild chimpanzees, with special reference to Imanishi's concept of kaluchua. *Primates* 47:35–42.

Nakayama, K. 2004. Observing conspecifics scratching induces a contagion of scratching in Japanese monkeys *(Macaca fuscata)*. *Journal of Comparative Psychology* 118:20–24.

Nass, R., S. Baker, P. Speiser, R. Virdis, A. Balsamo, E. Cacciari, A. Loche, M. Dumic, and M. New. 1987. Hormones and handedness: Left-hand bias in female adrenal hyperplasia patients. *Neurology* 37:711–715.

Neiworth, J. J., M. A. Burman, B. M. Basile, and M. T. Lickteig. 2002. Use of experimenter-given cues in visual co-orienting and in an object-choice task by a new world monkey species, cotton top tamarins *(Saguinus oedipus)*. *Journal of Comparative Psychology* 116:3–11.

New, M. I. 1998. Diagnosis and management of congenital adrenal hyperplasia. *Annual Review of Medicine* 49:311–328.

Nicol, C. J., and S. J. Pope. 1994. Social learning in small flocks of laying hens. *Animal Behaviour* 47:1289–1296.

———. 1999. The effects of demonstrator social status and prior foraging success on social learning in laying hens. *Animal Behaviour* 57:163–171.

Niedenthal, P. M. 2007. Embodying emotion. *Science* 316:1002–1005.

Nielsen, M. 2006. Copying actions and copying outcomes: Social learning through the second year. *Developmental Psychology* 42:555–565.

Nielsen, M., E. Collier-Baker, J. M. Davis, and T. Suddendorf. 2005. Imitation recognition in a captive chimpanzee *(Pan troglodytes)*. *Animal Cognition* 8:31–36.

Nietzsche, F. 1881. *Daybreak.* Cambridge: Cambridge University Press.

Nishida, T. 1983. Alpha status and agonistic alliance in wild chimpanzees *(Pan troglodytes schweinfurthii)*. *Primates* 24:318–336.

Nishida, T., T. Hasegawa, H. Hayaki, Y. Takahata, and S. Uehara. 1992. Meat-sharing as a coalition strategy by an alpha male chimpanzee? In T. Nishida, W. C. McGrew, P. Marler, M. Pickford, and F. B. M. de Waal, eds. *Topics in primatology,* vol. 1: *Human origins,* 159–174. Tokyo: University of Tokyo Press.

Nishida, T., S. Uehara, and R. Nyundo. 1979. Predatory behavior among wild chimpanzees of the Mahale Mountains. *Primates* 20:1–20.

Nishitani, N., and R. Hari. 2000. Temporal dynamics of cortical representation for action. *Proceedings of the National Academy of Sciences USA* 97:913–918.

Nissen, H., and M. Crawford. 1932. A preliminary study of food-sharing behaviour in young chimpanzees. *Journal of Comparative Psychology* 22:383–419.

Nissen, H. W. 1931. A field study of the chimpanzee: Observations of chimpanzee behavior and environment in Western French Guinea. *Comparative Psychology Monographs* 8:1–122.

———. 1951. Phylogenetic comparison. In S. S. Stevens, ed. *Handbook of experimental psychology,* 347–386. New York: Wiley.

———. 1956. Individuality in the behavior of chimpanzees. *American Anthropologist* 58:407–413.

Noë, R. 2006. Cooperation experiments: Coordination through communication versus acting apart together. *Animal Behaviour* 71:1–18.

Nordenstrom, A., A. Servin, G. Bohlin, A. Larsson, and A. Wedell. 2002. Sex-typed toy play behavior correlates with the degree of prenatal androgen exposure assessed by the *CYP21* genotype in girls with congenital adrenal hyperplasia. *Journal of Clinical Endocrinology and Metabolism* 87 (11):5119–5124.

Nottebohm, F., A. Alvarez-Buylla, J. Cynx, C.-Y. Ling, M. Nottebohm, R. Suter, A. Tolles, and H. Williams. 1990. Song learning in birds: The relation between perception and production. *Philosophical Transactions of the Royal Society B: Biological Sciences* 329:115–124.

Novy, M. J., and J. A. Resko, eds. 1981. *Fetal endocrinology*. New York: Academic Press.

Nugent, J. K., B. M. Lester, and T. B. Brazelton. 1989. *The cultural context of infancy*. Norwood, NJ: Ablex.

Nummenmaa, L., and A. J. Calder. 2009. Neural mechanisms of social attention. *Trends in Cognitive Sciences* 13:135–143.

Obayashi, S., T. Suhara, K. Kawabe, T. Okauchi, J. Maeda, Y. Akine, H. Onoe, and A. Iriki. 2001. Functional brain mapping of monkey tool use. *Neuroimage* 14:853–861.

Oberman, L. M., and V. S. Ramachandran. 2007. The simulating social mind: The role of simulation in the social and communicative deficits of autism spectrum disorders. *Psychological Bulletin* 133:310–327.

O'Connell, S. M. 1995. Empathy in chimpanzees: Evidence for theory of mind? *Primates* 36:397–410.

Odling-Smee, F. J., K. N. Laland, and M. W. Feldman. 2003. *Niche construction: The neglected process in evolution*. Princeton, NJ: Princeton University Press.

Ogden, C. L., C. Fryar, M. Carroll, and F. Flegal. 2004. Mean body weight, height, and body mass index, United States 1960–2002. *Advance Data from Vital and Heath Statistics* 347:1–18.

Onishi, K. H., and R. Baillargeon. 2005. Do 15-month-old infants understand false beliefs? *Science* 308:255–258.

Oster, H. 2005. The repertoire of infant facial expressions: An ontogenetic perspective. In J. Nadel and D. Muir, eds. *Emotional development: Recent research advances*, 261–292. New York: Oxford University Press.

Osvath, M., and H. Osvath. 2008. Chimpanzee *(Pan troglodytes)* and orangutan *(Pongo abelii)* forethought: Self-control and pre-experience in the face of future tool use. *Animal Cognition* 11:661–674.

Ottoni, E. B., and P. Izar. 2008. Capuchin monkey tool use: Overview and implication. *Evolutionary Anthropology* 17:171–178.

Ottoni, E. B., B. D. Resende, and P. Izar. 2005. Watching the best nutcracking: What capuchin monkeys *(Cebus apella)* know about other's tool using skills. *Animal Cognition* 24:215–219.

Palagi, E. 2006. Social play in bonobos *(Pan paniscus)* and chimpanzees *(Pan troglodytes)*: Implications for natural social systems and interindividual relationships. *American Journal of Physical Anthropology* 129:418–426.

Palagi, E., and G. Cordoni. 2009. Postconflict third-party affiliation in *Canis lupus:* Do wolves share similarities with the great apes? *Animal Behaviour* 78:979–986.

Palagi, E., A. Leone, G. Mancini, and P. F. Ferrari. 2009. Contagious yawning in gelada baboons as a possible expression of empathy. *Proceedings of the National Academy of Sciences USA* 106:19262–19267.

Palagi, E., T. Paoli, and S. M. Borgognini-Tarli. 2004. Reconciliation and consolation in captive bonobos *(Pan paniscus)*. *American Journal of Primatology* 62:15–30.

————. 2006. Short-term benefits of play behavior and conflict prevention in *Pan paniscus*. *International Journal of Primatology* 27:1257–1270.

Palme, R., S. Rettenbacher, C. Touma, S. M. El-Bahr, and E. Möstl. 2005. Stress hormones in mammals and birds: Comparative aspects regarding metabolism, excretion, and noninvasive measurement in fecal samples. *Annals of the New York Academy of Sciences* 1040:162–171.

Pang, S., L. S. Levine, D. M. Chow, C. Faiman, and M. I. New. 1979. Serum androgen concentrations in neonates and young infants with congenital adrenal hyperplasia due to 21-hydroxylase deficiency. *Clinical Endocrinology* 11:575–584.

Panger, M. A. 2007. Tool use and cognition in primates. In C. J. Campbell, A. Fuentes, K. C. MacKinnon, S. K. Bearder, and R. M. Stumpf, eds. *Primates in perspective*, 665–677. Oxford: Oxford University Press.

Papousek, H., and M. Papousek. 1987. Intuitive parenting: A dialectic counterpart to the infant's integrative competence. In J. Osofsky, ed. *Handbook of infant development*, 2nd ed., 669–720. New York: Wiley.

Parish, A. R. 1994. Female relationships in bonobos *(Pan paniscus):* Evidence for bonding, cooperation, and female dominance in a male-philopatric species. *Human Nature* 7:61–96.

Parr, L. A., E. Hecht, S. K. Barks, T. M. Preuss, and J. R. Votaw. 2009. Face processing in the chimpanzee brain. *Current Biology* 19:50–53.

Parvizi, J. 2009. Corticocentric myopia: Old bias in new cognitive sciences. *Trends in Cognitive Sciences* 13:354–359.

Pastalkova, E., V. Itskov, A. Amarasingham, and G. Buzsaki. 2008. Internally generated cell assembly sequences in the rat hippocampus. *Science* 321:1322–1327.

Patterson, N., D. J. Richter, S. Gnerre, E. S. Lander, and D. Reich. 2006. Genetic evidence for complex speciation of humans and chimpanzees. *Nature* 441: 1103–452.

Paukner, A., J. R. Anderson, E. Visalberghi, E. Borelli, and P. F. Ferrari. 2005. Macaques *(Macaca nemestrina)* recognize when they are being imitated. *Biology Letters* 1:219–222.

Paukner, A., S. J. Suomi, E. Visalberghi, and P. F. Ferrari. 2009. Capuchin monkeys display affiliation toward humans who imitate them. *Science* 325 (5942):880–883.

Peignot, P., and J. R. Anderson. 1999. Use of experimenter-given manual and facial cues by gorillas *(Gorilla gorilla)* in an object-choice task. *Journal of Comparative Psychology* 113:253–260.

Pelé, M., V. Dufour, B. Thierry, and J. Call. 2009. Token transfers among great apes (*Gorilla gorilla, Pongo pygmaeus, Pan paniscus,* and *Pan troglodytes*): Species differences, gestural requests, and reciprocal exchange. *Journal of Comparative Psychology* 123:375–384.

Pelphrey, K., R. Adolphs, and J. P. Morris. 2004. Neuroanatomical substrates of social cognition dysfunction in autism. *Mental Retardation and Developmental Disabilities Research Review* 10:259–271.

Penn, D. C., and D. J. Povinelli. 2007a. Causal cognition in human and nonhuman animals: A comparative, critical review. *Annual Review of Psychology* 58:97–118.

———. 2007b. On the lack of evidence that non-human animals possess anything remotely resembling a "theory of mind." *Philosophical Transactions of the Royal Society B: Biological Sciences* 362:731–744.

Perrett, D. I., M. H. Harries, R. Bevan, S. Thomas, P. J. Benson, A. J. Mistlin, A. K. Chitty, J. K. Hietanen, and J. E. Ortega. 1989. Frameworks of analysis for the neural representation of animate objects and actions. *Journal of Experimental Biology* 146:87–113.

Perrett, D. I., J. K. Hietanen, M. W. Oram, and P. J. Benson. 1992. Organization and functions of cells responsive to faces in the temporal cortex. *Philosophical Transactions of the Royal Society B: Biological Sciences* 335:23–30.

Perrett, D. I., E. T. Rolls, and W. Caan. 1982. Visual neurons responsive to faces in the monkey temporal cortex. *Experimental Brain Research* 47:329–342.

Perrett, D. I., P. A. Smith, D. D. Potter, A. J. Mistlin, A. S. Head, A. D. Milner, and M. A. Jeeves. 1985. Visual cells in the temporal cortex sensitive to face view and gaze direction. *Proceedings of the Royal Society B: Biological Sciences* 223:293–317.

Perry, S. 2009. Conformism in the food processing techniques of white-faced capuchin monkeys *(Cebus capucinus). Animal Cognition* 12:705–716.

Perry, S., and J. C. Ordoñez-Jiménez. 2006. The effects of food size, rarity, and processing complexity on white-faced capuchins' visual attention to foraging conspecifics. In G. Hohmann, M. M. Robbins, and C. Boesch, eds. *Feeding ecology in apes and other primates,* 203–234. Cambridge: Cambridge University Press.

Perry, S., and L. M. Rose. 1994. Begging and transfer of coati meat by white-faced capuchin monkeys, *Cebus capucinus. Primates* 35:499–515.

Peters, Michael. 1991. Sex differences in human brain size and the general meaning of differences in brain size. *Canadian Journal of Psychology* 45 (4):507–522.

Petit, O., C. Desportes, and B. Thierry. 1992. Differential probability of "coproduction" in two species of macaque *(Macaca tonkeana, M. mulatta). Ethology* 90:107–120.

Pfeifer, J. H., M. Iacoboni, J. C. Mazziotta, and M. Dapretto. 2008. Mirroring others' emotions relates to empathy and interpersonal competence in children. *Neuroimage* 39:2076–2085.

Phillips, K. A., B. W. Grafton, and M. E. Haas. 2003. Tap-scanning for invertebrates by capuchins *(Cebus apella). Folia Primatologica* 74:162–164.

Phoenix, C. H., R. W. Goy, A. A. Gerall, and W. C. Young. 1959. Organizing action of prenatally administered testosterone propionate on the tissues mediating mating behavior in the female guinea pig. *Endocrinology* 65:369–382.

Piaget, J. 1952. *The origins of intelligence in children.* New York: Basic Books.

Pico-Alfonso, M. A., F. Mastorci, G. Ceresini, G. P. Ceda, M. Manghi, O. Pino, A. Troisi, and A. Sgoifo. 2007. Acute psychosocial challenge and cardiac autonomic response in women: The role of estrogens, corticosteroids, and behavioral coping styles. *Psychoneuroendocrinology* 32:451–463.

Pika, S. 2008. Gestures of apes and prelinguistic human children: Similar or different? *First Language* 28:116–140.

Pitman, C., and R. Shumaker. 2009. Does early care affect joint attention in great apes *(Pan troglodytes, Pan paniscus, Pongo abelii, Pongo pygmaesu, Gorilla gorilla)? Journal of Comparative Psychology* 123:334–341.

Platek, S. M., F. B. Mohamed, and G. G. Gallup. 2005. Contagious yawning and the brain. *Cognitive Brain Research* 23:448–452.

Platt, M. L., and A. A. Ghazanfar. 2009. *Primate Neuroethology.* Oxford: Oxford University Press.

Plomin, R., R. N. Emde, J. M. Braungart, J. Campos, R. Corley, D. W. Fulker, J. Kagan, J. S. Reznick, J. Robinson, C. Zahn-Waxler, and J. C. DeFries. 1993. Genetic change and continuity from fourteen to twenty months: The MacArthur longitudinal twin study. *Child Development* 64:1354–1376.

Plooij, F. 1984. *The behavioral development of free-living chimpanzee babies and infants.* Norwood, NJ: Ablex.

Plotnik, J. M., F. B. M. de Waal, and D. Reiss. 2006. Self-recognition in an Asian elephant. *Proceedings of the National Academy of Sciences USA* 103:17053–17057.

Plutchik, R. 1987. Evolutionary bases of empathy. In N. Eisenberg and J. Strayer, eds. *Empathy and its development,* 3–46. Cambridge: Cambridge University Press.

Poe, E. A. 1982. *The tell-tale heart and other writings.* New York: Bantam Books.

Pongrácz, P., Á. Miklósi, E. Kubinyi, K. Gurobi, J. Topál, and V. Csányi. 2001. Social learning in dogs: The effect of a human demonstrator on the performance of dogs in a detour task. *Animal Behaviour* 62:1109–1117.

Pongrácz, P., Á. Miklósi, E. Kubinyi, J. Topál, and V. Csányi. 2003. Interaction between individual experience and social learning in dogs. *Animal Behaviour* 65:595–603.

Pongrácz, P., V. Vida, P. Bánhegyi, and Á. Miklósi. 2008. How does dominance rank status affect individual and social learning performance in the dog *(Canis familiaris)? Animal Cognition* 11:75–82.

Popper, K. R. 1972. *Objective knowledge.* Oxford: Oxford University Press.

Povinelli, D. J. 1989. Failure to find self-recognition in Asian elephants in contrast to their use of mirror cues to discover hidden food. *Journal of Comparative Psychology* 103:122–131.

———. 2000. *Folk physics for apes: The chimpanzee's theory of how the world works.* Oxford: Oxford University Press.

Povinelli, D. J., J. M. Bering, and S. Giambrone. 2000. Toward a science of other minds: Escaping the argument by analogy. *Cognitive Science* 24:509–541.

———. 2003. Chimpanzee's "pointing": Another error of the argument by analogy? In S. Kita, ed. *Pointing: Where language, culture, and cognition meet.* Mahwah, NJ: Lawrence Erlbaum.

Povinelli, D. J., and T. J. Eddy. 1996. What young chimpanzees know about seeing. *Monographs of the Society for Research in Child Development* 61:1–152.

Povinelli, D. J., and D. K. O'Neill. 2000. Do chimpanzees use their gestures to instruct each other? In S. Baron-Cohen, H. Tager-Flusberg, and D. J. Cohen, eds. *Understanding other minds: Perspectives from developmental cognitive neuroscience,* 459–487. New York: Oxford University Press.

Povinelli, D. J., and J. Vonk. 2003. Chimpanzee minds: Suspiciously human? *Trends in Cognitive Sciences* 7:157–160.

Prather, J. F., S. Nowicki, S., R. C. Anderson, S. Peters, and R. Mooney. 2009. Neural correlates of categorical perception in learned vocal communication. *Nature Neuroscience* 12:221–228.

Prather, J. F., S. Peters, S. Nowicki, and R. Mooney. 2008. Precise auditory-vocal mirroring in neurons for learned vocal communication. *Nature* 451: 305–310.

Prato-Previde, E., S. Marshall-Pescini, and P. Valsecchi. 2008. Is your choice my choice? The owners' effect on pet dogs' *(Canis lupus familiaris)* performance in a food choice task. *Animal Cognition* 11:167–174.

Premack, D. 1971a. Language in chimpanzee? *Science* 172:808–822.

———. 1971b. On the assessment of language competence in the chimpanzee. In A. M. Schrier and F. Stollnitz, eds. *Behavior of nonhuman primates.* New York: Academic Press.

———. 2004. Is language the key to human intelligence? *Science* 303:318–320.

———. 2007. Human and animal cognition: Continuity and discontinuity. *Proceeding of the National Academy of Sciences USA* 104:13861–13867.

Premack, D., and G. Woodruff. 1978a. Chimpanzee problem-solving: A test for comprehension. *Science* 202:532–535.

———. 1978b. Does the chimpanzee have a theory of mind? *Behavioral and Brain Sciences* 1:515–526.

Preston, S. D., and F. B. M. de Waal. 2002. Empathy: Its ultimate and proximate bases. *Behavioral and Brain Sciences* 25:1–72.

Preuschoft, S., and J. A. R. A. M. van Hooff. 1995. Homologizing primate facial displays: A critical review of methods. *Folia Primatologica* 65:121–137.

Price, E. E., C. A. Caldwell, and A. Whiten. 2010. Comparative cultural cognition. *Wiley Interdisciplinary Reviews* 1:23–31.

Price, E. E., S. P. Lambeth, S. J. Schapiro, and A. Whiten. 2009. A potent effect of observational learning in chimpanzee construction. *Proceedings of the Royal Society B: Biological Sciences* 276:3377–3383.

Prinz, W. 1997. Perception and action planning. *European Journal of Cognitive Psychology* 9:129–154.

————. 2002. Experimental approaches to imitation. In W. Prinz and A. N. Melt-zoff, eds. *The imitative mind: Development, evolution and brain bases*, 143–162. Cambridge: Cambridge University Press.

Pylshyn, Z. W. 1978. When is attribution of beliefs justified? *Behavioral and Brain Sciences* 4:492–593.

Quadagno, D. M., R. Briscoe, and J. S. Quadagno. 1977. Effects of perinatal gonadal hormones on selected nonsexual behavior patterns: A critical assessment of the nonhuman and human literature. *Psychological Bulletin* 84:62–80.

Raby, C. R., and N. S. Clayton. 2009. Prospective cognition in animals. *Behavioural Processes* 80:314–324.

Racine, T. P., D. A. Leavens, N. Susswein, and T. J. Wereha, 2008. Conceptual and methodological issues in the investigation of primate intersubjectivity. In F. Morganti, A. Carassa, and G. Riva, eds. *Enacting intersubjectivity*. Amsterdam: IOS Press.

Radford, A. N. 2008. Duration and outcome of intergroup conflict influences intragroup affiliative behaviour. *Proceedings of the Royal Society B: Biological Sciences* 275:2787.

Radford, A. N., and M. A. Du Plessis. 2006. Dual function of allopreening in the cooperatively breeding green woodhoopoe, *Phoeniculus purpureus*. *Behavioral Ecology and Sociobiology* 61:221–230.

Ramos da Silva, E. D. 2008. Escolha de alvos coespecíficos na observação do uso de ferramentas por macacos-prego *(Cebus libidinosus)* selvagens. Master's thesis, Instituto de Psicologia, Universidade de São Paulo.

Range, F., L. Horn, T. Bugnyar, G. K. Gajdon, and L. Huber. 2009. Social attention in keas, dogs, and human children. *Animal Cognition* 12:181–192.

Range, F., and L. Huber. 2007. Attention in common marmosets: Implications for social-learning experiments. *Animal Behaviour* 73:1033–1041.

Range, F., Z. Viranyi, and L. Huber. 2007. Selective imitation in domestic dogs. *Current Biology* 17:868–872.

Rapaport, L. G., and G. R. Brown. 2008. Social influences on foraging behaviour in young nonhuman primates: Learning what, where and how to eat. *Evolutionary Anthropology* 17:189–201.

Ratcliffe, D. 1997. *The raven*. San Diego: Academic Press.

Reader, S. M., and L. N. Laland. 2003. *Animal innovation*. New York: Oxford University Press.

Reaux, J. E., L. A. Theall, and D. J. Povinelli. 1999. A longitudinal investigation of chimpanzees' understanding of visual perception. *Child Development* 70:275–290.

Reiss, D., and L. Marino. 2001. Mirror self-recognition in the bottlenose dolphin: A case of cognitive convergence. *Proceedings of the National Academy of Sciences USA* 98:5937–5942.

Reissen, A. H., and E. F. Kinder. 1952. *Postural development of infant chimpanzees*. New Haven, CT: Yale University Press.

Resende, B. D., E. B. Ottoni, and D. M. Fragaszy. 2008. Ontogeny of manipulative behavior and nut-cracking in young tufted capuchin monkeys

(Cebus apella): A perception-action perspective. *Developmental Science* 11:828–840.

Resnick, S. M., S. A. Berenbaum, I. I. Gottesman, and T. J. Bouchard. 1986. Early hormonal influences on cognitive functioning in congenital adrenal hyperplasia. *Developmental Psychology* 22 (2):191–198.

Rilling, J. K., S. K. Barks, L. A. Parr, T. M. Preuss, T. L. Faber, G. Pagnoni, J. D. Bremmer, and J. R. Votaw. 2007. A comparison of resting-state brain activity in humans and chimpanzees. *Proceedings of the National Academy of Sciences USA* 104:17146–17151.

Rizzolatti, G., and M. A. Arbib. 1998. Language within our grasp. *Trends in Neurosciences* 21:188–194.

Rizzolatti, G., R. Camarda, M. Fogassi, M. Gentilucci, G. Luppino, and M. Matelli. 1988. Functional organization of inferior area 6 in the macaque monkey: II. Area F5 and the control of distal movements. *Experimental Brain Research* 71:491–507.

Rizzolatti, G., and L. Craighero. 2004. The mirror-neuron system. *Annual Review of Neuroscience* 27:169–192.

Rizzolatti, G., L. Fadiga, V. Gallese, and L. Fogassi. 1996. Premotor cortex and the recognition of motor actions. *Cognitive Brain Research* 3:131–141.

Rizzolatti, G., L. Fadiga, M. Matelli, V. Bettinardi, E. Paulesu, D. Perani, and F. Fazio. 1996. Localization of grasp representations in humans by PET: 1. Observation versus execution. *Experimental Brain Research* 111:246–252.

Rizzolatti, G., L. Fogassi, and V. Gallese. 2001. Neurophysiological mechanisms underlying the understanding and imitation of action. *Nature Reviews Neuroscience* 2:661–670.

Rizzolatti, G., and G. Luppino. 2001. The cortical motor system. *Neuron* 31:889–901.

Rizzolatti, G., and C. Sinigaglia. 2006. *Mirrors in the brain: How our minds share actions and emotions.* Oxford: Oxford University Press.

Rochat, M. J., F. Caruana, A. Jezzini, L. Escola, I. Intskirveli, F. Grammont, V. Gallese, G. Rizzolatti, and M. A. Umiltà. 2010. Responses of mirror neurons in area F5 to hand and tool grasping observation. *Experimental Brain Research* 204 (4):605–616.

Rochat, M. J., E. Serra, L. Fadiga, and V. Gallese. 2008. The evolution of social cognition: Goal familiarity shapes monkeys' action understanding. *Current Biology* 18:227–232.

Rochat, P. 2003. Five levels of self-awareness as they unfold early in life. *Consciousness and Cognition* 12:717–731.

Rodeck, C. H., D. Gill, D. A. Rosenberg, and W. P. Collins. 1985. Testosterone levels in midtrimester maternal and fetal plasma and amniotic fluid. *Prenatal Diagnosis* 5 (3):175–181.

Roell, A. 1978. Social behaviour of the jackdaw in relation to its niche. *Behaviour* 64:1–12.

Romero, T., M. A. Castellanos, and F. B. M. de Waal. 2010. Consolation as possible expression of sympathetic concern among chimpanzees. *Proceedings of the National Academy of Sciences USA* 107:12110–12115.

Romero, T., F. Colmenares, and F. Aureli. 2009. Testing the function of reconciliation and third-party affiliation for aggressors in hamadryas baboons *(Papio hamadryas hamadryas)*. *American Journal of Primatology* 71:60–69.

Romero, T., and F. B. M. de Waal. 2010. Chimpanzee *(Pan troglodytes)* consolation: Third party identity as a window on possible function. *Journal of Comparative Psychology* 124:278–286.

Rosati, A., J. Stevens, B. Hare, and M. Hauser. 2007. The evolutionary origins of human patience: Temporal preferences in chimpanzees, bonobos, and human adults. *Current Biology* 17:1663–1668.

Rosenbaum, D. A., R. G. Cohen, S. A. Jax, D. J. Weiss, and R. van der Wel. 2007. The problem of serial order in behavior: Lashley's legacy. *Human Movement Science* 26:525–554.

Ross, C. 2002. Park or ride? Evolution of infant carrying in primates. *International Journal of Primatology* 22:749–771.

Rozzi, S., R. Calzavara, A. Belmalih, E. Borra, G. G. Gregoriou, M. Matelli, and G. Luppino. 2006. Cortical connections of the inferior parietal cortical convexity of the macaque monkey. *Cerebral Cortex* 16:1389–1417.

Rozzi, S., P. F. Ferrari, L. Bonini, G. Rizzolatti, and L. Fogassi. 2008. Functional organization of inferior parietal lobule convexity in the macaque monkey: Electrophysiological characterization of motor, sensory and mirror responses and their correlation with cytoarchitectonic areas. *European Journal of Neuroscience* 28:1569–1588.

Ruiz, A. M., J. C. Gómez, J. J. Roeder, and R. W. Byrne. 2009. Gaze following and gaze priming in lemurs. *Animal Cognition* 12:427–434.

Rumbaugh, D. M. 1977. *Language learning by a chimpanzee.* New York: Academic Press.

Rumbaugh, D. M., T. V. Gill, and E. C. von Glasersfeld. 1973. Reading and sentence completion by a chimpanzee (Pan). *Science* 182:731–733.

Russell, C. L., K. A. Bard, and L. B. Adamson. 1997. Social referencing by young chimpanzees *(Pan troglodytes)*. *Journal of Comparative Psychology* 111:185–191.

Russon, A. 2003. Developmental perspectives on great ape traditions. In D. M. Fragaszy and S. Perry, eds. *The biology of traditions: Models and evidence,* 329–364. Cambridge: Cambridge University Press.

Rutter, M., A. Caspi, and T. E. Moffitt. 2003. Using sex differences in psychopathology to study causal mechanisms: Unifying issues and research strategies. *Journal of Child Psychology and Psychiatry* 44 (8):1092–1115.

Saarela, M. V., Y. Hlushchuk, A. C. Williams, M. Schurmann, E. Kalso, and R. Hari. 2007. The compassionate brain: Humans detect intensity of pain from another's face. *Cerebral Cortex* 17:230–237.

Sagi, A., and M. L. Hoffman. 1976. Empathic distress in newborns. *Developmental Psychology* 12:175–176.

Saito, N., H. Mushiake, K. Sakamoto, Y. Itoyama, and J. Tanji. 2005. Representation of immediate and final behavioral goals in the monkey prefrontal cortex during an instructed delay period. *Cerebral Cortex* 15:1535–1546.

Saitoh, A. 2008. Creativity emerges from imagination? [in Japanese]. *Kagaku*. December.

Salvador, A., and R. Costa. 2009. Coping with competition: Neuroendocrine responses and cognitive variables. *Neuroscience and Biobehavioral Reviews* 33:160–170.

Sanfey, A. G., R. Hastie, M. K. Colvin, and J. Grafman. 2003. Phineas gauged: Decision-making and the human prefrontal cortex. *Neuropsychologia* 41:1218–1229.

Santos, L. R., and M. D. Hauser. 1999. How monkeys see the eyes: Cotton-top tamarins' reaction to changes in visual attention and action. *Animal Cognition* 2:131–139.

Santos, L. R., A. G. Nissen, and J. Ferrugia. 2006. Rhesus monkeys *(Macaca mulatta)* know what others can and cannot hear. *Animal Behaviour* 71:1175–1181.

Sanz, C., J. Call, and D. Morgan. 2009. Design complexity in termite-fishing tools of chimpanzees *(Pan troglodytes)*. *Biology Letters* 5:293–296.

Sanz, C., D. Morgan, and S. Gulick. 2004. New insights into chimpanzees, tools, and termites from the Congo Basin. *American Naturalist* 164:567–581.

Sanz, C., C. Schoning, and D. B. Morgan. 2009. Chimpanzees prey on army ants with specialized tool set. *American Journal of Primatology* 71:1–8.

Sapolsky, R. 1998. *Why zebras don't get ulcers: A guide to stress, stress-related diseases and coping.* 2nd ed. New York: W. H. Freeman.

Satz, P., H. Soper, D. Orsini, R. Henry, and J. Zvi. 1985. Handedness subtypes in autism. *Psychiatric Annals* 15:447–451.

Savage-Rumbaugh, E. S. 1986. *Ape language: From conditioned response to symbol.* New York: Columbia University Press.

Savage-Rumbaugh, E. S., and R. Lewin. 1994. *Kanzi: The ape at the brink of the human mind.* New York: Wiley.

Savage-Rumbaugh, E. S., J. Murphy, R. A. Sevcik, K. E. Brakke, S. L. Williams, and D. M. Rumbaugh. 1993. Language comprehension in ape and child. *Monographs of the Society for Research in Child Development* 58:1–256.

Savage-Rumbaugh, E. S., D. M. Rumbaugh, and S. Boysen. 1978. Symbolic communication between two chimpanzees *(Pan troglodytes)*. *Science* 201:641–644.

Sawyer, S. C., and M. M. Robbins. 2009. A novel food processing technique by a wild mountain gorilla *(Gorilla beringei beringei)*. *Folia Primatologica* 80:83–88.

Saxe, R., and S. Baron-Cohen. 2006. The neuroscience of theory of mind. *Social Neuroscience* 1:i–ix.

Saxe, R., S. Carey, and N. Kanwisher. 2004. Understanding other minds: Linking developmental psychology and functional neuroimaging. *Annual Review of Psychology* 55:87–124.

Saxe, R., and N. Kanwisher. 2003. People thinking about thinking people: The role of the temporo-parietal junction in "theory of mind." *Neuroimage* 19:1835–1842.

Saxe, R., and A. Wexler. 2005. Making sense of another mind: The role of the right temporo-parietal junction. *Neuropsychologia* 43:1391–1399.

Scerif, G., J. C. Gómez, and R. W. Byrne. 2004. What do Diana monkeys know about the focus of attention of a conspecific? *Animal Behaviour* 68:1239–1247.

Schacter, D. L., D. R. Addis, and R. L. Buckner. 2007. Remembering the past to imagine the future: The prospective brain. *Nature Reviews Neuroscience* 8:657–661.

Scheid, C., and R. Noë. 2010. The performance of rooks in a cooperative task depends on their temperament. *Animal Cognition* 13:545–553.

Scheid, C., F. Range, and T. Bugnyar. 2007. When, what, and whom to watch? Quantifying attention in ravens *(Corvus corax)* and jackdaws *(Corvus monedula)*. *Journal of Comparative Psychology* 121:380–386.

Schenker, N. M., W. D. Hopkins, M. A. Spocter, A. Garrison, C. D. Stimpson, J. M. Erwin, P. R. Hof, and C. C. Sherwood. 2010. Broca's area homologue in chimpanzees *(Pan troglodytes)*: Probabilistic mapping, asymmetry, and comparison to humans. *Cerebral Cortex* 20 (3): 730–742.

Scheumann, M., and J. Call. 2004. The use of experimenter-given cues by South African fur seals *(Arctocephalus pusillus)*. *Animal Cognition* 7:224–230.

Schiel, N., and L. Huber. 2006. Social influences on the development of foraging behavior in free-living common marmosets *(Callithrix jacchus)*. *American Journal of Primatology* 68:1–11.

Schindler, A. E. 1982. Hormones in human amniotic fluid. *Monographs on Endocrinology* 21:1–158.

Schino, G. 2007. Grooming and agonistic support: A meta-analysis of primate reciprocal altruism. *Behavioral Ecology* 18:115–120.

Schino, G., and F. Aureli. 2008. Grooming reciprocation among female primates: A meta-analysis. *Biology Letters* 4:9–11.

Schino, G., S. Geminiani, L. Rosati, and F. Aureli. 2004. Behavioral and emotional response of Japanese macaque *(Macaca fuscata)* mothers after their offspring receive an aggression. *Journal of Comparative Psychology* 118:340–346.

Schino, G., G. Perretta, A. Taglioni, V. Monaco, and A. Troisi. 1996. Primate displacement activities as an ethopharmacological model of anxiety. *Anxiety* 2:186–191.

Schino, G., L. Rosati, S. Geminiani, and F. Aureli. 2007. Post-conflict anxiety in Japanese macaques *(Macaca fuscata)*: Aggressor's and victim's perspectives. *Ethology* 113:1081–1088.

Schino, G., S. Scucchi, D. Maestripieri, and P. G. Turillazzi. 1988. Allogrooming as a tension-reduction mechanism: A behavioral approach. *American Journal of Primatology* 16:43–50.

Scott, F. J., S. Baron-Cohen, P. Bolton, and C. Brayne. 2002. The CAST (Childhood Asperger Syndrome Test): Preliminary development of a UK screen for mainstream primary-school-age children. *Autism* 6 (1):9–13.

Seed, A. M., N. S. Clayton, and N. J. Emery. 2007. Postconflict third-party affiliation in rooks. *Current Biology* 17:152–158.

———. 2008. Cooperative problem solving in rooks *(Corvus frugilegus)*. *Proceedings of the Royal Society B: Biological Sciences* 275:1421–1429.

Selye, H. 1974. *Stress without distress*. New York: New American Library.

Seyama, J., and R. S. Nagayama. 2005. The effect of torso direction on the judgement of eye direction. *Visual Cognition* 12:103–116.

Shah, A., and C. Frith. 1993. Why do autistic individuals show superior performance on the block design task? *Journal of Child Psychology and Psychiatry* 34:1351–1364.

Shah, A., and U. Frith. 1983. An islet of ability in autistic children: A research note. *Journal of Child Psychology and Psychiatry* 24 (4):613–620.

Shapleske, J., S. L. Rossell, P. W. Woodruff, and A. S. David. 1999. The planum temporale: A systematic, quantitative review of its structural, functional and clinical significance. *Brain Research Reviews* 29:26–49.

Shepherd, S. V. 2010. Following gaze: Gaze-following behavior as a window into social cognition. *Frontiers in Integrative Neuroscience* 4 (5):1–13.

Shepherd, S. V., J. T. Klein, R. O. Deaner, and M. L. Platt. 2009. Mirroring of attention by neurons in macaque parietal cortex. *Proceedings of the National Academy of Sciences USA* 106:9489–9494.

Shepherd, S. V., and M. L. Platt. 2008. Spontaneous social orienting and gaze following in ringtailed lemurs *(Lemur catta)*. *Animal Cognition* 11:13–20.

Sherry, D. F., and B. G. Galef. 1984. Cultural transmission without imitation: Milk bottle opening by birds. *Animal Behaviour* 32:937–938.

———. 1990. Social learning without imitation: More about milk bottle opening by birds. *Animal Behaviour* 40:987–989.

Sherwood, C. C., D. C. Broadfield, R. L. Holloway, P. J. Gannon, and P. R. Hof. 2003. Variability of Broca's area homologue in great apes: Implication for language evolution. *Anatomical Record* 217A:276–285.

Shettleworth, S. J. 2010. *Cognition, evolution, and behavior*. 2nd ed. New York: Oxford University Press.

Shillito, D. J., R. W. Shumaker, G. G. Gallup, and B. B. Beck. 2005. Understanding visual barriers: Evidence for Level 1 perspective taking in an orang-utan, *Pongo pygmaeus*. *Animal Behaviour* 69:679–687.

Shubin, N., C. Tabin, and S. Carroll. 2009. Deep homology and the origins of evolutionary novelty. *Nature* 457:818–823.

Shutt, K., A. MacLarnon, M. Heistermann, and S. Semple. 2007. Grooming in Barbary macaques: Better to give than to receive? *Biology Letters* 3:231–233.

Silani, G., G. Bird, R. Brindley, T. Singer, C. Frith, and U. Frith. 2008. Levels of emotional awareness and autism: An fMRI study. *Social Neuroscience* 3:97–112.

Silk, J. B. 2008. Social preferences in primates. In P. W. Glimcher, C. F. Camerer, E. Fehr, and R. A. Poldrack, eds. *Neuroeconomics: Decision making and the brain*, 269–284. Amsterdam: Elsevier Academic Press.

Silk, J. B., S. F. Brosnan, J. Vonk, J. Henrich, D. J. Povinelli, A. S. Richardson, S. P. Lambeth, J. Mascaro, and S. J. Schapiro. 2005. Chimpanzees are indifferent to the welfare of unrelated group members. *Nature* 437:1357–1359.

Silk, J. B., D. Rendall, D. L. Cheney, and R. M. Seyfarth. 2003. Natal attraction in adult female baboons *(Papio cynocephalus ursinus)* in the Moremi Reserve, Botswana. *Ethology* 109:627–644.

Singer, T. 2006. The neuronal basis and ontogeny of empathy and mind reading: Review of literature and implications for future research. *Neuroscience and Biobehavioral Reviews* 30:855–863.

Singer, T., H. D. Critchley, and K. Preuschoff. 2009. A common role of insula in feelings, empathy, and uncertainty. *Trends in Cognitive Sciences* 13:334–340.

Singer, T., and C. Lamm. 2009. The social neuroscience of empathy. *The Year in Cognitive Neuroscience 2009: Annals of the New York Academy of Sciences* 1156:81–96.

Singer, T., and S. Leiberg. 2009. Sharing the emotions of others: The neural bases of empathy. In M. S. Gazzaniga, ed. *The cognitive neurosciences*, 4:971–984. Cambridge, MA: MIT Press.

Singer, T., B. Seymour, J. O'Doherty, H. Kaube, R. J. Dolan, and C. D. Frith. 2004. Empathy for pain involves the affective but not the sensory components of pain. *Science* 303:1157–1162.

Singer, T., B. Seymour, J. P. O'Doherty, K. E. Stephan, R. J. Dolan, and C. D. Frith. 2006. Empathic neural responses are modulated by the perceived fairness of others. *Nature* 439:466–469.

Singer, T., R. Snozzi, G. Bird, P. Petrovic, G. Silani, M. Heinrichs, and R. J. Dolan. 2008. Effects of oxytocin and prosocial behavior on brain responses to direct and vicariously experienced pain. *Emotion* 8:781–791.

Singer, T., and N. Steinbeis. 2009. Differential roles of fairness- and compassion-based motivations for cooperation, defection, and punishment. *Annals of the New York Academy of Sciences* 1167:41–50.

Siu, A. M. H., and D. T. L. Shek. 2005. Validation of the Interpersonal Reactivity Index in a Chinese context. *Research on Social Work Practice* 15:118–126.

Skinner, B. F. 1953. *Science and human behavior*. New York: Macmillan.

Smail, P. J., F. I. Reyes, J. S. D. Winter, and C. Faiman. 1981. The fetal hormonal environment and its effect on the morphogenesis of the genital system. In S. J. Kogan and E. S. E. Hafez, eds. *Pediatric andrology*. Boston: Martinus Nijhoff.

Smith, O. A., C. A. Astley, F. A. Spelman, E. V. Golanov, D. M. Bowden, M. A. Chesney, and V. Chalyan. 2000. Cardiovascular responses in anticipation of changes in posture and locomotion. *Brain Research Bulletin* 53:69–76.

Smucny, D. A., C. S. Price, and E. A. Byrne. 1997. Post-conflict affiliation and stress reduction in captive rhesus macaques. *Advances in Ethology*, 32, 157.

Smuts, B. B. 1985. *Sex and friendship in baboons*. Cambridge, MA: Harvard University Press.

Sommer, I., N. Ramsey, and R. Kahn. 2001. Handedness, language lateralisation and anatomical asymmetry in schizophrenia: Meta-analysis. *British Journal of Psychiatry* 178:344–351.

Soper, H., P. Satz, D. Orsini, R. Henry, J. Zvi, and M. Schulman. 1986. Handedness patterns in autism suggest subtypes. *Journal of Autism and Developmental Disorders* 16:155–167.

Spagnoletti, N., E. Visalberghi, E. Ottoni, P. Izar, and D. Fragaszy. 2011. Stone tool use by adult wild bearded capuchin monkeys *(Cebus libidinosus)*: Frequency, efficiency and tool selectivity. *Journal of Human Evolution* 61, 97–107.

Spinozzi, G. 1989. Early sensorimotor development in cebus *(Cebus apella)*. In F. Antinucci, ed. *Cognitive structure and development in nonhuman primates,* 55–66. Hillsdale, NJ: Lawrence Erlbaum.

Steiper, M. E., and N. M. Young. 2006. Primate molecular divergence dates. *Molecular Phylogenetics and Evolution* 41:384–394.

Stevens, J. M. G., H. Vervaecke, H. Vries, and L. van Elsacker. 2005. The influence of the steepness of dominance hierarchies on reciprocity and interchange in captive groups of bonobos *(Pan paniscus)*. *Behaviour* 142:941–960.

Stevens, J. R. 2004. The selfish nature of generosity: Harassment and food sharing in primates. *Proceedings of the Royal Society B: Biological Sciences* 21:451–456.

Stevens, J. R., and M. D. Hauser. 2004. Why be nice? Psychological constraints on the evolution of cooperation. *Trends in Cognitive Sciences* 8:60–65.

Stevens, J. R., and D. W. Stephens. 2002. Food sharing: A model of manipulation by harassment. *Behavioral Ecology* 13:393–400.

Strange, B. A., R. N. Henson, K. J. Friston, and R. J. Dolan. 2001. Anterior prefrontal cortex mediates rule learning in humans. *Cerebral Cortex* 11:1040–1046.

Subiaul, F. 2007. The imitation faculty in monkeys: Evaluating its features, distribution and evolution. *Journal of Anthropological Science* 85:35–62.

Subiaul, F., H. Lurie, K. Romansky, T. Klein, D. Holmes, and H. Terrace. 2007. Cognitive imitation in typically-developing 3- and 4-year olds and individuals with autism. *Cognitive Development* 22:230–243.

Suddendorf, T., and M. C. Corballis. 1997. Mental time travel and the evolution of the human mind. *Genetic, Social, and General Psychology Monographs* 123:133–167.

Sulloway, F. J. 1979. *Freud: Biologist of the mind; Beyond the psychoanalytic legend.* New York: Basic Books.

Suomi, S. J. 2006. Risk, resilience, and gene X environment interactions in rhesus monkeys. *Annals of the New York Academy of Science* 1094:52–62.

Susswein, N., and T. Racine. In press. Sharing mental states: Causal and definitional issues in intersubjectivity. In J. Zlatev, T. P. Racine, C. Sinha, and E. Itkonen, eds. *The shared mind: Perspectives on intersubjectivity.* Amsterdam: John Benjamins.

Swettenham, J., S. Baron-Cohen, T. Charman, A. Cox, G. Baird, A. Drew, L. Rees, and S. Wheelwright. 1998. The frequency and distribution of spontaneous attention shifts between social and non-social stimuli in autistic, typically developing, and non-autistic developmentally delayed infants. *Journal of Child Psychology and Psychiatry* 9:747–753.

Taglialatela, J. P., J. L. Russell, J. A. Schaeffer, and W. D. Hopkins. 2008. Communicative signaling activates "Broca's" homologue in chimpanzees. *Current Biology* 18:343–348.

Takeshita, H., M. Myowa-Yamakoshi, and S. Hirata. 2009. The supine position of postnatal human infants: Implications for the development of cognitive intelligence. *Interaction Studies* 10:252–268.

Tanji, J., and E. Hoshi. 2001. Behavioral planning in the prefrontal cortex. *Current Opinion in Neurobiology* 11:164–170.

————. 2008. Role of the lateral prefrontal cortex in executive behavioral control. *Physiological Reviews* 88:37–57.

Tebbich, S., M. Taborsky, and H. Winkler. 1996. Social manipulation causes cooperation in keas. *Animal Behaviour* 52:1–10.

Tennie, C., J. Call, and M. Tomasello. 2009. Ratcheting up the ratchet: On the evolution of cumulative culture. *Philosophical Transactions of the Royal Society B: Biological Sciences* 364:2405–2415.

Tennie, C., D. Hedwig, J. Call, and M. Tomasello. 2008. An experimental study of nettle feeding in captive gorillas. *American Journal of Primatology* 70:584–593.

Terkel, J. 1996. Cultural transmission of feeding behavior in the black rat *(Rattus rattus)*. In C. M. Heyes and B. G. Galef, eds. *Social learning in animals: The roots of culture*, 17–47. San Diego: Academic Press.

Terrace, H. S. 1979. *Nim*. New York: Knopf.

Terry, R. 1970. Primate grooming as a tension reduction mechanism. *Journal of Psychology* 76:129–136.

Thierry, B., D. Wunderlich, and C. Gueth. 1989. Possession and transfer of objects in a group of brown capuchins *(Cebus apella)*. *Behaviour* 110:294–305.

Thorndike, E. L. 1898. Animal intelligence: An experimental study of the associative processes in animals. *Psychological Review Monographs* 2 (4, whole number 8).

————. 1911. Animal intelligence. New York: Macmillan.

Thornton, A. N., and N. J. Raihani. 2008. The evolution of teaching. *Animal Behaviour* 75:1823–1836.

Thorpe, W. H. 1963. *Learning and instinct in animals*. Cambridge, MA: Harvard University Press.

Thorsteinsson, K., and K. A. Bard. 2009. Coding infant chimpanzee facial expressions of joy. In E. Banninger-Huber and D. Peham, eds. *Current and future perspectives in facial expression research: Topics and methodological questions. Proceedings of the International Meeting at the Institute of Psychology, University of Innsbruck/Austria*, 54–61. Innsbruck: Innsbruck University Press.

Tinbergen, N. 1973. Ethology and stress diseases. Nobel Lecture, December 12.

Tkach, D., J. Reimer, and N. G. Hatsopoulos. 2007. Congruent activity during action and action observation in motor cortex. *Journal of Neuroscience* 27:13241–13250.

Tolman, E. C. 1938. Determiners of behavior at a choice point. *Psychological Review* 45:1–41.

Tomaiuolo, F., J. D. Macdonald, Z. Caramanos, G. Posner, M. Chiavaras, A. C. Evans, and M. Petrides. 1999. Morphology, morphometry and probability mapping of the pars opercularis of the inferior frontal gyrus: An *in vivo* MRI analysis. *European Journal of Neuroscience* 11:3033–3046.

Tomasello, M. 1990. Cultural transmission in the tool use and communicatory signaling of chimpanzees. In S. T. Parker and K. R. Gibson, eds. *"Language" and intelligence in monkeys and apes: Comparative developmental perspectives*, 274–311. Cambridge: Cambridge University Press.

———. 1995. Joint attention as social cognition. In C. Moore and P. J. Dunham, eds. *Joint attention: Its origins and role in development,* 103–130. Hillsdale, NJ: Lawrence Erlbaum.

———. 1996. Do apes ape? In C. M. Heyes and B. G. Galef, eds. *Social learning in animals: The roots of culture,* 319–346. San Diego: Academic Press.

———. 1999. *The cultural origins of human cognition.* Cambridge, MA: Harvard University Press.

———. 2007. If they're so good at grammar, then why don't they talk? Hints from apes' and humans' use of gestures. *Language Learning and Development* 3:133–156.

Tomasello, M., and J. Call. 1997. *Primate cognition.* Oxford: Oxford University Press.

Tomasello, M., J. Call, and B. Hare. 1998. Five primate species follow the visual gaze of conspecifics. *Animal Behaviour* 55:1063–1069.

———. 2003. Chimpanzees versus humans: It's not that simple. *Trends in Cognitive Sciences* 7:239–240.

Tomasello, M., J. Call, K. Nagell, R. Olguin, and M. Carpenter. 1994. The learning and use of gestural signals by young chimpanzees: A trans-generational study. *Primates* 35:137–154.

Tomasello, M., and M. Carpenter. 2005. The emergence of social cognition in three young chimpanzees. *Monographs of the Society for Research in Child Development* 70:1–136.

Tomasello, M., M. Carpenter, J. Call, T. Behne, and H. Moll. 2005. Understanding and sharing intentions: The origins of cultural cognition. *Behavioral and Brain Sciences* 28:675–691.

Tomasello, M., M. Davis-Dasilva, L. Camak, and K. Bard. 1987. Observational learning of tool-use by young chimpanzees. *Human Evolution* 2:175–183.

Tomasello, M., A. C. Kruger, and H. H. Ratner. 1993. Cultural learning. *Behavioral and Brain Sciences* 16:495–552.

Tomasello, M., E. S. Savage-Rumbaugh, and A. C. Kruger. 1993. Imitative learning of actions on objects by children, chimpanzees, and enculturated chimpanzees. *Child Development* 64:1688–1705.

Tomonaga, M. 2006. Development of chimpanzee social cognition in the first 2 years of life. In T. Matsuzawa, M. Tomonaga, and M. Tanaka, eds. *Cognitive development in chimpanzees,* 182–197. Tokyo: Springer-Verlag.

Tomonaga, M., M. Tanaka, T. Matsuzawa, M. Myowa-Yamakoshi, D. Kosugi, Y. Mizuno, S. Okamoto, M. Yamaguchi, and K. A. Bard. 2004. Development of social cognition in infant chimpanzees *(Pan troglodytes):* Face recognition, smiling, gaze, and the lack of triadic interactions. *Japanese Psychological Research* 46:227–235.

Topal, J., G. Gergely, Á. Míklósi, A. Erdohegyi, and G. Csibra. 2008. Infants' perseverative search errors are induced by pragmatic misinterpretation. *Science* 321:1831–1834.

Tordjman, S., P. Ferrari, V. Sulmont, M. Duyme, and P. Roubertoux. 1997. Androgenic activity in autism. *American Journal of Psychiatry* 154 (11): 1626–1627.

Trevarthen, C. 1979. Communication and cooperation in early infancy: A descriptive of primary intersubjectivity. In M. Bullowa, ed. *Before speech: The beginnings of interpersonal communication.* New York: Cambridge University Press.

Trevarthen, C., and K. J. Aitken. 2001. Infant intersubjectivity: Research, theory, and clinical applications. *Journal of Child Psychology and Psychiatry* 42: 3–48.

Trivers, R. L. 1971. The evolution of reciprocal altruism. *Quarterly Review of Biology* 46:35–57.

Troisi, A. 2002. Displacement activities as a behavioural measure of stress in nonhuman primates and human subjects. *Stress* 5:47–54.

Tronick, E. Z., and S. A. Winn. 1992. The neurobehavioral organization of Efe (Pygmy) infants. *Developmental and Behavioral Pediatrics* 13:421–424.

Tsao, D. Y., S. Moeller, and W. A. Freiwald. 2008. Comparing face patch systems in macaques and humans. *Proceedings of the National Academy of Sciences USA* 105:19514–19519.

Tschudin, A., J. Call, R. I. M. Dunbar, G. Harris, and C. van der Elst. 2001. Comprehension of signs by dolphins *(Tursiops truncatus). Journal of Comparative Psychology* 115:100–105.

Tulchinsky, D., and A. B. Little. 1994. *Maternal-fetal endocrinology.* 2nd ed. Philadelphia: W. B. Saunders.

Tulving, E. 2005. Episodic memory and autonoesis: Uniquely human? In H. S. Terrace and J. Metcalfe, eds. *The missing link in cognition: Origins of self-reflective consciousness,* 3–56. Oxford: Oxford University Press.

Udry, J. R., N. M. Morris, and J. Kovenock. 1995. Androgen effects on women's gendered behaviour. *Journal of Biosocial Science* 27:359–368.

Umiltà, M. A., L. Escola, I. Intskirveli, F. Grammont, M. Rochat, F. Caruana, A. Jezzini, V. Gallese, and G. Rizzolatti. 2008. How pliers become fingers in the monkey motor system. *Proceedings of the National Academy of Sciences USA* 105:2209–2213.

Umiltà, M. A., E. Kohler, V. Gallese, L. Fogassi, L. Fadiga, C. Kaysers, and G. Rizzolatti. 2001. I know what you are doing: A neurophysiological study. *Neuron* 31:155–165.

Valsecchi, P., E. Choleris, A. Moles, C. Guo, and D. Mainardi. 1996. Kinship and familiarity as factors affecting social transfer of food preferences in adult Mongolian gerbils *(Meriones unguiculatus). Journal of Comparative Psychology* 110:243–251.

van Baaren, R. B., R. W. Holland, B. Steenaert, and A. van Knippenberg. 2003. Mimicry for money: Behavioral consequences of imitation. *Journal of Experimental Social Psychology* 39:393–398.

van de Beek, C., J. H. H. Thijssen, P. T. Cohen-Kettenis, S. H. van Goozen, and J. K. Buitelaar. 2004. Relationships between sex hormones assessed in amniotic fluid, and maternal and umbilical cord blood: What is the best source of information to investigate the effects of fetal hormonal exposure? *Hormones and Behavior* 46:663–669.

van de Rijt-Plooij, H. H. C., and F. X. Plooij. 1987. Growing independence, conflict and learning in mother-infant relations in free-ranging chimpanzees. *Behaviour* 101:1–86.

van Hooff, J. 1967. The facial displays of the catarrhine monkeys and apes. In T. Morris, ed. *Primate ethology,* 7–35. London: Weidenfeld & Nicolson.

van IJzendoorn, M. H., K. A. Bard, M. J. Bakermans-Kranenburg, and K. Ivan. 2009. Enhancement of attachment and cognitive development of young nursery-reared chimpanzees in responsive versus standard care. *Developmental Psychobiology* 51:173–185.

van Lawick-Goodall, J. 1968. The behavior of free-living chimpanzees of the Gombe Stream Reserve. *Animal Behavior Monographs* 1:161–311.

van Schaik, C. P., M. Ancrenaz, G. Borgen, B. Galdikas, C. D. Knott, I. Singleton, A. Suzuki, S. S. Utami, and M. Merrill. 2003. Orangutan cultures and the evolution of material culture. *Science* 299:102–105.

van Schaik, C. P., R. O. Deaner, and M. Y. Merrill. 1999. The conditions for tool use in primates: Implications for the evolution of material culture. *Journal of Human Evolution* 36:719–741.

van Schaik, C. P., and R. I. M. Dunbar. 1990. The evolution of monogamy in large primates: A new hypothesis and some crucial tests. *Behaviour* 115:30–62.

van Wolkenten, M. L., J. M. Davis, M. L. Gong, and F. B. M. de Waal. 2006. Coping with acute crowding by *Cebus apella. International Journal of Primatology* 27:1241–1256.

Vauclair, J., and K. A. Bard. 1983. Development of manipulations with objects in ape and human infants. *Journal of Human Evolution* 12:631–645.

Veit, R., H. Flor, M. Erb, C. Hermann, M. Lotze, W. Grodd, and N. Birbaumer. 2002. Brain circuits involved in emotional learning in antisocial behavior and social phobia in humans. *Neuroscience Letters* 328:233–236.

Verderane, M. P. 2010. Socioecologia de macacos-prego *(Cebus libidinosus)* em área de ecótono cerrado/caatinga. PhD dissertation, Instituto de Psicologia, Universidade de São Paulo, San Paolo, Brazil.

Vervaecke, H., H. de Vries, and L. van Elsacker. 2000. Function and distribution of coalitions in captive bonobos *(Pan paniscus). Primates* 41:249–265.

Vick, S-J., and J. R. Anderson. 2000. Learning and limits of use of eye gaze by capuchin monkeys *(Cebus apella)* in an object-choice task. *Journal of Comparative Psychology* 114:200–207.

Vick, S.-J., B. Waller, L. Parr, M. Smith Pasqualini, and K. A. Bard. 2007. A cross species comparison of facial morphology and movement in humans and chimpanzees using FACS. *Journal of Nonverbal Behavior* 31:1–20.

Virgin, C. E., and R. M. Sapolsky. 1997. Styles of male social behavior and their endocrine correlates among low-ranking baboons. *American Journal of Primatology* 42:25–39.

Visalberghi, E. 1987. Acquisition of nut-cracking behaviour by two capuchin monkeys *(Cebus apella). Folia Primatologica* 49:168–181.

Visalberghi, E., and E. Addessi. 2003. Food for thought: Social learning about food in feeding capuchin monkeys. In D. M. Fragaszy and S. Perry, eds.

The biology of traditions, 187–212. Cambridge: Cambridge University Press.

Visalberghi, E., E. Addessi, N. Spagnoletti, V. Truppa, E. Ottoni, P. Izar, and D. M. Fragaszy. 2009. Selection of effective stone tools by wild capuchin monkeys. *Current Biology* 19:213–217.

Visalberghi, E., and D. M. Fragaszy. 2002. Do monkeys ape? Ten years after. In K. Dautenhahn and C. L. Nehaniv, eds. *Imitation in animals and artifacts,* 471–499. Cambridge, MA: MIT Press.

———. 2006. What is challenging about tool use? The capuchins' perspective. In E. A. Wasserman and T. R. Zentall, eds. *Comparative cognition: Experimental explorations of animal intelligence,* 529–552. New York: Oxford University Press.

Visalberghi, E., D. M. Fragaszy, E. Ottoni, P. Izar, M. G. de Oliveira, and F. R. D. Andrade. 2007. Characteristics of hammer stones and anvils used by wild bearded capuchin monkeys *(Cebus libidinosus)* to crack open palm nuts. *American Journal of Physical Anthropology* 132:426–444.

Visalberghi, E., and C. Néel. 2003. Tufted capuchins *(Cebus apella)* use weight and sound to choose between full and empty nuts. *Ecological Psychology* 15:215–228.

Visalberghi, E., B. P. Quarantotti, and F. Tranchida. 2000. Solving a cooperation task without taking into account the partner's behavior: The case of capuchin monkeys *(Cebus apella). Journal of Comparative Psychology* 114:297–301.

Visalberghi, E., N. Spagnoletti, E. D. R. da Silva, F. R. D. Andrade, E. Ottoni, P. Izar, and D. M. Fragaszy. 2009. Distribution of potential hammers and transport of hammer tools and nuts by wild capuchin monkeys. *Primates* 50:95–104.

Voelkl, B., and L. Huber. 2000. True imitation in marmosets. *Animal Behaviour* 60:195–202.

———. 2007. Imitation as faithful copying of a novel technique in marmoset monkeys. *PLoS ONE* 7:e611.

Volkmar, F. R., A. Klin, B. Siegel, P. Szatmari, C. Lord, M. Campbell, B. J. Freeman, D. V. Cicchetti, M. Rutter, and W. Kline. 1994. Field trial for autistic disorder in DSM-IV. *American Journal of Psychiatry* 151:1361–1367.

Vonk, J., S. F. Brosnan, J. B. Silk, J. Henrich, A. S. Richardson, S. P. Lambeth, S. J. Schapiro, and D. J. Povinelli. 2008. Chimpanzees do not take advantage of very low cost opportunities to deliver food to unrelated group members. *Animal Behaviour* 75:1757–1770.

Wallace, A. R. 1869. *The Malay Archipelago.* London: Macmillan.

Wallis, J. 1997. A survey of reproductive parameters in the free-ranging chimpanzees of Gombe National Park. *Journal of Reproduction and Fertility* 109:297–307.

Wallis, J. D., K. C. Anderson, and E. K. Miller. 2001. Single neurons in prefrontal cortex encode abstract rules. *Nature* 411:953–956.

Wallis, J. D., and E. K. Miller. 2003. From rule to response: Neuronal processes in the premotor and prefrontal cortex. *Journal of Neurophysiology* 90:1790–1806.

Want, S. C., and P. L. Harris. 2001. Learning from other people's mistakes: Causal understanding in learning to use a tool. *Child Development* 72:431–443.

———. 2002. How do children ape? Applying concepts from the study of non-human primates to the developmental study of "imitation" in children. *Developmental Science* 5:1–41.

Wapner, S., and L. Cirillo. 1968. Imitation of a model's hand movement: Age changes in transposition of left-right relations. *Child Development* 39:887–894.

Warneken, F., B. Hare, A. P. Melis, D. Hanus, and M. Tomasello. 2007. Spontaneous altruism by chimpanzees and young children. *PLoS Biology* 5:1414–1420.

Warneken, F., and M. Tomasello. 2006. Altruistic helping in human infants and young chimpanzees. *Science* 311:1301–1303.

Warren, W. H. 1984. Perceiving affordances: Visual guidance of stair climbing. *Journal of Experimental Psychology: Human Perception and Performance* 10:683–703.

Wascher, C. A. F., I. B. R. Scheiber, and K. Kotrschal. 2008. Heart rate modulation in bystanding geese watching social and non-social events. *Proceedings of the Royal Society B: Biological Sciences* 275:1653–1659.

Washburn, M. F. 1923. *The animal mind: A textbook of comparative psychology.* 2nd ed. New York: Macmillan.

Watanabe, M., and M. Sakagami. 2007. Integration of cognitive and motivational context information in the primate prefrontal cortex. *Cerebral Cortex* 17 Suppl. 1:i101–i109.

Watanabe, S., and K. Ono. 1986. An experimental analysis of "empathic" response: Effects of pain reactions of pigeon upon other pigeon's operant behavior. *Behavioural Processes* 13:269–277.

Watkins, K. E., T. Paus, J. P. Lerch, A. Zijdenbos, D. L. Collins, P. Neelin, J. Taylor, K. J. Worsley, and A. C. Evans. 2001. Structural asymmetries in the human brain: A voxel-based statistical analysis of 142 MRI scans. *Cerebral Cortex* 11:868–877.

Watts, D. P. 2002. Reciprocity and interchange in the social relationships of wild male chimpanzees. *Behaviour* 139:343–370.

———. 2006. Conflict resolution in chimpanzees and the valuable-relationships hypothesis. *International Journal of Primatology* 27:1337–1364.

Watts, D. P., F. Colmenares, and K. Arnold. 2000. Redirection, consolation and male policing: How targets of aggression interact with bystanders. In F. Aureli and F. B. M. de Waal, eds. *Natural conflict resolution,* 281–301. Berkeley: University of California Press.

Watts, D. P., and J. C. Mitani. 2001. Boundary patrols and intergroup encounters in wild chimpanzees. *Behaviour* 138:299–327.

———. 2002. Hunting behavior of chimpanzees at Ngogo, Kibale National Park, Uganda. *International Journal of Primatology* 23:1–28.

Wehner, R. 1996. Preface. *Journal of Experimental Biology* 199:i.

Wellman, H. M., D. Cross, and J. Watson. 2001. Meta-analysis of theory-of-mind development: The truth about false belief. *Child Development* 7:655–684.

Wellman, H. M., A. T. Phillips, and T. Rodriguez. 2000. Young children's understanding of perception, desire, and emotion. *Child Development* 71:895–912.

Werdenich, D., and L. Huber. 2002. Social factors determine cooperation in marmosets. *Animal Behaviour* 64:771–781.

Westergaard, G. C., and D. M. Fragaszy. 1987. The manufacture and use of tools by capuchin monkeys *(Cebus apella)*. *Journal of Comparative Psychology* 101:159–168.

Wheelwright, S., S. Baron-Cohen, N. Goldenfeld, J. Delaney, D. Fine, R. Smith, L. Weil, and A. Wakabayashi. 2006. Predicting Autism Spectrum Quotient (AQ) from the Systemizing Quotient-Revised (SQ-R) and Empathy Quotient (EQ). *Brain Research* 1079 (1):47–56.

White, T. 2009. *Ardipithecus ramidus* and the paleobiology of early hominids. *Science* 326:75–86.

Whitehouse, A. J., M. T. Maybery, R. Hart, E. Mattes, J. P. Newnham, D. M. Sloboda, F. J. Stanley, and M. Hickey. In press. Fetal androgen exposure and pragmatic language ability of girls in middle adulthood: Implications for the extreme male-brain theory of autism. *Psychoneuroendocrinology*.

Whiten, A. 1992. On the nature and evolution of imitation in the animal kingdom: Reappraisal of a century of research. *Advances in the Study of Behavior* 21:239–283.

———. 1998. Imitation of the sequential structure of actions by chimpanzees *(Pan troglodytes)*. *Journal of Comparative Psychology* 112:270–281.

———. 1999a. The evolution of deep social mind in humans. In M. Corballis and S. E. G. Lea, eds. *The descent of mind,* 155–175. Oxford: Oxford University Press.

———. 1999b. Parental encouragement in *Gorilla* in comparative perspective: Implications for social cognition. In S. T. Parker, H. L. Miles, and R. W. Mitchell, eds. *The mentality of gorillas and orangutans,* 342–366. Cambridge: Cambridge University Press.

———. 2006. The place of "deep social mind" in the evolution of human nature. In M. A. Jeeves, ed. *Human nature,* 207–222. Edinburgh: Royal Society of Edinburgh.

———. 2011. The scope of culture in chimpanzees, humans and ancestral apes. *Philosophical Transactions of the Royal Society B,* 366: 997–1007.

———. In press. Primate social learning, traditions and culture. In J. C. Mitani, J. Call, P. Kappeler, R. Palombit, and J. B. Silk, eds. *The evolution of primate societies.* Chicago: University of Chicago Press.

Whiten, A., and R. W. Byrne. 1988. The manipulation of attention in primate tactical deception. In R. Byrne and A. Whiten, eds. *Machiavellian intelligence: Social expertise and the evolution of intellect in monkeys, apes, and humans,* 211–223. Oxford: Oxford University Press.

———. 1997. *Machiavellian intelligence II.* Cambridge: Cambridge University Press.

Whiten, A., D. M. Custance, J. C. Gómez., P. Teixidor, and K. A. Bard. 1996. Imitative learning of artificial fruit processing in children *(Homo sapiens)* and chimpanzees *(Pan troglodytes)*. *Journal of Comparative Psychology* 110:3–14.

Whiten, A., J. Goodall, W. C. McGrew, T. Nishida, V. Reynolds, Y. Sugiyama, C. E. G. Tutin, R. W. Wrangham, and C. Boesch. 1999. Cultures in chimpanzees. *Nature* 399:682–685.

Whiten, A., and R. Ham. 1992. On the nature and evolution of imitation in the animal kingdom: Reappraisal of a century of research. *Advances in the Study of Behaviour* 21:239–283.

Whiten, A., V. Horner, and F. B. M. de Waal. 2005. Conformity to cultural norms of tool use in chimpanzees. *Nature* 437:737–740.

Whiten, A., V. Horner, C. A. Litchfield, and S. Marshall-Pescini. 2004. How do apes ape? *Learning and Behavior* 32:36–52.

Whiten, A., N. McGuigan, S. Marshall-Pescini, and L. M. Hopper. 2009. Emulation, imitation, over-imitation and the scope of culture for child and chimpanzee. *Proceedings of the Royal Society B: Biological Sciences* 364:2417–2428.

Whiten, A., and A. Mesoudi. 2008. Establishing an experimental science of culture: Animal social diffusion experiments. *Philosophical Transactions of the Royal Society B: Biological Sciences* 363:3477–3488.

Whiten, A., A. Spiteri, V. Horner, K. E. Bonnie, S. P. Lambeth, S. J. Schapiro, and F. B. M. de Waal. 2007. Transmission of multiple experimentally-seeded traditions within and between groups of chimpanzees. *Current Biology* 17:1038–1043.

Whiten, A., and C. P. van Schaik. 2007. The evolution of animal "cultures" and social intelligence. *Philosophical Transactions of the Royal Society B: Biological Sciences* 362:603–620.

Wicker, B., C. Keysers, J. Plailly, J. P. Royet, V. Gallese, and G. Rizzolatti. 2003. Both of us disgusted in my insula: The common neural basis of seeing and feeling disgust. *Neuron* 40:655–664.

Wilkinson, A., K. Kuenstner, J. Mueller, and L. Huber. 2010. Social learning in a non-social reptile *(Geochelone carbonaria)*. *Biology Letters,* DOI: 10.1098/rsbl.2010.0092.

Williams, H., and F. Nottebohm, 1985. Auditory responses in avian cortical motor neurons: A motor theory for song perception in birds. *Science* 229:279–282.

Williams, J., F. Scott, C. Stott, C. Allison, P. Bolton, S. Baron-Cohen, and C. Brayne. 2005. The CAST (Childhood Asperger Syndrome Test): Test accuracy. *Autism* 9 (1):45–68.

Williams, J. G., C. Allison, F. J. Scott, P. F. Bolton, S. Baron-Cohen, F. E. Matthews, and C. Brayne. 2008. The Childhood Autism Spectrum Test (CAST): Sex differences. *Journal of Autism and Developmental Disorders* 38 (9):1731–1739.

Williams, J. H. G., A. Whiten, T. Suddendorf, and D. I. Perrett. 2001. Imitation, mirror neurons and autism. *Neuroscience and Biobehavioral Reviews* 25: 287–295.

Williams, J. M., G. W. Oehlert, J. V. Carlis, and A. E. Pusey. 2004. Why do male chimpanzees defend a group range? *Animal Behaviour* 68:523–532.

Wilson, E. O. 1975. *Sociobiology.* Cambridge, MA: Harvard University Press.

Wilson, S. M., and M. Iacoboni. 2006. Neural responses to non-native phonemes varying in producibility: Evidence for the sensorimotor nature of speech perception. *Neuroimage* 33:316–325.

Wilson, S. M., I. Molnar-Szakacs, and M. Iacoboni. 2008. Beyond superior temporal cortex: Intersubject correlations in narrative speech comprehension. *Cerebral Cortex* 18:230–242.

Wilson, S. M., A. P. Saygin, M. I. Sereno, and M. Iacoboni. 2004. Listening to speech activates motor areas involved in speech production. *Nature Neuroscience* 7:701–702.

Wimmer, H., and J. Perner. 1983. Beliefs about beliefs: Representation and constraining function of wrong beliefs in young children's understanding of deception. *Cognition* 13:103–128.

Wise, S. P., E. A. Murray, and C. R. Gerfen. 1996. The frontal cortex-basal ganglia system in primates. *Critical Reviews in Neurobiology* 10:317–356.

Witelson, S. F. 1991. Sex-differences in neuroanatomical changes with aging. *New England Journal of Medical* 325 (3):211–212.

Wittig, R. M., and C. Boesch. 2005. How to repair relationships—Reconciliation in wild chimpanzees *(Pan troglodytes)*. *Ethology* 111:736–763.

Wittig, R. M., C. Crockford, J. Lehmann, P. L. Whitten, R. M. Seyfarth, and D. L. Cheney. 2008. Focused grooming networks and stress alleviation in wild female baboons. *Hormones and Behavior* 54:170–177.

Wobber, V., B. Hare, J. Maboto, S. Lipson, R. Wrangham, and P. T. Ellison. 2010. Differential changes in steroid hormones before competition in bonobos and chimpanzees. *Proceedings of the National Academy of Sciences USA* 107:12457–12462.

Wobber, V., R. Wrangham, and B. Hare. 2010. Bonobos exhibit delayed development of social behavior and cognition relative to chimpanzees. *Current Biology* 20:226–230.

Wohlschläger, A., and H. Bekkering. 2002. Is human imitation based on a mirror-neuron system? Some behavioural evidence. *Experimental Brain Research* 143:335–341.

Wohlschläger, A., M. Gattis, and H. Bekkering. 2003. Action generation and action perception in imitation: An instance of the ideomotor principle. *Philosophical Transactions of the Royal Society B: Biological Sciences* 358:501–515.

Wood, D. 1989. Social interaction as tutoring. In M. H. Bornstein and J. S. Bruner, eds. *Interaction in human development*, 59–80. Hillsdale, NJ: Lawrence Erlbaum.

Wood, J. N. 2003. Social cognition and the prefrontal cortex. *Behavioral Cognitive Neuroscience Review* 2:97–114.

Woodward, A. L. 1998. Infants encode the goal object of an actor's reach. *Cognition* 69:1–34.

Wrangham, R. 1999. The evolution of coalitionary killing. *Yearbook of Physical Anthropology* 42:1–30.

Wright, B. W., K. A. Wright, J. Chalk, M. P. Verderane, D. M. Fragaszy, E. Visalberghi, P. Izar, E. B. Ottoni, P. Constantino, and C. Vinyard. 2009. Fallback foraging as a way of life: Using dietary toughness to compare the fallback signal among capuchins and implications for interpreting morphological variation. *American Journal of Physical Anthropology* 140:687–699.

Yamamoto, S., T. Humle, and M. Tanaka. 2009. Chimpanzees help each other upon request. *PLoS One* 4:e7416.

Yamamoto, S., and M. Tanaka. 2009a. Do chimpanzees spontaneously take turns in a reciprocal cooperation task? *Journal of Comparative Psychology* 123:242–249.

————. 2009b. Selfish strategies develop in social problem situations in chimpanzee *(Pan troglodytes)* mother-infant pairs. *Animal Cognition* 12:S27–S36.

————. 2010. The influence of kin relationship and reciprocal context on chimpanzees' other-regarding preferences. *Animal Behaviour* 79:595–602.

Yamazaki, Y., H. Yokochi, M. Tanaka, K. Okanoya, and A. Iriki. 2010. Potential role of monkey inferior parietal neurons coding action semantic equivalences as precursors of parts of speech. *Social Neuroscience* 5:105–117.

Yarnold, P. R., F. B. Bryant, S. N. Nightingale, and G. J. Martin. 1996. Assessing physician empathy using the interpersonal reactivity index: A measurement model and cross-sectional analysis. *Psychology, Health, and Medicine* 1:207–221.

Yerkes, R. M. 1925. *Almost human.* New York: Century.

————. 1943. *Chimpanzees: A laboratory colony.* New Haven, CT: Yale University Press.

Yerkes, R. M., and M. I. Tomilin. 1935. Mother-infant relations in chimpanzee. *Journal of Comparative Psychology* 20:321–359.

Yerkes, R. M., and A. Yerkes. 1929. *The great apes: A study of anthropoid life.* New Haven, CT: Yale University Press.

Yokochi, H., M. Tanaka, M. Kumashiro, and A. Iriki. 2003. Inferior parietal somatosensory neurons coding face-hand coordination in Japanese macaques. *Somatosensory Motor Research* 20:115–125.

Zahavi, D. 2008. Simulation, projection and empathy. *Consciousness and Cognition* 17:514–522.

Zahn-Waxler, C., and M. Radke-Yarrow. 1990. The origins of empathic concern. *Motivation and Emotion* 14:107–130.

Zahn-Waxler, C., M. Radke-Yarrow, E. Wagner, and M. Chapman. 1992. Development of concern for others. *Developmental Psychology* 28:126–136.

Zamma, K. 2002. Grooming site preferences determined by lice infection among Japanese macaques in Arashiyama. *Primates* 43:41–49.

Zanolie, K., S. Teng, S. E. Donohue, A. C. Van Duijvenvoorde, G. P. Band, S. A. Rombouts, and E. A. Crone. 2008. Switching between colors and shapes on the basis of positive and negative feedback: An fMRI and EEG study on feedback-based learning. *Cortex* 44:537–547.

Zentall, T. R. 2001. Imitation in animals: Evidence, function, and mechanisms. *Cybernetics and Systems* 32:53–96.

————. 2003. Imitation by animals: How do they do it? *Current Directions in Psychological Science* 12:91–95.

————. 2006. Imitation: Definitions, evidence, and mechanisms. *Animal Cognition* 9:335–353.

Zohar, O., and J. Terkel. 1991. Acquisition of pine cone stripping behaviour in black rats *(Rattus rattus). International Journal of Comparative Psychology* 5:1–6.

Contributors

FILIPPO AURELI, Research Centre in Evolutionary Anthropology and Palaeoecology, John Moores University

BONNIE AUYEUNG, Autism Research Centre, Department of Psychiatry, University of Cambridge

KIM A. BARD, Centre for Comparative and Evolutionary Psychology, Department of Psychology, University of Portsmouth

SIMON BARON-COHEN, Autism Research Centre, Department of Psychiatry, University of Cambridge

FRANS B. M. DE WAAL, Living Links, Yerkes National Primate Research Center, and Psychology Department, Emory University

PIER FRANCESCO FERRARI, Department of Neurosciences and Department of Evolutionary and Functional Biology, University of Parma; Dipartimento di Biologia and Italian Institute of Technology, University of Parma

LEONARDO FOGASSI, Dipartimento di Psicologia, Dipartimento di Neuroscienze and Italian Institute of Technology, University of Parma

DOROTHY FRAGASZY, Psychology Department, University of Georgia

ORLAITH N. FRASER, Department of Cognitive Biology, University of Vienna

NAOTAKA FUJII, Laboratory for Adaptive Intelligence, Brain Science Institute, Riken

BRIAN HARE, Department of Evolutionary Anthropology and Center for Cognitive Neuroscience, Duke University

GRIT HEIN, Laboratory for Social and Neural Systems Research, Institute for Empirical Research in Economics, University of Zurich

LYDIA M. HOPPER, Language Research Center, Georgia State University, Atlanta, Georgia; and Centre for Social Learning and Cognitive Evolution, and Scottish Primate Research Group, School of Psychology, University of St. Andrews

WILLIAM D. HOPKINS, Department of Psychology and Neuroscience, Agnes Scott College; and Division of Developmental and Cognitive Neuroscience, Yerkes National Primate Research Center

LUDWIG HUBER, Department of Cognitive Biology, University of Vienna

MARCO IACOBONI, Ahmanson-Lovelace Brain Mapping Center, Department of Psychiatry and Biobehavioral Sciences, Semel Institute for Neuroscience and Social Behavior, Brain Research Institute, David Geffen School of Medicine at UCLA

ATSUSHI IRIKI, Laboratory for Symbolic Cognitive Development, Brain Science Institute, Riken

SARAH MARSHALL-PESCINI, Department of Biological Science and Technology, University of Milan, Milan, Italy; and Centre for Social Learning and Cognitive Evolution, and Scottish Primate Research Group, School of Psychology, University of St. Andrews

TETSURO MATSUZAWA, Primate Research Institute, Kyoto University

APRIL M. RUIZ, Yale University

CHARLES R. MENZEL, Language Research Center, Georgia State University

EMIL W. MENZEL JR., Professor Emeritus, Department of Psychology, Stony Brook University

LAURIE R. SANTOS, Yale University

TANIA SINGER, Laboratory for Social and Neural Systems Research, Institute for Empirical Research in Economics, University of Zurich, and Department of Social Neuroscience, Max Planck Institute for Human Cognitive and Brain Sciences, Leipzig

JINGZHI TAN, Department of Evolutionary Anthropology, Duke University

JARED P. TAGLIALATELA, Division of Developmental and Cognitive Neuroscience, Yerkes National Primate Research Center, and Department of Biology and Physics, Kennesaw State University

ELISABETTA VISALBERGHI, Istituto di Scienze e Tecnologie della Cognizione, Consiglio Nazionale delle Ricerche, Rome

ANDREW WHITEN, Centre for Social Learning and Cognitive Evolution, and Scottish Primate Research Group, School of Psychology, University of St Andrews

Index